Technologies for Fingermark Age Estimations: A Step Forward

Josep De Alcaraz-Fossoul
Editor

Technologies for Fingermark Age Estimations: A Step Forward

Springer

Editor
Josep De Alcaraz-Fossoul
Henry C. Lee College of Criminal Justice
and Forensic Science
Forensic Science Department
University of New Haven
West Haven, CT, USA

ISBN 978-3-030-69336-7 ISBN 978-3-030-69337-4 (eBook)
https://doi.org/10.1007/978-3-030-69337-4

This Springer imprint is published by the registered company Springer Nature Switzerland AG
The registered company address is: Gewerbestrasse 11, 6330 Cham, Switzerland

Foreword

When Joe (Josep De Alcaraz-Fossoul) asked me to participate in the writing of a book about age determination of fingermarks back in 2018, during the American Academy of Forensic Sciences (AAFS) annual conference in Seattle, I was immediately seduced by the idea for two main reasons: (a) there has been a general revival of interest for this subject over the last 10–15 years, mainly triggered by the evolution of technical and analytical capacities which has resulted in many publications in various scientific journals; and (b) questions about the age of fingermarks collected at crime scenes are quite regularly asked by/to police officers, fingerprint examiners, and other members of the justice system. It was thus time to gather past and current knowledge on this topic in one reference book to provide a compact overview of the different approaches and visions that have been explored to date and help researchers and practitioners to understand the challenges hidden behind this simple question: "*How old is this fingermark*?"

Let me use this brief introduction to put this book back to its forensic context.

Forensic science, in particular criminalistics, is commonly defined as the application of scientific methods and techniques to search, collect, analyse, and interpret traces left by criminal actions to reconstruct crimes, support investigations, and help the judiciary to make decisions about someone's involvement in a crime. The trace (or the mark) is thus the central element of forensic science. Whether the question is academic or operational, investigative or evaluative, forensic science strives to make traces "talk".

During investigations or court proceedings, one of the first questions asked to experts about traces mainly concern their sources: "*Who left the trace*?" or "*What object is at the source of the trace*?" However, the answer to the questions "*Who*?" or "*What*?" does not always suffice. In fact, what if the suspect/s does/do not deny being on the scene but claim/s they were at the location for legitimate reasons and, perhaps, even at another time than that of the crime? In such cases, the answers to the questions "*How*?" and "*When*?" are of uppermost importance in order to reconstruct the past activities leading to the traces collected. Forensic scientists will recognize here the *hierarchy of propositions* structuring the information given by traces in three distinct levels, that is, *source*, *activity,* and *crime*. While questions related to the *source* have already been the topic of several research and publications, the issues related to the *activity*-level, including time-related questions, have only begun to be explored in depth more recently.

Of course, in our modern society, the questions related to the timeframe of past events could be answered by CCTV-cameras, messages left on social media platforms, connections of a cell phone to specific cell towers, etc. However, when such data are not available, physical traces ought to be used to this endeavour. Research on this matter has been conducted in different disciplines of forensic science, as for example gunshot residues, inks, blood stains, and fingermarks. In fact, fingermarks are usually very good at answering the "*Who*?" and, to some extent, the "*How*?" questions based on their position and orientation on an item, but fingermarks still represent a challenge regarding the "*When*?" question; the one relevant question no scientist can claim having the solution for yet. This is mostly due to the fact that fingermarks are characterized by physical (i.e., general pattern, minutiae, ridge flow), biological, and chemical (composition) parameters that may be modified/degraded over time and influenced by many different endogenous and exogenous factors. The complexity of fingermarks makes it very challenging to understand and model their aging and degradation processes. Certainly, this opens the possibility to numerous research approaches based on the measurement of many potential aging indicators which can be monitored using a variety of analytical techniques and statistical models. The abundance of possibilities explains the richness of research articles on the subject, each focusing on a different aspect of a fingermark's aging according to the specialty of research groups.

This book reflects the richness of the past and current literature/research, aiming at opening a discussion and providing possible answers and approaches to the topic of age estimation of fingermarks. It starts with an introductory chapter on the past, present, and future of the forensic use of fingermarks in order to set the framework of this book (Chap. 1); it continues with an overview into their complex structure, covering their physical, chemical, and microbiological composition (Chap. 2); and it follows with the relevance of estimating the age of fingermarks, highlighting the difficulty of this endeavour and proposing a formal approach to address it in a systematic manner (Chap. 3). After these general chapters, the book contains six chapters focused on the two main research approaches identified so far: on physical (visual) characteristics and on chemical and biological parameters. Results obtained using contactless technologies and digital methods are discussed, including 2D imaging approaches for topographical analyses; optical profilometry as a 3D imaging technology; analytical chemistry; chemical imaging; and electrochemistry, respectively, in Chaps. 4, 5, 6, 7, 8, and 9. As a complex structure, a fingermark may also contain other sources of information such as DNA, microbes, and blood traces, whose analysis and impact on the aging processes are discussed, respectively, in Chaps. 10, 11, and 12. Lastly, statistics applied to fingermarks are the object of Chap. 13.

This book does not contain any ready-to-use methodologies or technologies that can be immediately applied to routine casework, since such standards do not yet exist. The age estimation of fingermarks is a forensic topic of interest and concern, and, as such, only an interdisciplinary approach will allow to identify the most promising solution. This is the reason why a wide range of techniques and methodologies are presented, even if these processes are currently at different developmental stages – from academic research to pseudo-operational applications. This book allows the reader to take a critical look at the advancements already achieved and those to be aimed at. It represents a good basis for any forensic science students, forensic practitioners, forensic scientists, crime investigators, police officers, lawyers, and members of the justice system who wish to get a thorough overview on the topic. This book is a very relevant first step in combining the interdisciplinary knowledge needed to solve one of the most interesting forensic challenges of the twenty-first century.

Vienna, Austria Aline Girod-Frais

Acknowledgments

As editor of this book, I would like to deeply thank all of you who believed in this inspiring project and have helped to make it a reality. First, to the internationally renowned contributors and reviewers whose valuable work has been key in the realization of this publication. Also, to all those who indirectly have encouraged and challenged me for nearly a decade to research in the stimulating, and sometimes controversial, endeavor of investigating a reliable methodology that could help determine the age of fingermarks. Above all, I would like to mention the Catalonia Police – *Mossos d'Esquadra*, and especially Antoni Balaciart Muntaner, Lisard Idalgo, Cristina Mestres Patris, Lourdes Puigbarraca, and Joan Guinovart; at the University of Barcelona, Carme Barrot Feixat, Manel Gené Badia, and Jaume Turbany Oset; at the University of California–Los Angeles, Katherine A. Roberts; and at Arizona State University, Michelle Mancenido, Emmanuel Soignard, and Jennifer Broatch. A special thanks to all forensic scientists and professionals involved, including Donna Brandelli, Luke McGarr, and Karen Stow. Last but not least, to all undergraduate and graduate students who have been involved in this work, directly or indirectly. Thank you.

Chapter External Reviewers[1]

David Cunningham Department of Chemistry and Physics, Southeast Missouri State University, Cape Girardeau, MO, USA

Koby Kizzire Henry C. Lee College of Criminal Justice and Forensic Science, Forensic Science Department, University of New Haven, West Haven, CT, USA

Luke McGarr Forensic Focus Limited, Ripley

Kevin Parisi Forensic Science Examiner 1 – Latent Print Unit, CT DESPP Forensic Laboratory, Meriden, CT

John Reffner John Jay College of Criminal Justice, New York, NY, USA

Robert Ramotowski National Institute of Standards and Technology, Special Programs Office, Gaithersburg, MD, USA

Nikolas M. Stasulli Department of Biology and Environmental Sciences, University of New Haven, West Haven, CT, USA

Karen Stow Forensic Focus Limited, UK

Ryan Tully-Doyle Department of Mathematics, California Polytechnic State University, California, USA

Annemieke van Dam Department of Biomedical Engineering & Physics, Amsterdam UMC, Amsterdam, the Netherlands

David Zakon Los Angeles, CA, USA

Silvia Zoppis INPS – UOC 5, Rome, Italy

[1]The Editor and all contributors of this book acknowledge and appreciate the expert and independent revisions provided by (in alphabetical order).

Contents

Contributors[1]

Andy Bécue École des Sciences Criminelles, Faculty of Law, Criminal Justice and Public Administration, University of Lausanne, Lausanne, Switzerland

Aline Girod-Frais Crime Scene Unit, Criminal Intelligence Service (Bundeskriminalamt), Vienna, Austria

Department of Criminal Law and Criminology, University of Vienna, Vienna, Austria

Josep De Alcaraz-Fossoul Henry C. Lee College of Criminal Justice and Forensic Science, Forensic Science Department, University of New Haven, West Haven, CT, USA

Céline Weyermann École des Sciences Criminelles, University of Lausanne, Lausanne, Switzerland

Ronny Merkel Otto-von-Guericke-University Magdeburg, Magdeburg, Germany

Jennifer Chisholm Henry C. Lee College of Criminal Justice and Forensic Science, Forensic Science Department, University of New Haven, West Haven, CT, USA

Meredith A. Narowski Raleigh/Wake City-County Bureau of Identification, Raleigh, NC, USA

Amanda A. Frick École des Sciences Criminelles, University of Lausanne, Lausanne, Switzerland

Department of Applied Sciences, Northumbria University, Newcastle upon Tyne, UK

Ana Moraleda École des Sciences Criminelles, University of Lausanne, Lausanne, Switzerland

---

[1] This book has become a reality thanks to the combined effort of (by Chapter order).

Simona Francese Biomolecular Science Research Centre, Centre for Mass Spectrometry Imaging, Sheffield Hallam University, Sheffield, WB, UK

Robert Bradshaw Biomolecular Science Research Centre, Centre for Mass Spectrometry Imaging, Sheffield Hallam University, Sheffield, WB, UK

Roberto Rosa Department of Sciences and Methods for Engineering, University of Modena and Reggio Emilia, Reggio Emilia, Italy

Consuelo Mugoni Department of Engineering "Enzo Ferrari", University of Modena and Reggio Emilia, Modena, Italy

Massimiliano Bononi Department of Engineering "Enzo Ferrari", University of Modena and Reggio Emilia, Modena, Italy

Roberto Giovanardi Department of Engineering "Enzo Ferrari", University of Modena and Reggio Emilia, Modena, Italy

Sara C. Zapico International Forensic Research Institute, Chemistry and Biochemistry Department, Florida International University, Miami, FL, USA

Anthropology Department, Smithsonian Institution, National Museum of Natural History, Washington, DC, USA

Joe Adserias-Garriga Department of Applied Forensic Sciences, Mercyhurst University, Erie, PA, USA

Jesús L. Garcia-Gil Department of Biology, Campus Montilivi, University of Girona, Girona, Spain

Alexis R. Weber Department of Chemistry, University at Albany, State University of New York, Albany, NY, USA

Igor K. Lednev Department of Chemistry, University at Albany, State University of New York, Albany, NY, USA

Brooke Weinger Kammrath Henry C. Lee College of Criminal Justice and Forensic Science, Forensic Science Department, University of New Haven, West Haven, CT, USA

Henry C. Lee Institute of Forensic Science, West Haven, CT, UK

Chapter 1
Past, Present, and Future of the Forensic Use of Fingermarks

Aline Girod-Frais and Andy Bécue

Abstract The focus of this chapter is to describe the forensic uses of fingermarks and fingerprints over the years and their awaited evolution. To do this, past conceptions are combined with the current knowledge status to highlight the overall potential of fingermarks and friction ridge skin patterns. For each application, the main academic and operational challenges are described.

Therefore, this chapter is not aimed at providing a comprehensive literature review on the historical and technological developments regarding the exploitation of fingermarks and fingerprints. It is rather built to set the framework for the next book chapters, as well as to promote a fruitful and relevant future for research and operational applications.

Keywords Fingerprint · Ridge pattern · Composition · Identification · Detection · Profiling · Age

Acronyms and Definitions

ACE-V	Analysis, Comparison, Evaluation, and Verification (four-step process)
AFIS	Automated Fingerprint Identification System
Dactyloscopy	Etymologically, the word dactyloscopy is derived from two old Greek terms *dactylos* and *skopein*, meaning *finger* and *to consider*, *to observe*, or *to study*, respectively. Strictly speaking, dactyloscopy is the study/observation of friction ridges present on the surface of the fingers

A. Girod-Frais (✉)
Criminal Intelligence Service (Bundeskriminalamt), Crime Scene Unit, Vienna, Austria

Department of Criminal Law and Criminology, University of Vienna, Vienna, Austria
e-mail: aline.girod-frais@bmi.gv.at; aline.girod-frais@univie.ac.at

A. Bécue
École des Sciences Criminelles, Faculty of Law, Criminal Justice and Public Administration, University of Lausanne, Lausanne, Switzerland
e-mail: andy.becue@unil.ch

J. De Alcaraz-Fossoul (ed.), *Technologies for Fingermark Age Estimations: A Step Forward*, https://doi.org/10.1007/978-3-030-69337-4_1

	only and does not include those on the palms, feet soles, or toes. To refer to friction ridges independently from their location on the body, the term *lophoscopy* should be used. However, this term is not very popular, and so the word dactyloscopy is now commonly used to refer to the study of friction ridges in general, regardless of their origin. This general meaning of the word dactyloscopy is used in this chapter.
Detection	In this chapter, the term "detection" is used to describe the whole process aiming at making (latent) forensic marks visible so that they can be recorded. In the case of fingermark detection, it encompasses the application of dedicated techniques (e.g., reagents), the visualization of the results by optical means (e.g., coaxial episcopy, luminescence), or the application of an analytical process aiming at the same goal (e.g., chemical imaging). In some countries, such as the UK, the term "visualization" alone is used to describe the whole process.
DNA	Deoxyribonucleic acid
ENFSI	European Network of Forensic Science Institutes
Fingerprint/Fingermark	In this chapter, we use the convention applied by UK practitioners defining the residue of friction ridge recovered from crime scenes as a mark (i.e., *fingermark*) and that collected under controlled conditions from an individual as a print (i.e., *fingerprint*). As a consequence of this differentiation, fingermarks are more likely to be of a lower quality compared to fingerprints, which tend to be "… a near perfect representation of the friction ridge skin" if collected correctly [25]. Furthermore, for the sake of language homogeneity, we will use the English term *Fingerprint Officer* to designate practitioners engaged in the three existing comparison processes: (1) between fingerprints (print to print), (2) between fingermarks and fingerprints and inversely (mark to print/print to mark), and (3) between fingermarks (mark to mark). This chapter will mainly focus on the comparison and evaluation processes, as they are prone to controversies in the dedicated literature.
IAI	International Association for Identification
Identification/ Individualization	In this chapter, we use the term *identification* synonymously with *individualization*. It represents an opinion expressed by an examiner that a particular fingermark was left by the friction ridge skin of a particular person.

A Fingerprint Officer will thus generally provide an identification opinion "... when the mark shows sufficient quality (clarity) of friction ridges in agreement with the print so that the probability for such a match to happen if a print from another source is submitted is deemed to be impossible" [25]. Furthermore, it is important to add that there must not be any unexplained characteristics in disagreement between mark and print and that a certain quantity of characteristics has to be observed in some countries in order for Fingerprint Officers to provide an identification.

LR — Likelihood Ratio A likelihood ratio is a measure of the relative strength of support that particular findings give to one proposition against a stated alternative. It therefore provides a logically defensible way to deal with inferential reasoning [5]. LR is defined in terms of the ratio between two conditional probabilities: (i) the probability of the findings given that one proposition is true and given the conditioning information and (ii) the probability of the findings given that the other proposition is true and given the conditioning information.

SWGFAST — Scientific Working Group on Friction Ridge Analysis, Study, and Technology

1.1 Introduction

While the first use of friction ridge skin patterns as ways to prove legitimacy date back thousands of years [67], their usage by law enforcement for classification and identification purposes was not established internationally until the beginning of the twentieth century [12, 27] when the science of friction ridge skin patterns – generally called dactyloscopy (see section Acronyms and Definitions) – was born. At that time, the classification and comparison tasks were conducted manually and could take from a few minutes to many hours. While the classification of a fingerprint was a long but quite well-defined process, the comparison between a fingermark and a fingerprint left more room for interpretation by the specialist and did not necessarily lead to an identification. The notion of uncertainty was consequently an integral part of the dactyloscopy, as described by pioneers such as Locard in his tripartite rule [91]. However, over time, dactyloscopy mainly became a law enforcement tool, whose results had to be quickly obtained and associated with absolute certainty and hence generally reduced to the decisions of identification or exclusion.

This binary vision of the information potential of traces is not shared by the academic community defining forensic science as the science of the trace [118]. As stated by Margot [93]: "Forensic science is looking at the least likely, fragmented, imperfect, uncontrolled element in an event: the trace. It is the remnant (the memory) of the source and the activity that produced it. It has to be decoded and understood to elicit some knowledge about the event. The study of its relation to other traces as well as their environment provides many, and sometimes unsuspected, clues about the event and is a rich source of hypotheses to be tested as well as providing knowledge about reality." Behind this interpretation lay major principles linked to a trace: its incomplete and imperfect nature, its materiality, and the fact that once detected it bears information that can be measured, compared, or assessed. The traces left by friction ridge skin at crime scenes, i.e., the fingermarks, are thus incomplete and imperfect. The information provided by the comparison of fingermarks with fingerprints can be insufficient to achieve any certainty; the concept of probability is therefore crucial in the field of dactyloscopy. Furthermore, fingermarks are material. They are "made of something." They have a composition, and this composition bears information that can be very relevant for forensic purposes. The position and orientation of fingermarks in crime scenes and on items are also relevant parameters able to provide information linked to the reconstruction of the events and the significance of the marks present. Thus, forensically speaking, ridge details are only one source of information a fingermark contains, as depicted in Fig. 1.1.

The aim of this chapter is to introduce and describe the forensic use of fingermarks and place the major scientific developments and challenges in their respective contexts: research or operational application. The use of the morphological characteristics contained in the friction ridge skin patterns (Sect. 1.2) as well as the ways the chemical composition can be exploited (Sect. 1.3) will be covered. For readers interested in a thorough review of these topics, we refer to the latest dedicated Interpol reports [14, 15, 49].

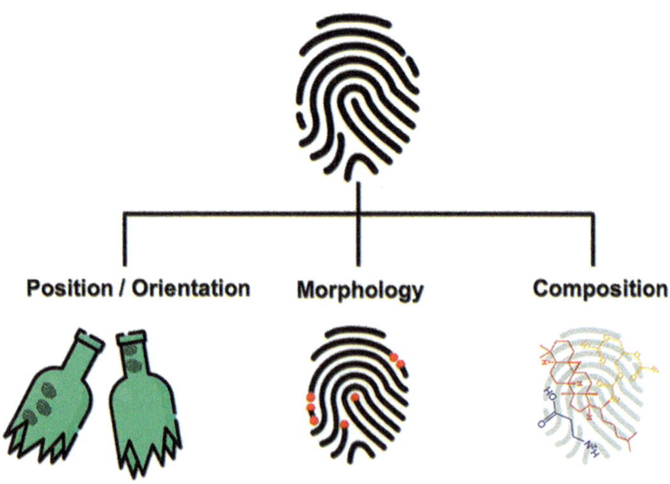

Fig. 1.1 Illustration of the different sources of information a fingermark contains when placed in its context. Fingermark icon from www.pngfind.com (last consultation: 19th of March 2020)

1.2 Morphological Characteristics

1.2.1 About the Past: Historical Milestones

As mentioned in the introduction, the morphological characteristics of fingerprints were already used more than a century ago for identification purposes linked with law enforcement duties. It is not our aim to propose a detailed historical overview of these developments, as the interested readers can find more information on this topic elsewhere [16, 25, 27, 74], but some elements of the past should be mentioned in order to understand the present and future challenges linked with the use of the morphological characteristics of friction ridge skin patterns in a forensic context.

William Herschel, working within the British Empire in Bengal (India), was most likely the first European to identify the potential of fingerprints for individualization purposes [74]. In the second half of the nineteenth century, while in charge of the Department for Registration of Deeds in the Hooghly District, Herschel encountered numerous cases of identity fraud (unrelated cadavers were brought to the British authorities and presented as relatives in order to obtain deceased persons rent). In that context, Herschel started studying fingerprints and discovered their stability over time and their extreme variability among people [68]. He introduced this method of individualization for civil and criminal matters but failed to extend its use to the whole of Bengal, India, or abroad, mostly due to a lack of support from his hierarchy. Shortly after Herschel, Henry Faulds, a medical doctor working within the British Empire in Indian and Japanese colonies, made similar observations concerning the individualization potential of fingerprints due to their variability among individuals. He was also one of the first to mention the potential of fingerprints to identify criminals [52]. At the end of the nineteenth century, Francis Galton became aware of Faulds' research and conducted his own large empirical test series to prove the immutability and the extreme variance of fingerprints based on statistical data. He also developed a classification system to categorize the ten-print cards (paper template used to record a person's fingerprints, mostly through inked fingers) and quickly retrieved them when searching a suspect. Furthermore, Galton also published the first detailed descriptions of minutia/ridge characteristics appearing in the fingerprints, in particular bifurcations and ridge endings [58]. Finally, Edward Henry, Inspector General of the Bengal District, came across Galton's book and decided to optimize his classification system to make it more efficient, with the aim of convincing European countries to use friction ridge skin patterns as standard classification and identification means. With the major contribution of two local police officers, Qazi Azizul Haque and Hem Chandra Bose, Henry proposed a new classification model in 1896. The so-called "Galton-Henry" system allowed the classification of several millions of ten-print cards. As a result, the British Empire abandoned the anthropometry system (Bertillon's identification methodology) in favor of dactyloscopy to classify and identify persons based on their fingerprints in 1901 [12].

However, the identification of fingermarks found on crime scenes by comparing them with fingerprints collected on ten-print cards was not yet widespread, even if the idea had already been mentioned by Faulds. The developments leading to the first documented identification of a criminal based on fingermarks collected from a crime scene are attributed to Juan Vucetich, who worked for a police department in Argentina. Vucetich, inspired by Galton's publication, created his own fingerprint classification method and exposed interested police inspectors in the potential of dactyloscopy. When the two children of Francisca Rojas were murdered in 1892, one of these inspectors, Alvarez, had the idea to compare one bloody fingermark found at the crime scene with the fingerprints of Francisca, which lead to her identification. Confronted by this evidence, Francisca admitted her guilt [12]. This first success highlighted the potential of the comparison of unknown fingermarks with known fingerprints for law enforcement purposes and emphasized the need for more research in this particular field of dactyloscopy.

In this context, French authors published different articles about the use of a defined number of corresponding minutiae or characteristics between a mark and a fingerprint in order for an identification to be established. Victor Balthazard, coroner and professor at the Faculty of Medicine, Paris, conducted his own statistical research on the morphological characteristics of friction ridge skin. In 1911, he suggested the use of 16 corresponding minutia, with this standard adjusted to 11 or 12 if the population of interest was reduced to France or Europe [11]. However, his conclusions were based on somewhat rather dubious assumptions. In 1912, Alphonse Bertillon, Police Clerk in Paris and creator of the anthropometry methodology applied to criminals, highlighted that the incomplete nature of fingermarks could hide potential differences in the marks when compared to fingerprints, thus leading to wrong decisions or outcomes [17]. To support his claims, Bertillon presented 2 friction ridge patterns (voluntarily partially masked) bearing 16 corresponding minutiae despite these 2 patterns originating from 2 different sources. Two years later, Edmond Locard, a medical doctor practicing legal medicine in Lyon, dedicated a large part of his work to dactyloscopy and, in 1914, described the comparison process between a fingermark and a reference fingerprint in the form of a tripartite rule [91]. The strength of Locard's publications lay in the consideration of a combination of quantitative and qualitative aspects for the identification process. The tripartite rule can be translated as follows, citing [27]:

> *Rule #1: If more than 12 concurring points are present and the fingermark is sharp, then the certainty of identity is beyond debate. (The imperative requirement for the absence of significant differences is implicit.)*

> *Rule #2: If 8 to 12 concurring points are involved, then the case is borderline, and the certainty of identity will depend on: (a) the sharpness of the fingermark, (b) the rarity of its type (general pattern), (c) the presence of the center of the figure (core) and the triangle (delta) in the exploitable part of the mark, (d) the presence of pores, (e) the perfect and obvious identity regarding the width of the papillary ridges and valleys, the direction of the lines, and the angular value of the bifurcations. In these instances, certainty can only be established following discussion of the case by at least two competent and experienced specialists.*

Rule #3: If a limited number of characteristic points are present, the fingermark cannot provide certainty for an identification, but only a corroborative element of a strength proportional to the number of points available and their clarity.

The first statement of this tripartite rule was (and still is) used by numerous police services to define a numerical standard which *ensures* the identification of an individual. In Europe, the most common numerical standard is set on 12 points, generally meaning that when fewer than 12 corresponding minutiae are found between a mark and a print, the comparison is to be considered of no value for identification purposes. However, this interpretation deviates from Locard's tripartite rule. Indeed, in the first statement, the quality (i.e., sharpness) of the mark is explicitly cited, implying that an individualization conclusion should not be reached based on a number of corresponding minutiae solely. In the second statement, it is clearly stated that an individualization conclusion is possible with 8–12 corresponding minutiae, depending on different factors – of rather qualitative nature – and based on a discussion between at least 2 experts. Finally, in the third statement, the certainty of source is replaced by corroborative elements of variable strength with fewer than eight corresponding minutiae. To summarize, Locard's tripartite rule should not be reduced to the observation of a specific number of minutiae to conclude with certainty about the source of a fingermark (i.e., identification or exclusion). It simply indicated how many corresponding minutiae and which qualitative parameters had to be observed to consider an identification as certain (statements 1 and 2) or probable (statement 3). Locard's tripartite rule actually offered the Fingerprint Officers quite a lot of leeway to reach nuanced and probabilistic decisions based on quantitative and qualitative parameters.

Among others, the publications of Balthazard, Bertillon, and Locard gave rise to a debate about the minimum number of corresponding minutiae required to conclude that a fingermark and a fingerprint originated from the same source, particularly in Europe. However, due to the outbreak of the First and Second World Wars, scientific research on this subject stagnated. Therefore, it was not until the late 1960s that scientific research was conducted on the relevance of a numerical standard for identification decisions in the field of dactyloscopy. In this context, the IAI organization was set up in 1971 with the aim to evaluate the relevance of a numerical standard. This resulted in the publication of a report in 1973 [75] concluding that "No valid basis exists at this time for requiring that a pre-determined minimum number of friction ridge characteristics must be present in two impressions in order to establish positive identification." In 1995, this declaration was slightly updated by the 28 participants (representing 11 countries) of the International Symposium on Fingerprint Detection and Identification, organized in Ne'urim (Israel) which stated: "No scientific basis exists for requiring that a pre-determined minimum number of friction ridge features must be present in two impressions in order to establish a positive identification" [94]. This update was mainly influenced by the study of Evett and William regarding the relevance of the 16-point Standard in England and Wales [50]. This study allowed experts to realize that "… point counts do not make a lot of sense," as this count is highly dependent on the fingerprint experts' personal opinions. For the authors, training, certification testing, performance testing of experts, file audits, and blind trials are the keys to quality results in the field of dactyloscopy, and not "…a rigid numerical points count." Furthermore, the empirical studies conducted by Herschel, Faulds, Galton, Balthazard, or even Locard cannot support the 12-point standards, as they were all based on a data set far too small to allow generalizations. Moreover, most of the studies principally considered fingerprints (i.e., nearly perfect impressions) and not fingermarks (i.e., imperfect and partial trace elements). Thus, a numerical standard used as a generic rule for the three types of comparison conducted in the field of dactyloscopy (print-print, mark-print/print-mark, and mark-mark) cannot be justified.

Based on the IAI and Ne'urim decisions, the so-called holistic approach emerged in the field of dactyloscopy [69–72]. This standard is based on the three levels of information that can be extracted from fingermarks:

- Level I: General patterns (ridge flow), characterized by their ridge form, direction, and position
- Level II: Minutiae, characterized by their type, form, and position as constellations
- Level III: Pores and other specificities, characterized by their form and position

Therefore, the holistic approach considers the number of minutiae in combination with qualitative parameters. Fingerprint Officers following a holistic approach must decide whether their observations are sufficient to provide an identification, based on a detailed documentation of the ACE-V process [7]. The holistic approach shares commonality with Locard's tripartite rule. However, among the fingerprint expert's community, practitioners criticized the lack of clarity of this approach and the difficulty in communicating the results to the court. Furthermore, it was (and still is) considered that the personal opinion of Fingerprint Officers takes far too much importance in the holistic approach, which could lead to wrong decisions and undermine the confidence of the judiciary in dactyloscopy (personal communications from different fingerprint experts in FR, CH, AUT, and DE). Nonetheless, some countries, such as the USA and Canada, adopted this standard very soon after the publication of the IAI declaration.

Besides the research dedicated to identification processes, in the 1950s researchers started to collect statistical data about the morphological characteristics of friction ridge skin patterns (e.g., frequency of appearance of general patterns or minutia in a population) [64, 108, 122]. These first models were mostly based on data sets which were too small and included oversimplified assumptions concerning the interdependency of the studied parameters. In the 1980s, researchers started to consider the fact that the distribution of minutiae among the friction ridge skin patterns was not even and that some areas (e.g., core or delta) or patterns (e.g., whorls or double loops) were particularly rich in minutiae [89, 116, 130].

In summary, towards the end of the twentieth century, while the forensic literature was starting to call for more scientific and transparent knowledge to justify decisions made in the field of dactyloscopy, two main identification standards coexisted: the numerical standard, based mainly on the statistical data obtained in the early twentieth century, and the holistic approach, promoted by the IAI declaration. Each system had its supporters and its detractors.

1.2.2 About the Present: Identification Standards

Several European countries still use a numerical standard with either little or no room for interpretation to make decisions following dactyloscopic comparisons: Albania, Austria, Bulgaria, Belgium, France, Germany, Greece, Italy, Poland, Portugal, Romania, Slovenia, Spain, and Czech Republic, with nuanced application schemes. Indeed, some countries with their standard defined at 12 corresponding minutia will not declare a fingerprint as a potential source of a fingermark if fewer than 12 corresponding minutiae are found (a questionable interpretation of Locard's first rule), while other countries still allow identification with 8–12 corresponding minutia based on qualitative aspects (a rather correct interpretation of Locard's first and second rules). Furthermore, the use of different numerical standards has also been reported by other countries, for example, 8–10 (based on different interpretations of Locard's rule) and even 16–17 (based on Balthazard's work and an incorrect interpretation of Bertillon's publication) [27]. This disparity in approach illustrates the controversy surrounding the so-called numerical standard. Additionally, given that countries are free to define their own national standards, the outcome of the comparison process between a given fingermark and a fingerprint may drastically differ, with outcomes ranging from exclusion to identification. This raises concerns with regard to the ground truth, which remains the same for a given comparison (being from the same source or not).

Furthermore, the use of a numerical standard is often associated with the risk of focusing on a number of minutiae to be reached in order to be able to identify a mark with a print, without considering other relevant quantitative data (e.g., frequency of appearance of minutiae constellations) or qualitative characteristics (e.g., sharpness, ridge shape, presence of pores). All quantitative and qualitative information evident in marks and prints should be considered during dactyloscopic comparisons. Otherwise, the process could result in wrong decisions [25]. Different studies conducted in Europe also showed that numerical standards influenced how many corresponding minutiae were found by Fingerprint Officers. Studies performed among fingerprint experts emphasized the fact that no participant (or a limited number of them) reported *n-1* minutiae when referring to a numerical standard set to *n* minutiae [27, 50] (personal communications based on internal studies conducted in different police units). In addition, large-scale studies conducted on American Fingerprint Officers highlighted a large variability with regard to the number of annotated minutiae between experts (i.e., inter-variability) and even for the same expert on different days (i.e., intra-variability) [81–83]. As a result, an increasing number of forensic scientists denounced the false premise of certainty purported by a numerical standard and called on academics and practitioners for more transparent and scientifically correct workflows [26, 50, 119].

Due to the increasing awareness of the limits of a numerical standard, some countries joined the USA and Canada by adopting a holistic approach (i.e., England and Wales, Scotland, Australia and New Zealand, Switzerland, and the Scandinavian countries) [27], thus leading to more publications on processes to formalize the holistic approach, such as the quantity-quality graphic proposed by SWGFAST in 2013 [123]. Often, the adoption of such an approach by Fingerprint Officers is closely linked to a general awareness by the Justice Systems of the need to develop documented strategies based on transparent and robust scientific workflows. In fact, as the forensic community had started to question the relevance of the numerical standard and appealed for further developments in the field, the judiciary also started to ask for more comprehensive information on the rules governing the field of dactyloscopy, especially in adversarial systems. As a result, the National Academy of Sciences published a report in 2009 in which the authors criticized the practices linked to some forensic traces (aka NAS report) [101].

While the fingerprint community struggled to find appropriate answers to the expressed criticisms and discipline gaps concerning the numerical and holistic standards, the rapid development in the field of genetics provided assistance. In the early 2000s, the routine use of DNA analyses for forensic purposes was introduced. This emerging technology, particularly the methods used to interpret the acquired data, inspired some within the fingerprint community. Even during the introduction of forensic DNA analysis, calculations were conducted to statistically assess the rarity of the obtained DNA profiles. These calculations were based on the collection of objective data related to allele frequencies in different populations. The potential of such statistics and probability in the field of dactyloscopy was therefore recognized by numerous forensic scientists and research groups focused on this promising endeavor [27]. These studies started the development of the third standard in the field of dactyloscopy, the probabilistic standard.

One of the main research directions regarding the application of statistical data to dactyloscopic comparisons has been inspired by the pioneering works mentioned in the previous section. This can be defined as "feature-based," as the researchers are collecting data about the random correspondence of feature configurations (note: the term features encompasses level I, II, and/or III characteristics). Researchers attempted to tackle the drawbacks of the first studies (i.e., inadequate data sets and erroneous assumptions concerning the interdependency of morphological characteristics) by considering larger data sets of fingerprints and fingermarks to assess the frequency of incidence of level I (i.e., ridge flow) [3, 27, 38, 65], level II (e.g., minutia) [24, 28, 29, 46, 65, 66, 102, 103], or level III (e.g., pores) [6, 117] characteristics.

The most recent among these studies, performed by de Jongh et al. on level I patterns [38], is worth mentioning. Based on a collection of 24,104 fingerprints, the authors observed the same trend as in earlier studies with approximately 5% of arches (mainly on index/forefingers), 60–70% of loops (mainly on middle and little fingers), and 25–35% of whorls (mainly on thumbs). This study goes further by proposing to divide the classical pattern types into 35 subclasses: 8 for arches, 8 for loops, and 19 for whorls. The frequencies of 35 different patterns were obtained and are currently used by Dutch Fingerprint Officers as a basis to assign evidential weight to fingerprint general patterns when they are visible in marks. This study also reported the development of a software tool called Pattern Encoding Tool (PET) to classify questioned marks in the 35 general pattern classes. Furthermore, the relationship that can be observed between some general patterns and the fingers was also highlighted, particularly for arch patterns, which are mainly present on the index/fore and middle fingers. To conclude, the authors emphasize that the data were collected to support the work of fingerprint experts using a probabilistic approach but could also support the holistic process by providing experts with objective data to evaluate their observations.

The frequency of level II characteristics was the subject of a study by Gutierrez-Redomero et al. conducted on 2000 fingerprints [65]. The authors statistically measured the distribution of 20 types of minutiae in 8 different sectors of the fingerprints (i.e., 4 sectors around the core and 4 others by the edges) according to the pattern types (e.g., arches, loops, and whorls) and associated fingers. It was observed that 55–65% of minutiae were ridge endings, 13–18% bifurcations or convergences, and 0.1–2% other types (e.g., points, crossbars, bridges, or opposed bifurcations, as referred to by the authors). Furthermore, the authors observed a greater frequency of ridge endings on the thumbs and index/forefingers in the periphery of the fingertips. Bifurcations and convergences were principally observed in the middle, ring, and little fingers, in the central area of the fingertips and in the distal sectors. Differences between gender based on general pattern and ridge density were also observed in this study. The authors concluded that in the process of comparison between fingermarks and fingerprints using morphological characteristics, the different discriminatory value of the minutiae has to be taken into account, as follows "[…] features will be less indicative if they are more frequent in the population […]." They also highlighted that "[…] a numeric standard that assigns the same value of evidence for purposes of identification to each of these minutiae should not be used in the expert evidence as the sole criterion […]" [65].

The latest developments proposing models that consider the distribution of morphological characteristics as a function of the general patterns were achieved during the last decade. Numerous efforts were made to develop statistical models able to account for distortion and clarity issues of the general pattern and to capture the spatial relationship between minutiae. Such works offered solutions which have already been applied in practice [1, 104]. However, full validation processes and developments towards user-friendly application still need to be conducted.

In addition to these statistical studies, actual research also aims at assessing the result of a dactyloscopic comparison from the scores or similarity measurements provided, for example, by an AFIS algorithm. Given that numerous law enforcement agencies rely on AFIS, such an approach would probably benefit from an easier acceptance by the practitioners. The studies and models corresponding to this category can be defined as "similarity metric models," as they calculate the probability of the observations based on the distributions of similarity scores.

Swofford et al. proposed a review on this topic and also described the development of an evaluation tool called *FRStat* [133]. This software is able to calculate the similarity between two configurations of minutiae on friction ridge skin patterns and the probability of occurrence of this similarity metric (called GSS) when the two configurations come from a common source or from different sources. This software is currently used by the US Department of Defense. The authors of the study also offer the possibility for people interested to test this software by contacting them.

Numerous studies were also conducted on an AFIS score-based likelihood ratio [47, 48, 85, 86, 98, 113, 114]. Among other results, a software combining a user interface called PiAnoS (Picture Annotation System) and an AFIS-based calculator (called XENA) was developed by the École des Sciences Criminelles of the University of Lausanne. This software, from a technical point of view, offers a way to document the analysis, comparison, and evaluation steps of the ACE-V methodology. It also contains a likelihood ratio (LR) calculation tool [131]. To do this, the software relies on the Swiss Fingerprints Database for the years 1930–1940 containing 963,710 fingerprints and considers about 500–700 distortion possibilities. After the analysis of the questioned mark by an expert (including minutia annotation), the mark is compared with the print potentially at its source in order to obtain a score based on the annotated minutia constellation, similarly to the process taking place in an AFIS search. This minutia constellation is then compared to the 963,710 fingerprints available in the database to obtain the distribution of scores when mark and print are not from the same source. The distribution of scores when the mark and the print were left by the same finger is obtained by comparing the minutia constellation annotated on the questioned mark with its distorted self (applying distortion models). Based on these data, a LR can be calculated to evaluate two alternative hypotheses: the questioned mark was left by the same source as the print (hypothesis 1) or by a different source (hypothesis 2). It is expected that the integration of such a calculation tool in an AFIS process would encourage its use by practitioners.

In 2015, the ENFSI Fingerprint Working Group included this approach in its Best Practice Manual [49]. Thus, three approaches are currently accepted for the reporting of fingerprint evidence among ENFSI: the numerical, the holistic, and the probabilistic approaches. These three methodologies also co-exist in international practice, with countries currently using either a numerical standard (most of the European countries), a holistic approach (USA, Canada, England/Wales, and Australia), or a probabilistic process (some Swiss State Police, the Netherlands Forensic Institute, and some US forensic laboratories). The latter approach, however, has currently been adopted by the fewest number of forensic laboratories (Fig. 1.2).

Based on the most recent literature and in our opinion, of the three approaches, the probabilistic approach is the only one which guarantees a scientifically balanced and transparent consideration of the morphological characteristics of friction ridge skin patterns. Technological tools allowing the calculation of LRs to weight two alternative hypotheses are already available. However, these tools are not perfectly adapted to the routine use of major police institutions, and the majority of Fingerprint Officers and police bureaus/labs in Europe still rely on a numerical standard, mainly

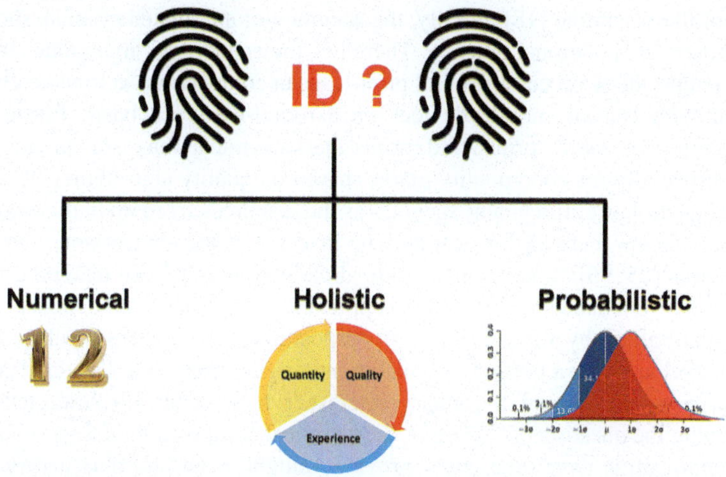

Fig. 1.2 Illustration of the three identification standards currently used in practice. Fingermark icon from www.pngfind.com (last consultation: 19th of March 2020)

because of its ease of use and the false security provided by the notion of *certainty*. Statistics and probabilities are mainly considered as useful for academic purposes but not for operational work, and probabilistic models are currently mainly not compatible with police structure. Thus, with regard to identification standards in the field of dactyloscopy, we could say that a debate is underway between the conception praised by the scientific literature and the position defended by numerous practitioners.

1.2.3 About the Future: Challenges

A very well-known way of structuring the questions emerging during forensic processes relies on what is known as the "Hierarchy of propositions" [31]. It describes the different levels of information that the analysis, comparison, and evaluation of marks produce. These levels are (1) the Source (i.e., who, what), (2) the Activity (e.g., how, when, where), and (3) the Crime.

Based on this hierarchy, the different identification standards described in the previous section concerned the Source level. The main challenge for the future of dactyloscopy at this level mainly lies, in our opinion, in the development of probabilistic models dedicated to dactyloscopic comparison and useable by any practitioners working in the field for routine applications. While these developments could fall within the remit of the academic community, the practitioners and judiciary should start considering probabilistic models as a valuable solution for operational practice. The development of international reporting standards governing dactyloscopic comparison and based on evaluative reasoning should also be part of the future of the discipline [18]. Finally, the accuracy of automated systems should be further used in operational practice. There lies the concept of "lights-out" systems. Such systems allow processing of all print-to-print comparisons automatically, even coupled with various mobile devices on a round-the-clock basis. Furthermore, applications for marks-prints and prints-marks comparisons are already being tested, especially for good quality marks (based on quality algorithms). This could clearly benefit law enforcement agencies as the resources freed up in this way (budget and personnel) could, for example, be invested in the comparison of complex fingermarks [25, 96]. Some countries are already more advanced than others in the use of these systems, but a generalized application is still far from being achieved.

As highlighted by the hierarchy of propositions, the questions related to the Source level are only a portion of what an investigator may ask. Time and Activity level aspects are also of a primary interest, which opens up new challenges. Concerning the question of "When?", the chemical composition of fingermarks and their modification over time could provide valuable answers, thus justifying the particular efforts set on this topic by the forensic community (addressed in more detail in the second part of this chapter). Furthermore, as mentioned by Champod and Chamberlain [25]: "an area of particular interest in the detection of criminal activities is the interpretation of mark placement." Such information can indeed help to answer the questions related to the Activity level and hence help reconstruct what happened at a crime scene [41]. Research efforts should be focused on helping crime scene investigators and other forensic specialists to correctly note, evaluate, and interpret such elements.

The last level of the hierarchy of propositions concerns the Crime question, which is the question that should be answered by the court. However, in order for the judiciary to be able to correctly assess the information provided by fingermarks, the forensic community has to clearly demonstrate and explain the methodologies used and to provide transparent documentation. The communication between Fingerprint Officers and the judiciary must be further developed and include appropriate training for both communities.

Finally, a very significant challenge in the field of forensic science in general, including dactyloscopy, is the gap between the scientific developments and their adoption by practitioners. The priorities currently set in operational conditions in the field of dactyloscopy aim mostly to accelerate the processing of comparisons in order to provide certainty of identity or exclusion as quickly as possible. This aim is being fulfilled for print-to-print comparisons such as those conducted for security purposes at the borders or in custody suites to control whether a person has already been registered in a database with "lights-out" systems in some countries. However, this endeavor is not realistic when talking about low-quality prints or fingermarks as these require manual intervention. Thus, practitioners have to be trained to make this distinction and prepared to process comparisons differently based on the quality of the material to be compared. This implies the use of an ACE-V process including a thorough documentation of the quality of fingerprints or fingermarks in order to be able to decide how the comparison and evaluation should be conducted ("easy" or "complex" case). These accurate and defined working processes may be followed in some countries but still need expanding in others. Furthermore, the conception that dactyloscopy is a tool for law enforcement officers leading to certain results and being enclosed in a police environment has to evolve. Dactyloscopy is a scientific endeavor and, as such, should evolve with the latest scientific developments. Practitioners and the academic community should therefore communicate and work together to ensure that new practices, if relevant, are fully validated and implemented as soon as possible, if proved to be of probative value. The judiciary community would benefit from such developments. Such a process may take place by introducing national expert networks or by establishing effective communication channels between the academic community, the fingerprint practitioners, and the judiciary.

1.3 Chemical Composition

In addition to the ridge pattern, the chemical composition of fingermarks is also of primary interest in a forensic context. Well before the development of advanced analytical means, general knowledge about the secretion (residue) composition became the basis for the development of the first detection techniques. In the last few decades, the forensic community has witnessed an increase in the amount of data that can be gathered from fingermark composition and in the number of studies in that field, principally due to the development of numerous advanced analytical techniques. It is crucial to clarify what lies behind the "composition of fingermarks" and for which purposes it can be used, to comprehend the usefulness of fingermark composition for forensic science.

First, it is crucial to understand that the molecular composition of the secretion residue (i.e., endogenous, exogenous, and semi-exogenous compounds) is only part of the chemistry of fingermarks. Indeed, a fingermark is one element of a complex system that encompasses its environment and the events having led to its creation. In addition to the molecular composition, other relevant characteristics need to be covered and understood: the structure and physicochemical properties of the secretion residue (i.e., emulsion), the sources of variability (e.g., donors, pressure, environmental conditions), the interactions with the underlying substrates (e.g., adsorption, absorption, and diffusion mechanisms), the evolution of composition over time, and the modifications caused by external factors post-deposition (e.g., light, wind, temperature, humidity, detection techniques) [60]. While our understanding of the chemistry linked to fingermarks is continuously increasing, the interaction mechanisms with their surroundings and their evolution over time remain poorly understood.

Three main outcomes derived from the chemistry of fingermarks can be identified (Fig. 1.3): (1) improving the detection capabilities, (2) estimating the age of a fingermark, and (3) providing information about the donor, *aka* donor profiling. The detection of fingermarks has been carried out for more than a century; age estimation and donor profiling are more recent challenges.

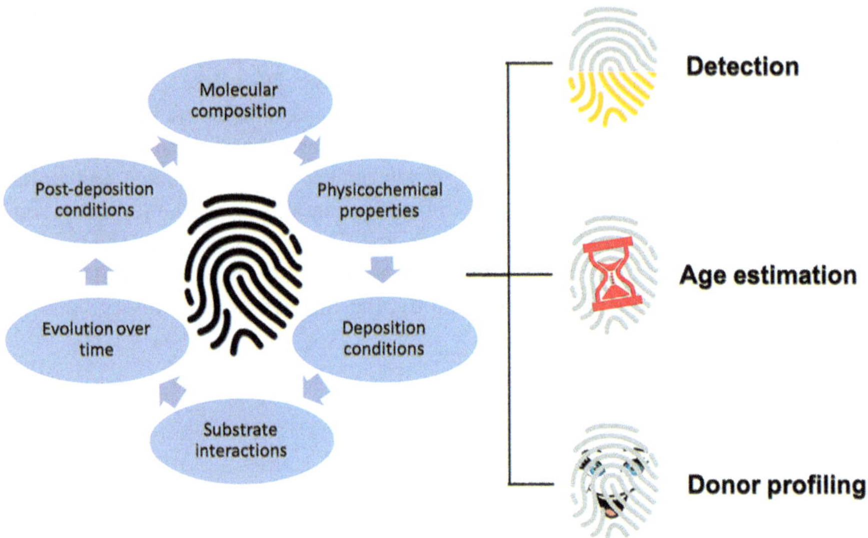

Fig. 1.3 Illustration of the complexity of a (latent) fingermark when placed in its surrounding and of the three goals derived from its composition. Fingermark icon from www.pngfind.com (last consultation: 19th of March 2020)

1.3.1 Detection Capabilities

As a preliminary remark, it is important to state that this section sets generalities with regard to fingermark detection. It aims to describe the most common practices and to define the relationship between fingermark composition and detection strategies. The authors are aware that each case is specific and that many exceptions may apply. It is not the aim of this section to cover the topic in an extensive way.

Overall, fingermark detection strategies rely on two fundamental principles:

- Detecting as many (relevant) marks as possible on a given item or on a surface of interest
- Generating a recordable contrast between the marks and the underlying substrate

With regard to the first principle, the most common method relies on targeting different molecular species contained in the fingermarks of interest. To achieve this goal, different fingermark detection techniques are usually combined and applied sequentially on an item. A typical sequential processing starts with optical methods, followed by techniques that target the most fragile fraction (e.g., water-soluble fraction) and eventually those interacting with the most robust one (i.e., non-water-soluble fraction). Sequential processing relies mostly on the empirical characterization of the substrate of interest to associate it with a common class (e.g., porous, non-porous, semi-porous, adhesives, to cite a few) [27, 73]. In some cases, knowledge of the mark contaminant can help refine the choice of sequence used, for example, by introducing specific techniques (e.g., blood stains). In other cases, knowledge of environmental conditions or other intelligence information about the substrate and/or the marks can prevent the application of unsuitable techniques (e.g., detrimental event, exposure to contaminants or to outdoor conditions). In addition, the recovery of other forensic traces (e.g., DNA, ink, footwear marks) may compete with the detection of fingermarks, some detection processes in a sequence adversely affecting these traces due to the use of solvents, heat, or moisture.

It is the responsibility either of the crime scene investigator or of the forensic expert to know how and when these traces have to be collected in an overall evidence recovery sequence including these other evidence types as well as fingermarks. This is still the common practice, and it will most certainly remain as such for the future as it is quick, low cost, and independent of the applied detection means. The weakest point of this practice, however, is the characterization of the surface which relies mostly on empirical observations and on the experience and competence of the expert rather than on actual surface characterization. Therefore, issues may emerge with some substrates. For example, plastics are characterized by a wide range of chemical compositions and properties. They may be processed by following a standard "non-porous substrate" sequence by a laboratory, due to a lack of knowledge about their chemical composition or about existing specific subsequential processes. Consequently, the processing may not be optimal and yield poor results. This is another proof of how the knowledge about the substrate composition and the interaction with secretion residue is as important as the chemical composition of fingermarks.

With regard to the second principle, all detection techniques rely on optical, physical, or (physico) chemical properties of the mark and its substrate to generate a recordable contrast. Except for the optical observation of latent (unprocessed) marks, most of the current detection techniques involve the application of a reagent to the substrate (being under atomic form, powder, solution, or fumes). This is followed by the optical recording of the contrast that has been generated between the marks and the underlying substrate. Dozens of techniques following that principle are commonly used worldwide by forensic laboratories: 1,2-indanedione, ninhydrin, cyanoacrylate fumes combined with dye staining, physical developer, vacuum metal deposition, dry powders and powder suspensions, and blood protein dyes, to cite the most common ones [27, 73].

More recently, the combination of analytical techniques with visualization capabilities, known as chemical imaging, has opened up new possibilities [10, 13, 54]. Chemical imaging is a false-color reconstruction depicting the qualitative and quantitative distribution of a specific molecule (or a range of chemical species) across a scanned area. It can be applied to fingermark detection if the molecular composition differs sufficiently between the secretion residue and the underlying substrate, with regard to the molecule(s) of interest. Many studies based on chemical imaging are preliminary or pilots, and only few have thoroughly assessed the possibility of introducing chemical imaging-based techniques into an operational workflow [54, 84] or reported its use on casework-related items [21]. For this reason, the operational application of chemical imaging to casework currently remains anecdotal, mostly because of the need for advanced analytical equipment, experienced users, and the technical impossibility to quickly scan a large area. It is unlikely chemical imaging will ever be used for routine detection purposes but could be considered in uncommon cases or to assist in better understanding the chemistry behind fingermarks (e.g., structure of the emulsion and spatial distribution of chemical constituents along the ridges) (discussed in detail in Chap. 8).

As previously stated, fingermark detection strategies are closely bound to fingermark composition. In this context, several publications describe the average composition of endogenous compounds excreted from eccrine, sebaceous, and apocrine glands [60]. Exogenous and semi-exogenous compounds must also be added to these [54]. Exogenous compounds are present on the fingertips by contact with a foreign substance/contaminant (e.g., narcotics, explosives, blood, grease). Semi-exogenous compounds are directly related to the lifestyle of the individual (e.g., diet, medication, smoking habits) and encompass mostly the metabolites secreted by the body conjointly with the common endogenous compounds. Due to the evolution of analytical devices, it is now quite easy to gather information about the exact composition of fingermarks in a controlled environment, as witnessed by the numerous publications in this field. While each of the chemicals contained in fingermarks could theoretically be targeted or imaged for detection purposes, it is much more relevant to consider the average composition of secretion residue and to define global molecular fractions rather than knowing the exact composition of each fingermark under examination (which is technically and fundamentally impossible given that most of these marks are latent and yet to be visualized). Consequently, current detection strategies rely on targeting broad molecular fractions, mostly amino acids, lipids, non-water-soluble fraction, and major contaminants (such as blood) – if relevant. This is simply an outcome of the detection objective, which is to locate as many marks as possible on an item. Therefore, an effective sequential process should combine detection techniques targeting two or three different molecular fractions, which is usually the case. This also explains why effective detection techniques have been developed well before the availability of analytical devices and detailed compositional information. As examples, we could cite:

- The development of dry powders early in the twentieth century, which relied on the knowledge that secretion residue contains lipids and a certain amount of moisture [127]
- The use of silver nitrate, targeting the chloride ions contained in human sweat [8]
- The transition of ninhydrin from biology to forensic science, which has been made possible by the awareness that human sweat contains amino acids [106]

With regard to the endogenous secretions, qualitative and quantitative variations are observed between individuals, as well as for a same individual over time, affecting the performance of the detection techniques [57]. This variability of composition however does not prevent the successful application of detection techniques for casework. On the other hand, if the molecular target continuously diffuses on the substrate over time, it will most likely cause a blurriness of the ridges as the mark ages. Hence, the detection techniques used to target these particular molecules will become obsolete operationally, as was the case for silver nitrate, for example.

The targeting of a specific molecule in fingermarks is mostly counterproductive, even if it is technically possible. For example, researchers demonstrated the possibility of targeting cotinine (a metabolite of nicotine) using functionalized nanoparticles and hence detecting the fingermarks of smokers [87] or the targeting of cannabis using Fast Blue B [142]. If it makes sense from a chemical perspective, it is often counterproductive from a forensic viewpoint due to the high selectivity of these processes, and therefore, such developments have not been adopted in the field. The only exception is the targeting of specific contextual contaminations such as blood, grease, soot, etc.

What are the future challenges related to fingermark detection?

- A better characterization of the emulsion behavior, including the spatial distribution of eccrine and sebaceous materials [44] combined with a better understanding of the interaction mechanisms with the underlying substrate over time [100, 112]. Such studies are crucial because they emphasize micron-scale phenomena that could help to improve the effectiveness of detection techniques: water-in-oil emulsion, migration of some components in the inter-ridge area post-deposition, absorption into the plasticizer layer, etc. On the same basis, some developments in surface engineering are also of interest for forensic scientists: resistance of secretion residue to cleaning [129], behavior of anti-fingerprint coatings [45, 53, 79], and visibility of secretion residue [92]. However, much is yet to be discovered with regard to the fundamentals linked to fingermark secretion residue and its relationship with detection effectiveness.
- A better knowledge of how fingermark components evolve over time (see Age Estimation section). For detection purposes, this may help to explain the variations of performance of detection techniques as the secretion residue ages [19, 120]. It may also provide information about relevant molecular targets, either those remaining stable over time or those degrading quickly into other species [55]. Such studies are particularly useful to verify or invalidate preconceptions that emerged from practice, for example, the fact that a powdered mark of good quality has to be fresh. This has been proven false, as aged marks can show good levels of clarity upon detection [23, 95].
- A better understanding of the detection mechanisms involving the non-water-soluble fraction (i.e., physical developer and single/multi-metal deposition) [39, 40, 99]. This is crucial in order to adapt or improve a technique when it performs worse than expected, as it has been shown that the performance can be linked to specific elements such as the chemical reagent provider or the water quality.
- The consideration of artificial secretions as a complement to natural fingermarks for research and proficiency testing purposes. This requires rethinking entirely the use of artificial secretions to go beyond the uses of commercial artificial secretion pads [128]. Indeed, past studies have shown that such pads are not adapted to the research linked to fingermark detection [141], mostly because of simplistic and unrealistic compositions resulting in erratic behavior with detection techniques. The publication of complex emulsions reacting normally with

conventional detection techniques opens new perspective with regard to artificial secretions [125].

- A stronger interaction/collaboration between crime scene/lab practitioners and researchers (e.g., forensic scientists, chemists, biologists). Far too often, publications describing a *revolutionary* detection technique are not followed by any additional experimentation/validation nor adopted by the field practitioners. In most cases, the reason is a lack of forensic awareness linked to fingermarks (their complexity and the limitations that are induced by the investigation) or a lack of consideration of the investigators' actual needs or constraints.

To summarize, the composition of fingermarks plays a major role in detection capability. However, this is coupled with additional mechanisms such as the interaction with the underlying substrate which has an impact on the effectiveness of detection techniques. Therefore, it is desirable that future developments focus on strengthening the foundations or basics: interaction of the fingermarks with the underlying substrates and understanding the complex detection mechanisms. The development of detection techniques targeting a specific range of molecules will most likely continue, but the possibility of new techniques replacing well-established ones is quite low as the overview of fingermark detection history showed.

1.3.2 Age Estimation

Age estimation aims to answer the question that crime investigators and judiciary may ask: "When was this fingermark left?" or "Is this fingermark actually linked to the case?". Some may be interested in getting accurate age estimation (in days, weeks, months, years), while others could be interested in obtaining an indication of anteriority or posteriority to a specific time reference. The interest, for crime investigators, is undeniable as it would help decide between two plausible scenarios or provide a timeline reference. However, answering this simple question constitutes one of the most difficult challenges with regard to the composition of fingermarks, mostly because the factors that influence this are numerous (e.g., donor, substrate, deposition pressure, relative humidity, environment) and the way they influence the fingermark composition remains unknown in a casework scenario, as opposed to the controlled conditions encountered in most of the published studies [59].

In that context, it is crucial to distinguish monitoring the evolution of secretion residue over time from the actual estimation of a fingermark age. The first step is currently quite easy to conduct, mostly due to the range of analytical devices available. It consists of monitoring the qualitative and quantitative evolution of a range of molecules contained in fingermarks as they are aging, mostly under controlled conditions. From such studies, it is possible to express conclusions about how molecular species degrade, appear, or remain constant over time. Age estimation requires designing a decision model encompassing the observations made during the first step and considering the fingermark in its context (e.g., range of substrates, exposure to external events). In that sense, building up an age estimation model is a goal significantly more complex and difficult to achieve. To evidence this, most of the available articles describe the evolution of compounds over time (e.g., lipids, proteins, water), whereas only a limited number actually present or validate an age estimation model. The confusion between both goals is often found in the literature, mostly because authors assume that an age model will be easily derived from data collected over time. Unfortunately, it is not, and some authors realize that eventually [105].

As age estimation is one of the main topics of the following chapters, we limit this section to different ways to monitor the evolution of fingermarks over time as described in the literature (non-exhaustive citations):

- Evolution of fingermark morphology and topology using optical means [32–37, 43, 77, 97, 100, 112]
- Evolution of the lipid fraction [55, 63], including a focus on unsaturated fatty acids [111], unsaturated triglycerides [110], and glycerides [56], mainly using mass spectrometry
- Evolution of fluorescent oxidation products using spectroscopic analyses [135]
- Evolution of proteins using high-resolution mass spectrometry-based proteomics approach [107]

From the list above, it can be seen that optical, physical, physicochemical, and chemical characteristics were investigated. The outcomes of monitoring fingermark evolution over time provide a better understanding of how fingermarks interact with their surroundings and how detection techniques perform on aged secretion residue (see above). In the context of age estimation, such studies should help to identify aging markers (e.g., molecules, molecular ratios, kinetic ratios) so that an age estimation model can be built. The challenge actually lies in building up a model that is then validated to ensure it provides accurate and relevant information in realistic conditions. In that sense, very few studies successfully reached that goal:

- Model built on luminescent oxidation products [135]
- Model built on lipid molecular ratio [61]
- Model built on image contrast [97]

Caution should therefore be exercised with regard to age estimation as some published "models" are limited in their application range (e.g., certain types of substrate, specific detection techniques) or require additional studies to be carried out before being reliable.

So what are the future challenges related to fingermark age estimation?

- A systematic use of natural fingermarks before promoting any operational use of an age determination model. As mentioned above, fingermarks are very complex matrices, and this complexity increases with the passage of time. Therefore, even if artificial secretions could provide some answers in the preliminary stages of research (or to monitor the efficiency of a detection technique with aging secretions), it would be invalid to develop an age determination model built on non-natural fingermarks (i.e., artificial or sebum-rich).

- An increasing use of realistic conditions, including the consideration of various substrates or the integration of the model in a detection sequence. In addition, another primary objective should be that *the technique should be easily deployed at the scene or in a forensic lab*. Here it is important to highlight again the difference between research aiming at gaining fundamental knowledge on the detailed composition of fingermarks – this may indeed need the use of uncommon substrates, non-enhanced fingermarks, and advanced techniques – and research aimed at proposing an age estimation model. The latter needs to focus on methodologies and processes that could be used in operational environments and take the usual workflows into account (e.g., before the application of fingermark detection techniques); otherwise there is a risk of developing a method confined to niche cases.
- A monitoring of target compounds other than lipids, such as those found in eccrine sweat (e.g., amino acids, proteins). The degradation of these compounds is rather slow and could make possible the development of models able to date fingermarks over a longer timeframe (up to some years). Such extended timeframe may have limited impact in most cases, but it could help eliminate fingermarks on relevance if it is determined they are older than the investigated events [30].
- The consideration of a combination of aging parameters of different types, as, for example, morphological and chemical aspects. To the best of the authors' knowledge, the development of aging models based on mixed parameters has never been published but could lead to interesting results.
- The adoption of a research strategy encompassing the forensic context, highlighting the relevancy and application range of an age estimation model, and including realistic validation steps, as further detailed in Chap. 3.
- A detailed consideration of probabilistic approaches to model fingermark aging. Such approach would require taking into consideration numerous aging parameters (multivariable approach) and encompassing different influence factors such as temperature, humidity, light conditions, and the alleged activity at the crime scene. We think that aiming at evaluating the age of a fingermark based on one aging parameter alone is not feasible, and we see more potential in the multivariate probabilistic evaluation of two alternative time hypotheses.

To summarize, estimating the age of a fingermark constitutes certainly the biggest challenge, mostly because it encompasses many limitations: complexity of the fingermark composition, consideration of intra- and inter-variability, impact of the environmental conditions on the fingermarks over time (mostly unknown), interaction with the substrates, and impact of the detection techniques applied beforehand (to visualize the marks). All those factors are known to impact the composition of fingermarks and may prevent the application of an age estimation model or broaden the uncertainty range. As stated, a multivariate approach may represent a way to circumvent such difficulties and widen the analytical capabilities while considering the operational limitations.

1.3.3 Donor Profiling

Donor profiling aims to provide personal information about the individual who has left a specific fingermark, other than through the comparison with reference fingerprints. In this context, numerous papers describe the attempts to infer the gender of a donor from morphological information, such as the ridge width and ridge densities [2, 9, 42, 121, 126, 134], to cite the most recent ones. More recently, the ability to collect and analyze the genetic material that may be contained in fingermarks (aka "touch DNA") has constituted a useful source of complementary information (i.e., genetic profile and gender). The increasing sensitivity of DNA analyses has permitted this practice to become a conventional way for investigators to obtain additional information from fingermarks or from surfaces of interest that may have been touched [88]. This approach is not without limitations: DNA degradation caused by the detection techniques applied beforehand or by the optical methods used for the observation, DNA recovery success rates varying with the substrates, and contamination risks (e.g., contaminated surface or DNA transferred on someone else's hands before fingermarks are left), to cite a few. Most recently, inferring biological or lifestyle-related information (e.g., gender, ethnicity, medical treatment, donor age, smoking habits, contact with illicit substances) from the detailed composition of a fingermark has emerged in the scientific literature [20, 136, 140]. This requires the molecular information to be processed through decision models built on identified lifestyle markers, whose presence enables conclusions about an alleged activity or enables comparison with known endogenous, semi-exogenous, and exogenous compounds on databases. Similarly to touch DNA, the analysis of fingermark composition to infer donor profiling suffers from major limitations: impact of the detection process and contamination material brought by another individual. Profiling an individual from the molecular composition of their fingermark is the most recent research approach linked to fingermarks and certainly the most questionable from a forensic point of view with regard to its investigative applicability and usefulness.

So far, molecular donor profiling has been attempted through amino acid profiling [4, 22, 138–140], lipid profiling [124, 143], specific chromosome targeting [137], or bacterial profiling [109]. An omics approach (i.e., consideration of the secretion residue as a whole) has been used to infer the donors' lifestyle from fingermarks found on mobile phones [20]. While acquiring physical or behavioral information from fingermarks seems appealing, it must be clear that donor profiling suffers from several limitations: first of all, this field is emerging and most of the published studies are pilot studies that require follow-up research; second, the inferences that can be drawn are characterized by a high range of uncertainty, hence impacting the strength of the information provided to the investigators; and third, databases containing numerous exogenous products encountered in our daily environment are required (in most cases, the development and updating of such databases is unrealistic). Finally, if we consider that a fingermark has first to be detected before being chemically analyzed, it is also expected that the detection techniques applied beforehand would dramatically impact the ability to perform any profiling from the modified composition. Studies addressing this question are consequently awaited in the literature, as it is also the case for touch DNA [90, 115, 132]. With regard to illicit substances found in secretion residue (due to contact or consumption), it has been shown that fingermark detection techniques applied beforehand have a detrimental impact and should be limited to the minimum required to detect the fingermarks of interest [80]. Alternative protocols were also proposed, involving the lifting of fingermarks on specific medium before their analysis [139]. The compatibility of such approaches with conventional investigative practices must be assessed as well. The conclusions of the first studies addressing the implementation of molecular donor profiling in a casework environment emphasize the proposition that compromises have to be made between the application of detection strategies and the use of molecular composition for other purposes. Such choice is closely linked to the context of the investigated case, the questions to answer, and the limits of each approach (e.g., impact of contaminations present on a given surface prior to fingermark deposition).

What are the future challenges related to donor profiling from fingermarks?

- For investigatory purposes, donor profiling is unlikely to be relevant if it is proposed as a way to provide investigators with lifestyle information and hoping it will help to find a suspect among the population. Indeed, so far, we do not live in a dystopian society in which information about the lifestyle of citizens is collected and centralized. The only exceptions are the genetic profile and the gender, as they both constitute strong investigative information that can be readily compared to existing databases. However, molecular donor profiling mostly suffers from the need to refer to databases of lifestyle markers and from high uncertainty degrees compared to classical approaches (e.g., gender determination from composition as opposed to DNA analysis). In some cases, it could help narrow down individuals among a group of suspects, if insufficient ridge detail was observed on the mark and if no genetic material was recovered (both conditions required). It is most likely molecular donor profiling will be used in specific contexts such as the handling of illicit substances (i.e., explosives, narcotics) or their consumption (i.e., narcotics). In that case, it is awaited that contamination is taken into account and that studies will be conducted to assess the impact of the fingermark detection techniques on the recovery of such information. Also, research should address the (in)compatibility of molecular profiling with the collection and analysis of touch DNA as it currently constitutes a well-established and effective investigative process. Touch DNA cannot be ignored anymore in this context, mostly due to its success and its strong investigative contribution. However, the limitations and constraints associated with it should be kept in mind.

- For research purposes, molecular donor profiling could definitely play a role, for example, by associating individuals at a donorship level (e.g., "good," "medium," or "poor"). This would constitute a helpful source of information for research projects linked to the development or assessment of detection techniques. Indeed, such projects require a pool of donors of varying donorship [76]. To the best of our knowledge, no study has been undertaken to address this so far, except for an attempt to classify donors based on their lipid profile which was published [62], but not verified by the application of detection techniques.
- With regard to donor profiling obtained from touch DNA, investigators are currently facing a fundamental issue: it is currently impossible to know if a fingermark or a surface of interest bears DNA (in a sufficient quantity). As a result, the investigators decide to "blindly" swab a fingermark or a surface of interest for DNA analysis. The development of an approach helping investigators emphasize target areas is thus of a primary interest. Preliminary studies can be found in the literature [51, 78], and it is anticipated that more will be published in the future. Providing such a tool to investigators would definitely increase the chances of isolating a DNA profile from a swabbed area. Additionally, it could also provide a way to decide between touch DNA collection and molecular donor profiling when facing a fingermark or a surface of interest.

To summarize, for donor profiling to become of forensic interest, it is essential to prove that the provided information could help lead to a suspect or narrow down a suspect from a group, beyond what current conventional practices allow already. So far, only touch DNA meets this criterion (genetic profile and information about the gender). With the increasing development of DNA phenotyping (e.g., skin, eye, and hair colors, freckles, ancestry) and its use for investigations, it is presumed that this way of providing information about the donor (e.g., from the DNA contained in fingermarks) will compete with donor profiling obtained from the fingermark composition. In both cases, it will be essential to assess their actual performance in different scenarios and how the composition of fingermarks (and hence the profiling of the individual) or the genetic material is impacted by the application of detection techniques beforehand.

1.4 Conclusions and Future Directions

In this chapter, we have discussed that, even if dactyloscopy emerged more than 100 years ago and that many (technological) developments have been witnessed since, some of the challenges identified at that time are still not addressed. The question concerning which identification standard should be used was asked at the beginning of the twentieth century and is currently still the focus on many publications and scientific meetings. In a time of international data exchange aimed at fighting crime, it is concerning to know that depending on a country's position regarding identification criteria, different decisions regarding comparison outcomes may be taken, ranging from identification to exclusion. With regard to the chemical composition of fingermarks, most of the recent developments are the result of the versatility of current analytical capabilities. By the number of publications in this field, it has been proven that it is technically possible to use the chemical information contained in fingermarks to improve detection techniques, to estimate their age, or to get information regarding the donor. However, despite such publications, there is evidence of an increasing gap between researchers' objectives and investigators' needs/constraints. To avoid wasting resources on forensically non-relevant topics, it is crucial to be aware of this gap and try to address it rather than ignoring it.

Both aspects of the science of friction ridge skin patterns (i.e., morphological and chemical) are interesting for research and for practical application, even if the needs of these communities may seem, at a first glance, very different. This highlights the need for the discipline to work towards bridging the gap between academics and practitioners. The first should realize that the more interesting achievements may not always be linked with the use of the most advanced technologies and the collection of more data. The latter should consider opening up to more novel approaches rather than solely relying on traditional positions and accept that forensic disciplines are scientific and require continuous development and training.

Thus, in our opinion, the better aspects of the past and present of the forensic use of fingermarks should combine to provide a forensically relevant pathway focused on three equally important elements, namely, research, training, and operational applicability.

References

1. Abraham J, Champod C, Lennard C, Roux C (2013) Modern statistical models for forensic fingerprint examinations: a critical review. Forensic Sci Int 232(1):131–150
2. Adamu LH, Taura MG (2017) Application of likelihood ratio and posterior probability density in sex estimation from level two fingerprint features among Hausa ethnic group. Egypt J Forensic Sci 7(1):25
3. Adda Neggaz L, Meroufel DN, Deba T, Bekada A, Hammadi M, Mediene Benchekor S, Rosa A, Benhamamouch S (2017) Digital dermatoglyphic study in three west Algerian populations: Reguibates, Zenata, Oran. J Can Soc Forensic Sci 50(4):164–174
4. Agudelo J, Privman V, Halámek J (2017) Promises and challenges in continuous tracking utilizing amino acids in skin secretions for active multi-factor biometric authentication for cybersecurity. ChemPhysChem 18(13):1714–1720
5. Aitken C, Taroni F (2004) Statistics and the evaluation of evidence for forensic scientists, 2nd edn. Wiley, Chichester
6. Anthonioz A, Champod C (2014) Integration of pore features into the evaluation of fingerprint evidence. J Forensic Sci 59(1):82–93
7. Ashbaugh DR (1991) Ridgeology. J Forensic Identif 41:16–64
8. Aubert P-F (1878) Des modifications subies par la sécrétion de la sueur dans les maladies de la peau. Ann Dermatol Venereol 9:359–373
9. Badiye A, Kapoor N, Mishra SD (2019) A novel approach for sex determination using palmar tri-radii: a pilot study. J Forensic Legal Med 65:22–26

10. Bailey MJ, Bright NJ, Croxton RS, Francese S, Ferguson LS, Hinder S, Jickells S, Jones BJ, Jones BN, Kazarian SG, Ojeda JJ, Webb RP, Wolstenholme R, Bleay S (2012) Chemical characterization of latent fingerprints by matrix-assisted laser desorption ionization, time-of-flight secondary ion mass spectrometry, mega electron volt secondary mass spectrometry, gas chromatography/mass spectrometry, X-ray photoelectron spectroscopy, and attenuated total reflection Fourier transform infrared spectroscopic imaging: an intercomparison. Anal Chem 84:8514–8523
11. Balthazard V (1911) De l'identification par les empreintes digitales. C R Séances Acad Sci 152:1862–1864
12. Barnes JG (2011) Chapter 1: history. In: Alan M (ed) The fingerprint sourcebook. National Institute of Justice (NIJ), Washington, DC, pp 1–22
13. Bécue A (2016) Emerging fields in fingermark (meta)detection – a critical review. Anal Methods 8:7983–8003
14. Bécue A, Champod C (2016) Fingermarks and other body impressions – a review (July 2013–July 2016). In: 18th Interpol forensic science symposium, Lyon (France), 2016
15. Bécue A, Eldridge H, Champod C (2020) Interpol review of fingermarks and other body impressions 2016–2019. Forensic Sci Int: Synergy 2: 442–480
16. Berry J, Stoney DA (2001) The history and development of fingerprinting. In: Lee HC, Gaensslen RE (eds) Lee and Gaensslen's advances in fingerprint technology, 2nd edn. CRC Press LLC, Boca Raton, pp 1–40
17. Bertillon A (1912) Notes et observations médico-légales – Les empreintes digitales. Archives d'Anthropologie Criminelle, de Médecine Légale et de Psychologie Normale et Pathologique 27:36–52
18. Biedermann A, Vuille J, Taroni F, Champod C (2015) The need for reporting standards in forensic science. Law Probab Risk 14(2):169–173
19. Boudreault A, Beaudoin A (2017) Pseudo-operational study on the efficiency of various fingermark development techniques during the aging process. J Forensic Identif 67(1):85–117
20. Bouslimani A, Melnik AV, Xu Z, Amir A, da Silva RR, Wang M, Bandeira N, Alexandrov T, Knight R, Dorrestein PC (2016) Lifestyle chemistries from phones for individual profiling. Proc Natl Acad Sci 113(48):E7645–E7654
21. Bradshaw R, Denison N, Francese S (2017) Implementation of MALDI MS profiling and imaging methods for the analysis of real crime scene fingermarks. Analyst 142:1581–1590
22. Brunelle E, Le AM, Huynh C, Wingfield K, Halámková L, Agudelo J, Halámek J (2017) Coomassie brilliant blue G-250 dye: an application for forensic fingerprint analysis. Anal Chem 89(7):4314–4319
23. Bunter S (2014) How long can an identifiable fingerprint persist on an exterior surface? CSEye (April)
24. Champod C (1996) Reconnaissance automatique et analyse statistique des minuties sur les empreintes digitales. PhD thesis. University of Lausanne, Lausanne-Dorigny (Switzerland)
25. Champod C, Chamberlain P (2009) Fingerprints. In: Fraser J, Williams R (eds) Handbook of forensic science. Willan Publishing, Cullompton, pp 57–83
26. Champod C, Lennard C, Margot P, Stoilovic M (2004) Fingerprints and other ridge skin impressions. CRC Press LLC, Boca Raton
27. Champod C, Lennard C, Margot P, Stoilovic M (2016) Fingerprints and other ridge skin impressions, 2nd edn. CRC Press LLC, Boca Raton
28. Champod C, Margot P (1997) Analysis of minutiæ occurrences in fingerprints – the search for noncombined minutiæ. In: Takasu TTaA (ed) Current topics in forensic science – proceedings of the 14th meeting of the International Association of Forensic Sciences, vol 1. Shunderson Communications, Ottawa, pp 55–58
29. Champod C, Margot P (1996) Computer assisted analysis of minutiæ occurrences on fingerprints. In: Springer JAAE (ed) Proceedings of the international symposium on fingerprint detection and identification. Israel National Police, Ne'urim, pp 305–318

30. Cohen Y, Rozen E, Azoury M, Attias D, Gavrielli B, Levin Elad M (2012) Survivability of latent fingerprints part I: adhesion of latent fingerprints to smooth surfaces. J Forensic Identif 62(1):47–53
31. Cook R, Evett IW, Jackson G, Jones PJ, Lambert JA (1998) A hierarchy of propositions: deciding which level to address in casework. Sci Justice 38:231–239
32. De Alcaraz-Fossoul J, Barrot Feixat C, Zapico SC, McGarr L, Carreras-Marin C, Tasker J, Gené Badia M (2019) Latent fingermark aging patterns (part IV): ridge width as one indicator of degradation. J Forensic Sci 64(4):1057–1066
33. De Alcaraz-Fossoul J, Barrot Feixat C, Carreras-Marin C, Tasker J, Zapico SC, Badia MG (2017) Latent fingermark aging patterns (part III): discontinuity index as one indicator of degradation. J Forensic Sci 62:1180
34. De Alcaraz-Fossoul J, Barrot Feixat C, Tasker J, McGarr L, Stow K, Carreras-Marin C, Turbany Oset J, Badia MG (2016) Latent fingermark aging patterns (part II): color contrast between ridges and furrows as one Indicator of degradation. J Forensic Sci 61(2):322–333
35. De Alcaraz-Fossoul J, Mancenido M, Soignard E, Silverman N (2019) Application of 3D imaging technology to latent Fingermark aging studies. J Forensic Sci 64(2):570–576
36. De Alcaraz-Fossoul J, Mestres Paris C, Barrot Feixat C, McGarr L, Brandelli D, Stow K, Badia MG (2016) Latent fingermark aging patterns (part I): minutiae count as one Indicator of degradation. J Forensic Sci 61(2):322–333
37. De Alcaraz-Fossoul J, Barrot-Feixat C, Zapico SC, Mancenido M, Broatch J, Roberts KA, Carreras-Marin C, Tasker J (2018) Ridge width correlations between inked prints and powdered latent fingerprints. J Forensic Sci 63(4):1085–1091
38. de Jongh A, Lubach AR, Kwie SLL, Alberink I (2019) Measuring the rarity of fingerprint Patterns in the Dutch population using an extended classification set. J Forensic Sci 64(1):108–119
39. De la Hunty M, Moret S, Chadwick S, Lennard C, Spindler X, Roux C (2015) Understanding physical developer (PD): part I – is PD targeting lipids? Forensic Sci Int 257:481–487
40. De la Hunty M, Moret S, Chadwick S, Lennard C, Spindler X, Roux C (2015) Understanding physical developer (PD): part II – is PD targeting eccrine constituents? Forensic Sci Int 257:488–495
41. de Ronde A, van Aken M, de Puit M, de Poot C (2019) A study into fingermarks at activity level on pillowcases. Forensic Sci Int 295:113–120
42. Dhall JK, Kapoor AK (2016) Fingerprint ridge density as a potential forensic anthropological tool for sex identification. J Forensic Sci 61(2):424–429
43. Dorakumbura BN, Becker T, Lewis SW (2016) Nanomechanical mapping of latent fingermarks: a preliminary investigation into the changes in surface interactions and topography over time. Forensic Sci Int 267:16–24
44. Dorakumbura BN, Boseley RE, Becker T, Martin DE, Richter A, Tobin MJ, van Bronswjik W, Vongsvivut J, Hackett MJ, Lewis SW (2018) Revealing the spatial distribution of chemical species within latent fingermarks using vibrational spectroscopy. Analyst 143(17):4027–4039
45. Druart M-E, Recloux I, Derouaux A, Van De Weerdt C, Olivier M-G (2018) Development of mesoporous sol–gel films for antifingerprint applications on glass. J Sol-Gel Sci Technol 88(2):334–344
46. Edgubov LG (1996) National standards of fingerprint identification. Fingerprint Whorld 22:153–160
47. Egli Anthonioz N (2009) Interpretation of partial fingermarks using an automated fingerprint identification system. PhD thesis. University of Lausanne, Lausanne-Dorigny (Switzerland)
48. Egli Anthonioz NM, Champod C (2014) Evidence evaluation in fingerprint comparison and automated fingerprint identification systems – modeling between finger variability. Forensic Sci Int 235:86–101
49. Egli N, Moret S, Bécue A, Champod C (2013) Fingermarks and other impressions – a review (August 2010–June 2013). In: 17th Interpol forensic science symposium, Lyon (France), 2013

50. Evett IW, Williams R (1996) A review of the sixteen points fingerprint standard in England and Wales. J Forensic Identif 46:49–73
51. Falkena K, Hoveling RJM, van Weert A, Lambrechts SAG, van Leeuwen TG, Aalders MCG, van Dam A (2019) Prediction of DNA concentration in fingermarks using autofluorescence properties. Forensic Sci Int 295:128–136
52. Faulds H (1880) On the skin-furrows of the hand. Nature 22:605
53. Forchelet S, Bécue A (2018) Impact of anti-fingerprint coatings on the detection of fingermarks. J Forensic Identif 68(3):348–368
54. Francese S, Bradshaw R, Denison N (2017) An update on MALDI mass spectrometry based technology for the analysis of fingermarks – stepping into operational deployment. Analyst 142:2518–2546
55. Frick AA, Chidlow G, Goodpaster JV, Lewis SW, Van Bronswijk W (2016) Monitoring compositional changes of the lipid fraction of fingermark residues deposited on paper during storage. Forensic Chem 2:29–36
56. Frick AA, Weyermann C (2019) An untargeted lipidomic approach for qualitative determination of latent fingermark glycerides using UPLC-IMS-QToF-MSE. Analyst 144(11):3590–3600
57. Fritz P, Frick AA, Van Bronswijk W, Beaudoin A, Bleay S, Lennard C, Lewis SW (2017) Investigations into the influence of donor traits on the performance of fingermark development reagents. Part 1: 1,2-Indanedione–zinc chloride. J Forensic Identif 67(3):410–425
58. Galton F (1892) Finger prints. Macmillan and Co, London
59. Girod A, Ramotowski RS, Lambrechts S, Misrielal P, Aalders M, Weyermann C (2016) Fingermark age determinations: legal considerations, review of the literature and practical propositions. Forensic Sci Int 262:212–226
60. Girod A, Ramotowski RS, Weyermann C (2012) Composition of fingermark residue: a qualitative and quantitative review. Forensic Sci Int 223:10–24
61. Girod A, Spyratou A, Holmes D, Weyermann C (2016) Aging of target lipid parameters in fingermark residue using GC/MS: effects of influence factors and perspectives for dating purposes. Sci Justice 56(3):165–180
62. Girod A, Weyermann C (2014) Lipid composition of fingermark residue and donor classification using GC/MS. Forensic Sci Int 238:68–82
63. Girod A, Xiao L, Reedy B, Roux C, Weyermann C (2015) Fingermark initial composition and aging using Fourier transform infrared microscopy (μ-FTIR). Forensic Sci Int 254:185–196
64. Gupta SR (1968) Statistical survey of ridge characteristics. Int Crim Police Rev 5(218):130–134
65. Gutiérrez-Redomero E, Alonso-Rodríguez C, Hernández-Hurtado LE, Rodríguez-Villalba JL (2011) Distribution of the minutiae in the fingerprints of a sample of the Spanish population. Forensic Sci Int 208:79–90
66. Gutiérrez E, Galera V, Martínez JM, Alonso C (2007) Biological variability of the minutiae in the fingerprints of a sample of the Spanish population. Forensic Sci Int 172(2):98–105
67. Heindl R (1927) System und Praxis der Daktyloskopie, 3rd edn. De Gruyter, Berlin
68. Herschel W (1880) Skin furrows of the hand. Nature 23:76
69. Hess J (1971) The reference shelf. Fingerprint Identif Mag 52(June):13
70. Hess J (1971) The reference shelf. Fingerprint Identif Mag 52(April):16–17 & 23
71. Hess J (1971) The reference shelf. Fingerprint Identif Mag 52(March):12–14 & 17
72. Hess J (1971) The reference shelf. Fingerprint Identif Mag 52(May):11–13
73. Home Office – Centre for Applied Science and Technology (CAST) (2014) In: Bandey H (ed) Fingermark visualization manual. Home Office, Centre for Applied Science and Technology, Sandridge
74. Hutchins LA (2011) Chapter 5: Systems of friction ridge classification. In: Alan M (ed) The fingerprint sourcebook. National Institute of Justice (NIJ), Washington, DC, pp 1–25
75. International Association for Identification (1973) Standardization committee report. FBI Law Enforcement Bull 42:7–8

76. International Fingerprint Research Group (IFRG) (2014) Guidelines for the assessment of Fingermark detection techniques. J Forensic Identif 64(2):174–200
77. Johnston A, Rogers K (2018) A study of the intermolecular interactions of lipid components from analogue fingerprint residues. Sci Justice 58(2):121–127
78. Kanokwongnuwut P, Kirkbride KP, Kobus H, Linacre A (2019) Enhancement of fingermarks and visualizing DNA. Forensic Sci Int 300:99–105
79. Kesmez Ö, Tamsü Selli N, Tunalı A, Akarsu E, Akarsu M, Arpaç E (2017) Fingerprint resistant coatings for stainless steel substrates. Prog Org Coat 112:51–56
80. King S, Benson S, Kelly T, Lennard C (2013) Determining the effects of routine fingermark detection techniques on the subsequent recovery and analysis of explosive residues on various substrates. Forensic Sci Int 233(1–3):257–264
81. Langenburg G (2012) A critical analysis and study of the ACE-V process. PhD thesis in forensic science. University of Lausanne, Lausanne-Dorigny (Switzerland)
82. Langenburg G, Champod C (2011) The GYRO system – a recommended approach to more transparent documentation. J Forensic Identif 61(4):373–384
83. Langenburg G, Champod C, Wertheim P (2009) Testing for potential contextual bias effects during the verification stage of the ACE-V methodology when conducting fingerprint comparisons. J Forensic Sci 54(3):571–582
84. Lauzon N, Dufresne M, Beaudoin A, Chaurand P (2017) Forensic analysis of latent fingermarks by silver-assisted LDI imaging MS on nonconductive surfaces. J Mass Spectrom 52(6):397–404
85. Lee R, Comber B, Abraham J, Wagner M, Lennard C, Spindler X, Roux C (2017) Supporting fingerprint identification assessments using a skin stretch model – a preliminary study. Forensic Sci Int 272:41–49
86. Leegwater AJ, Meuwly D, Sjerps M, Vergeer P, Alberink I (2017) Performance study of a score-based likelihood ratio system for forensic fingermark comparison. J Forensic Sci 62(3):626–640
87. Leggett R, Lee-Smith EE, Jickells SM, Russell DA (2007) "Intelligent" fingerprinting: simultaneous identification of drug metabolites and individuals by using antibody-functionalized nanoparticles. Angew Chem Int Ed Engl 46:4100–4103
88. Lim S, Subhani Z, Daniel B, Frascione N (2016) Touch DNA – the prospect of DNA profiles from cables. Sci Justice 56(3):210–215
89. Lin CH, Liu JH, Osterburg JW, Nicol JD (1982) Fingerprint comparison I: similarity of fingerprints. J Forensic Sci 27:290–304
90. Lin S-w, Ip SCY, Lam T-t, Tan T-f, Yeung W-l, Tam W-m (2017) Compatibility of DNA IQ™, QIAamp® DNA Investigator, and QIAsymphony® DNA Investigator® with various fingerprint treatments. Int J Legal Med 131(2):293–301
91. Locard E (1914) La Preuve judiciaire par les empreintes digitales. Archives d'anthropologie criminelle, de médecine légale et de psychologie normale et pathologique 29:321–348
92. Luda MP, Li Pira N, Trevisan D, Pau V (2018) Evaluation of antifingerprint properties of plastic surfaces used in automotive components. Int J Polym Sci 2018:11
93. Margot P (2011) Forensic science on trial – what is the law of the land? Aust J Forensic Sci 43(2–3):89–103
94. Margot P, German E (1996) Fingerprint identification breakout meeting. In: Springer JAAE (ed) International Symposium on fingerprint detection and identification. Israel National Police, Ne'Urim
95. Matuszewski S (2015) Age-dependent changes of contrast in fingermarks of various lipid content. Probl Forensic Sci 101:5–13
96. Meagher S, Dvornychenko V, Garris M (2014) Characterization of latent print "lights-out" modes for automated fingerprint identification systems. J Forensic Identif 64(3):255–284
97. Merkel R, Dittmann J, Vielhauer C (2016) Resource-efficient latent fingerprint age estimation for Adhoc crime scene forensics: quality assessment of flatbed scans and statistical features.

In: 2016 Sixth International Conference on Image Processing Theory, Tools and Applications (IPTA). IEEE, Oulu, pp 1–6
98. Meuwly D, Ramos D, Haraksim R (2017) A guideline for the validation of likelihood ratio methods used for forensic evidence evaluation. Forensic Sci Int 276:142–153
99. Moret S, Bécue A, Champod C (2014) Nanoparticles for fingermark detection: an insight into the reaction mechanism. Nanotechnology 425502:25
100. Moret S, Spindler X, Lennard C, Roux C (2015) Microscopic examination of fingermark residues: opportunities for fundamental studies. Forensic Sci Int 255:28–37
101. National Research Council (2009) Strengthening forensic science in the United States: a path forward. National Academies Press, Washington, DC
102. Neumann C, Champod C, Puch-Solis R, Egli N, Anthonioz A, Bromage-Griffiths A (2007) Computation of likelihood ratios in fingerprint identification for configurations of any number of Minutiæ. J Forensic Sci 52(1):54–64
103. Neumann C, Champod C, Puch-Solis R, Egli N, Anthonioz A, Meuwly D, Bromage-Griffiths A (2006) Computation of likelihood ratios in fingerprint identification for configurations of three Minutiæ. J Forensic Sci 51(6):1255–1266
104. Neumann C, Evett IW, Skerrett J (2012) Quantifying the weight of evidence from a forensic fingerprint comparison: a new paradigm. J R Stat Soc A Stat Soc 175(2):371–415
105. O'Neill KC, Lee YJ (2018) Effect of aging and surface interactions on the diffusion of endogenous compounds in latent fingerprints studied by mass spectrometry imaging. J Forensic Sci 63(3):708–713
106. Odén S, Von Hofsten B (1954) Detection of fingerprints by the Ninhydrin reaction. Nature 173:449–450
107. Oonk S, Schuurmans T, Pabst M, de Smet LCPM, de Puit M (2018) Proteomics as a new tool to study fingermark ageing in forensics. Sci Rep 8(1):16425. (16411 pp.)
108. Osterburg JW, Parthasarathy T, Raghavan TES, Sclove S (1977) Development of a mathematical formula for the calculation of fingerprint probabilities based on individual characteristics. J Am Stat Assoc 72(360):772–778
109. Phan K, Barash M, Spindler X, Gunn P, Roux C (2019) Retrieving forensic information about the donor through bacterial profiling. Int J Legal Med 134:21
110. Pleik S, Spengler B, Ram Bhandari D, Luhn S, Schäfer T, Urbach D, Kirsch D (2018) Ambient-air ozonolysis of triglycerides in aged fingerprint residues. Analyst 143(5):1197–1209
111. Pleik S, Spengler B, Schäfer T, Urbach D, Luhn S, Kirsch D (2016) Fatty acid structure and degradation analysis in fingerprint residues. J Am Soc Mass Spectrom 27(9):1565–1574
112. Popov KT, Sears VG, Jones BJ (2017) Migration of latent fingermarks on non-porous surfaces: observation technique and nanoscale variations. Forensic Sci Int 275:44–56
113. Ramos D, Haraksim R, Meuwly D (2017) Likelihood ratio data to report the validation of a forensic fingerprint evaluation method. Data Brief 10:75–92
114. Ramos D, Krish RP, Fierrez J, Meuwly D (2017) From biometric scores to forensic likelihood ratios. In: Tistarelli M, Champod C (eds) Handbook of biometrics for forensic science. Springer, Cham, pp 305–327
115. Raymond JJ, Roux C, Du Pasquier E, Sutton J, Lennard C (2004) The effect of common fingerprint detection techniques on the DNA typing of fingerprints deposited on different surfaces. J Forensic Identif 54(1):22–44
116. Rignell A, Sjöqvist KEA (1983) Swedish method of fingerprint classification. Hereditas 98(1):115–125
117. Roddy A.R. and Stosz J.D. Fingerprint feature processing techniques and poroscopy. In: L. C. Jain U. Halici, I. Hayashi, S. B. Lee, and S. Tsutsui (Ed), Intelligent biometric techniques in fingerprint and face recognition. CRC Press, Boca Raton, 1997, pp. 35–105
118. Roux C, Talbot-Wright B, Robertson J, Crispino F, Ribaux O (2015) The end of the (forensic science) world as we know it? The example of trace evidence. Philos Trans R Soc Lond B Biol Sci 370:20140260

119. Saks MJ (2010) Forensic identification: from a faith-based "science" to a scientific science. Forensic Sci Int 201(1–3):14–17
120. Salama J, Aumeer-Donovan S, Lennard C, Roux C (2008) Evaluation of the fingermark reagent oil red O as a possible replacement for physical developer. J Forensic Identif 58(2):203–237
121. Sánchez-Andrés A, Barea JA, Rivaldería N, Alonso-Rodríguez C, Gutiérrez-Redomero E (2018) Impact of aging on fingerprint ridge density: anthropometry and forensic implications in sex inference. Sci Justice 58(5):323–334
122. Santamaria Beltran FA (1955) A new method for evaluating ridge characteristics. Fingerprint Identif Mag 36:3–8 & 16–18
123. Scientific Working Group on Friction Ridge Analysis S.a.T.S (2013) Standards for examining friction ridge impressions and resulting conclusions, ver. 2.0, http://clpex.com/swgfast/documents/examinations-conclusions/130427_Examinations-Conclusions_2.0.pdf. Last accessed 10 Mar 2020
124. Shetage SS, Traynor MJ, Brown MB, Chilcott RP (2018) Sebomic identification of sex- and ethnicity-specific variations in residual skin surface components (RSSC) for bio-monitoring or forensic applications. Lipids Health Dis 17(1):194. (110 pp.)
125. Sisco E, Staymates J, Schilling K (2015) A chemically relevant artificial fingerprint material for the cross-comparison of mass spectrometry techniques. J Can Soc Forensic Sci 48(4):200–214
126. Soanboon P, Nanakorn S, Kutanan W (2016) Determination of sex difference from fingerprint ridge density in northeastern Thai teenagers. Egypt J Forensic Sci 6(2):185–193
127. Sodhi GS, Kaur J (2001) Powder method for detecting latent fingerprints: a review. Forensic Sci Int 120:172–176
128. Steiner R, Roux C, Moret S (2019) Controlling fingermark variability for research purposes: a review. Wiley Interdiscip Rev: Forensic Sci 1(4):e1338
129. Stoehr B, Hall C, Evans D, Murphy P (2016) Cleaning dirty surfaces: a three-body problem. ACS Appl Mater Interfaces 8(28):18534–18539
130. Stoney DA (1986) A critical analysis of quantitative fingerprint individuality models. J Forensic Sci 31(4):1187–1216
131. Stoney DA, De Donno M, Champod C, Wertheim PA, Stoney PL (2020) Occurrence and associative value of non-identifiable fingermarks. Forensic Sci Int 309:110219
132. Subhani Z, Daniel B, Frascione N (2019) DNA profiles from fingerprint lifts - enhancing the evidential value of fingermarks through successful DNA typing. J Forensic Sci 64(1):201–206
133. Swofford HJ, Koertner AJ, Zemp F, Ausdemore M, Liu A, Salyards MJ (2018) A method for the statistical interpretation of friction ridge skin impression evidence: method development and validation. Forensic Sci Int 287:113–126
134. Taduran RJO, Tadeo AKV, Escalona NAC, Townsend GC (2016) Sex determination from fingerprint ridge density and white line counts in Filipinos. HOMO 67(2):163–171
135. van Dam A, Schwarz JCV, de Vos J, Siebes M, Sijen T, van Leeuwen TG, Aalders MCG, Lambrechts SAG (2014) Oxidation monitoring by fluorescence spectroscopy reveals the age of fingermarks. Angew Chem Int Ed Engl 53:6272–6275
136. Van Dam A, van Beek FT, Aalders MCG, Van Leeuwen T, Lambrechts SAG (2016) Techniques that acquire donor profiling information from fingermarks – a review. Sci Justice 56(2):143–154
137. van Dam A, van Weert A, Falkena K, Weymans C, Kloosterman AD, Lambrechts SAG, van Leeuwen TG, Aalders MCG (2018) Sex determination from fingermarks using fluorescent *in situ* hybridization. Anal Methods 10(12):1413–1419
138. van Helmond W, Kuijpers C-J, van Diejen E, Spiering J, Maagdelijn B, de Puit M (2017) Amino acid profiling from fingerprints, novel methodology using UPLC-MS. Anal Methods 9(38):5561–5712
139. van Helmond W, O'Brien V, de Jong R, van Esch J, Oldenhof S, de Puit M (2018) Collection of amino acids and DNA from fingerprints using hydrogels. Analyst 143(4):900–905

140. van Helmond W, van Herwijnen AW, van Riemsdijk JJH, van Bochove MA, de Poot CJ, de Puit M (2019) Chemical profiling of fingerprints using mass spectrometry. Forensic Chem 16:100183
141. Zadnik S, van Bronswijk W, Frick AA, Fritz P, Lewis SW (2013) Fingermark simulants and their inherent problems: a comparison with latent fingermark deposits. J Forensic Identif 63(5):593–608
142. Zampa F, Furlan G, Furlan G, Bellizia M, Iuliano G, Ripani L (2014) New forensic perspective for fast blue B: from cannabinoid reagent in toxicology to latent fingerprint developer in drug cases. J Forensic Identif 64(6):523–535
143. Zhou Z, Zare RN (2017) Personal information from latent fingerprints using desorption electrospray ionization mass spectrometry and machine learning. Anal Chem 89(2):1369–1372

Chapter 2
Latent Fingermarks as Complex Physical, Chemical, and Biological Structures

Josep De Alcaraz-Fossoul

Abstract Latent fingermarks are complex and dynamic biochemical matrices that continuously change after deposition, affected by a myriad of influencing factors. In practical scenarios, a fingermark will never be immediately recovered from a scene; time will go by, and its physical, chemical, and biological composition will inevitably be altered. For many years, scientists have been searching for aging parameters or indicators that could be sensitive to and representative of the effects of donor, substrate, and environmental factors. In most cases, research on age determinations of fingermarks have taken place in laboratory controlled or monitored settings and have mostly been conducted to replicate common crime indoor conditions. In this chapter, the fundamentals of latent fingermarks will be reviewed and the potential role of age estimations in criminal investigations discussed, including an overview of some of the (known) factors that are involved in the natural degradation process and some potentially relevant aging parameters. This chapter will also introduce concepts that will be further discussed in other chapters of this book.

Keywords Fingermarks · Degradation · Aging · Parameters · Indicators · Predictors · Activity level propositions · 2D · 3D · Sweat · Eccrine · Sebaceous · Composition · Chemical · Biological · Optical · Environment · Substrate · Donor · Endogenous · Exogenous · Deposition · Microbiome

Acronyms

AFIS	Automated Fingerprint Identification System
DNA	Deoxyribonucleic acid
ULW	Universal Latent Workstation

J. De Alcaraz-Fossoul (✉)
Henry C. Lee College of Criminal Justice and Forensic Science,
Forensic Science Department, University of New Haven,
West Haven, CT, USA
e-mail: jfossoul@newhaven.edu

J. De Alcaraz-Fossoul (ed.), *Technologies for Fingermark Age Estimations: A Step Forward*, https://doi.org/10.1007/978-3-030-69337-4_2

2.1 Introduction

Human friction ridge skin pattern impressions, most commonly in the form of fingermarks,[1] have been used to identify persons for many decades, in criminal investigations as well as for non-crime-related purposes. In fact, this biometric is the current leader in identification markers [1]. Latent fingermarks, which result from touching an object with bare hands, are the most common type whose structure is subject to concurrent physical, chemical, and biological alterations post-deposition. These time-dependent modifications are the product of the natural degradation process, or aging, of the fingermark, which is influenced by several factors that could be generally classified as having donor, substrate (including contaminants), or environmental origin. It is still unknown to scientists how many factors are simultaneously involved in latent fingermark degradation at any given time, nor their relative influence neither which ones could be relevant or redundant in a hypothetical age determination model.

Despite certain skepticism among members of the forensic science community and little financial resources available for decades, it has been the goal of many researchers to seek and develop a reliable scientific method that could: 1) determine the effect of as many influencing factors as possible on fingermark aging; 2) detect and describe aging patterns in controlled and uncontrolled situations based on robust indicators or predictors of degradation; and 3) design mathematical models to infer fingermark age with high degree of certainty. The discovery and standardization of such method/s would revolutionize forensic science, more specifically friction ridge skin pattern analysis, and prove extremely valuable in revealing new information for crime scene reconstructions, for example, by placing a suspect within a timeframe of the alleged criminal event [2].

Research conducted to date has revealed several pre-deposition and peri-deposition factors that may influence the "life expectancy" of a latent fingermark, mostly related to donor, deposition, and surface characteristics, and the presence of certain contaminants [3]. It has also been demonstrated that microscopic modifications in the topography of the fingermark (ridges) will occur post-deposition, possibly as a result of changes in its chemical and microbial composition, induced by the permanent interaction with the surface and the environment. Topographical alterations can be detected through conventional visual analysis of enhanced (e.g., powdered and photographed) and unenhanced fingermarks [3] and may include a reduction in the count of viable minutiae, a loss in color contrast between ridges and furrows, an increase of the number of ridge discontinuities, and a fluctuation (loss or gain) of ridge width [4–8], among other possible modifications detected by optical means.

[1]This term refers to latent or invisible impressions (i.e., undetectable to the naked eye), plastic or molded impressions (e.g., prints on clay or wax), and patent impressions (i.e., visible prints) which are unintentionally made by friction ridge skin on a surface. The term *fingerprint* refers to exemplars obtained under controlled settings (e.g., inked or scanned impressions taken from detainees).

There is little doubt within the scientific community that the integrity of ridge impressions becomes compromised post-deposition which translates to a gradual decrease in the visual quality (clarity) of the fingermark over time. It is also known, and well documented, that fingermarks degrade following certain patterns under a specific combination of (laboratory controlled) conditions [3, 9]. Other naturally occurring topographical modifications may include *ridge drift* [10], defined as an apparent "move" of a single ridge, as well as other modifications that can only be captured using 3D technologies, such as the loss of height and volume of ridges [11]. Despite the successful observation of these time-sensitive features and the identification of some influencing aging factors, there is currently no standard methodology for estimating the time since deposition, or the age, of latent fingermarks, neither by optical, chemical, nor biological analytical approaches. Once a mathematical aging model is developed, fingerprinting may transition to *activity level propositions* (i.e., *when* and *how* was the ridge impression deposited) which may allow crime investigators to set a viable timeframe where suspects can be excluded, or included, from a criminal event [12]. In addition, the probative value of the evidence will increase and errors in identifications will be reduced while opening new opportunities to the forensic subdiscipline of fingermark age determinations in crime scene reconstruction.

In this chapter, the fundamentals of latent fingermarks will be briefly reviewed and the potential role of age estimations in criminal investigations discussed, including some of the (known) factors that are involved in the natural degradation process. This chapter will also introduce basic concepts that will be further discussed in Chaps. 5 and 6, such as the impact of aging factors from a 2D and 3D topographical point of view. Chapters 3 and 7 will focus on the chemical composition and degradation of fingermarks, whereas Chaps. 10 and 11 will explain degradation from a biological perspective.

2.2 What Are (Latent) Fingermarks?

A friction ridge skin impression could be defined as a physical pattern transferred from ridged skin of (mostly) uncontrollable quality, which may be visible or latent (invisible) to the unaided eye [13]. The most common type, latent fingermarks, results from touching a surface with friction ridge skin that is located at the ends of our fingers. Friction ridges are also located on the palms of our hands and soles of our feet and toes. The ridges (hills) refer to the raised part of a friction ridge skin whose pattern is transferred during the imprinting process, while the furrows (valleys) are the depressed areas. Fingermarks can be categorized by general patterns and by microscopical elements (i.e., commonly known as *minutiae*, Galton's details, or ridge characteristics) that are dependent on interrelated factors that occur in utero [1]. Friction skin ridge patterns are permanent and remain unchanged throughout a person's lifetime unless the basal layer of the dermis of the skin is damaged. This will result in permanent scars on the epidermis which will then alter the ridge patterns of a fingermark [14]. In rare circumstances, the intake of certain cell growth inhibitors by medication [15], work environments where abrasive products are used, or even genetic cases of adermatoglyphia [16] may inhibit the presence of skin ridges and the subsequent quality of the impression with few to no identifiable features.

2.2.1 Topography of a Latent Fingermark: The Basics

The general patterns of friction ridge skin impressions, in the case of fingertips, are classifiable and identifiable based on certain focal points. These are the core (i.e., the approximate center of the pattern), the flow of ridges (i.e., ridges that surround the core or travel side to side), and the deltas (i.e., the triangular shape ridge area/s where ridges diverge in three different directions) found in some patterns. The three main types of fingerprint patterns are arches (adeltas), loops (monodeltas), and whorls (bideltas) [1, 17–20]. This is also known as level one detail [20] or the general classification.

A simple or plain arch presents no delta nor an effective core but consists of ridges that enter on one side and rise and exit on the other side. One subtype of arch is a tented arch, where the arches near the central region are sharper and more angular and sometimes displaying a delta-like structure within the central region of the fingerprint. A loop type has one delta and at least one ridge that by turning back on itself forms a clearly discernible staple in appearance. Two subtypes exist. A "right loop", sloping to the right, also referred to as an ulnar loop if found in the right hand, presents the core to the right of the delta, where the ridges enter from the right side of the finger impression and recurve back on themselves to exit on the right side. A "left loop", sloping to the left, or radial loop if in the right hand, has the core to the left of the delta, where the ridges enter from the left side of the finger impression and recurve back on themselves to exit on the left side. A whorl type typically has two deltas but on rare occasions can have more than two. The cores are generally more circular in shape. There are four main subtypes of whorls: plain whorl, central pocket loop whorl, twinned or double loop whorl, and elongated whorls. Accidental whorls can also be noticed but are much rarer. A plain whorl and a central pocket loop (which is also a whorl pattern) both have a minimum of one fully recurving revolution of a ridge at the core which is essentially either circular or spiral but usually displays concentric spirals or recurves. A plain whorl's core is located between the two deltas which appear on opposite sides of the print but may not be equidistant from the core. A twinned or double loop whorl has two deltas, one on either side of the core with two interlinking loops (essentially "S" shaped), typically one with an upward tendency loop and the other downward. Finally, an elongated whorl is usually elliptical in shape and can be quite shallow or conversely quite deep. The elongated whorl again has a delta to each side of the finger. An accidental whorl may consist of more than two deltas and is a combination of two different types of patterns with the exception of the plain arch and cannot be classified as pertaining to any of the previous types alone [1, 17–20]. These general patterns have some degree of discriminating power as class characteristics whereby suspects can be excluded from an inquiry, but patterns alone lack the power to individualize.

Since its inception, fingerprinting has relied on microscopic friction ridge features called *minutiae* or level two detail (features) for individualization [1]. These minutiae (i.e., their relative location, orientation, frequency, and type) are the basis for comparison and identification between a dactylogram of unknown source recovered from a crime scene (or deceased person) and a known fingerprint (collected from an identified donor). There are three main types of minutiae: a ridge ending (an abrupt ending of a ridge line); a bifurcation (a splitting of two ridges); and a dot or island (a very short ridge section separated from the flanking ridges) [17]. Other known denominations are combinations of all these basic types. Lastly, there are level three features that are extremely fine such as sweat pores and ridge contours as well as random features including scars and other skin marks. These features may also be unique and identifiable but only used in conjunction with level two detail, not by themselves for identifications [17, 20, 21].

The "natural" fate of a latent fingermark, unless destroyed by an active mechanical force, is dependent on many factors since the moment of its deposition. There are two main groups of factors that may affect the chemical and biological composition of a fingermark and, therefore, its physical appearance: exogenous and endogenous [22, 23]. Exogenous factors are those that broadly relate to the environment, whereas endogenous factors originate from the donors themselves. Some examples of exogenous factors are cosmetics, blood, and dirt; they adhere to the surface of fingertips and can alter the initial chemical signature of the fingermark when it is transferred. These external factors can impact degradation post-deposition because of the way the constituents in cosmetics and dirt interact and decay over time. Other exogenous factors such as the nature of the receiving surface, exposure to natural light, relative humidity, and temperature, among many others, may also play a role in the degradation of latent fingermarks peri- and post-deposition (see Sect. 2.2.4). Examples of endogenous factors include relative water content, presence of drugs/medications, and all sorts of organic and inorganic substances that are present in the secretions of the donor. The detection of these compounds, and how they degrade overtime, is dependent on, and the result of, their concurrent interactions with the substrate and the environment. Chapters 3 and 7 explore in more detail the chemical composition and degradation of these compounds over time.

2.2.2 *Chemical Composition of a Latent Fingermark: The Basics*

Latent fingermarks are composed of sweat secretions and other contaminants that are transferred onto a substrate, resulting in an (initial) exact copy impression of the ridge skin pattern (frequently partial), under ideal conditions [23]. This pattern is on the epidermis, the outermost layer of the skin generated deeper in the skin within the dermis. The shape of the boundary between the epidermis and the dermis, made up of dermal papillae, determines the form and pattern of the ridges on the surface of the skin. The dermis is the bottom layer of the skin and contains secretory glands including apocrine, eccrine, and sebaceous, among other structures. These glands produce secretions that reach the skin surface through pores [17]. Apocrine glands are found mostly in the genital, groin, and armpit regions. Secretions from the apocrine glands include proteins, carbohydrates, and cholesterol, among others. Due to their location, they play a minor role in latent fingermark composition, and no study has directly discussed their contribution in enhancement methods or age determination. Eccrine glands are present all over the body, and their secretions consist mainly of water, proteins, urea, amino acids, uric acid, lactic acid, sugars, creatine, and choline [17]. They play an important role in fingermark composition and the focus of many scientific researches. Lastly, sebaceous glands are also found all over the body except the hands and feet. Their secretions are sebum generally made up of lipids consisting of glycerides, fatty acids, wax esters, cholesterol, phospholipids, and squalene [17, 22, 23]. These secretions are transferred onto the fingertips after they have had contact with other parts of the body and therefore contribute to the fingermark composition. In fingermark aging studies involving the examination of chemical compounds, it is important to identify the initial composition of the deposition and how each target organic compound degrades relative to other invariable (degradation-free) substances over time. Changes in relative concentrations of these compounds might be indicative of being directly involved on how the latent fingermark visually degrades. Endogenous sweat compounds were originally thought to be represented by about 95–99% water [17]. However, recent studies have argued that latent fingermarks are estimated to contain approximately 20% water at the instant of deposition due to evaporation and other processes [24].

From a research perspective, three chemically distinct types of latent fingermarks can be used. Eccrine-rich fingermarks may be considered ungroomed impressions. These are obtained after washing hands and allowing the person to continue with normal activities for a certain length of time without touching any other surfaces of the body or by wearing nitrile gloves for a set amount of time; thus, these impressions may potentially contain exogenous contaminants. Sebaceous-rich fingermarks (groomed) are the result of wiping the fingers, after washing hands, across a sebum-rich part of the skin, such as one's forehead or the sides of the nose (previously washed), and will contain a mix of secretions but no exogenous contaminants. These two secretion types are composed of a matrix of organic and inorganic substances and will degrade following different patterns [3, 23, 25, 26]. Over time, the components of the secretions undergo chemical changes like evaporation and oxidation [22, 23, 25, 27]. The third group, referred to as "natural" fingermarks, can also be used in experimental settings; however, the conclusions reached from this type of studies have to be carefully interpreted as this type of fingermark is prone to introducing a large number of unknown and uncontrolled variables that can inadvertently impact/skew the results. These fingermarks are ideal representations of "real" case scenarios, with no washing or grooming processes, but should only be used, for example, when validating a method or testing (training) a mathematical model that has been previously developed using a certain degree of control over the variables involved in fingermark composition, deposition, and degradation processes.

2.2.3 Biological Composition of a Latent Fingermark: The Basics

The widely popular Locard's exchange principle[2] states that "*whenever two objects come in contact, a transfer of material occurs*" [28]; in other words, it claims that "*every contact leaves a trace*". The source of a fingermark whose ridge pattern is beyond recognition can also be confirmed with the analysis of another trace element, DNA. In fact, deoxyribonucleic acid, or DNA, can also be detected, extracted, and identified from latent fingermarks, also referred as "touch DNA". Most optical and physical methods used to enhance a latent fingermark do not affect the recovery and identification of DNA profiles providing robust anti-contamination steps are taken. Once a fingermark has been properly documented but can no longer aid in identification, it can be swabbed and analyzed for its DNA content. This DNA may originate from human cells or as cell-free DNA that shed off of people at different rates [29]. There is a wide range of factors that can affect the recovery (i.e., quantity and quality) of DNA from friction skin ridge impressions. It includes the substrate where the fingermark was deposited on, the age and genetics of the donor and their personal habits, the duration/pressure of contact, the part of the hand/finger that is making contact, and the environmental conditions post-deposition, including the fingermark enhancement method (if applicable). The quantity detected will also depend on the sensitivity of the techniques applied and could vary between 15 pg and 200 pg per fingermark [29, 30]. Chapter 10 discusses the involvement of DNA as a biological trace in relation to latent fingermarks.

In addition to human DNA, humans harbor microbial organisms on the skin, as well as in other areas of the body, that may have multiple uses in forensics, including differentiating individuals from one another [31–33]. This topic is presented in more detail in Chap. 11. According to the literature, microbial communities on our skin develop as a result of how we are exposed, and for how long, to particular organisms and from other personal choices, such as diet and exposure to the environment [31]. Skin microbiomes are very identifiable, and it has been reported that microorganism signatures found on two people's hands can vary more than 80% from one another [33]. It is reasonable to hypothesize that microorganisms are also transferred and grow for a (short) period of time on surfaces after a fingermark is deposited [34]. The presence and succession of microbes on a specific surface over time could also be used as an indicator for latent fingermark aging; however, no

[2] Locard E (1920) L'enquête criminelle et les méthodes scientifiques, Ernest Flammarion, Paris pp. 129–130. Available: http://babel.hathitrust.org/cgi/pt?id=njp.32101068784956

studies have yet demonstrated this concept. In theory, the rich matrix of chemical compounds and skin cells within a fingermark (from endogenous or exogenous sources) may be a probable energy source for microorganisms. Therefore, it could be assumed that microbes are also a part of the biological structure of a latent fingermark and part involved in the chemical and physical (visual) degradation post-deposition.

2.2.4 Other Factors Involved in Latent Fingermark Matrix Degradation

As mentioned above, the chemical composition of latent fingermarks, especially post-deposition, will vary greatly between donors, and even within the same donor, due to a myriad of influencing factors. In fact, it has already been demonstrated that two different person's fingermarks are chemically distinct [35]. The main reason for this distinctiveness is the result of the unique interaction of donor characteristics with environmental and substrate conditions (pre-, peri-, and post-deposition) [22]. Although more scientific research is needed, some indicators of chemical degradation appear to be sufficiently robust, time-sensitive and discriminatory to become the foundation of future mathematical models for age estimates. This topic is covered in more detail in Chaps. 3 and 7. However, it is yet not clear whether these chemical differences over time translate to, or correlate with, any topographical changes observed during the aging process of latent fingermarks.

Donor-related factors involved in degradation may include age, biological sex, use of cosmetics/medication, and diet [22, 23, 25, 36–38]. In relation to the donor's age, children reportedly do not produce sebum until 7–10 years of age [25]. Children's fingermark chemical makeup includes three major classes of compounds: carboxylic acid, salts, esters, and secondary amides [37]. All of these compounds are also detected in adult fingermarks; however, at extremely different quantities [37]. Studies have shown that children's fingermarks degrade faster than adult fingermarks due to the absence of squalene, cholesterol, large fatty acid esters, wax esters, and triglycerides [22]. This could be explained by the fact that pre-pubescent donors contain more eccrine components (no sebum), and as the donor ages, they develop more sebaceous components [23]. Thus, there is a significant effect on enhancement and detection where children's fingermarks have degraded after 48 h, whereas impressions from adults can be successfully enhanced more than a week under the same conditions [23]. Biological sex is another source of chemical variability. Males tend to have a higher concentration of fatty acids within their fingermark residue, whereas sterols, sterol esters, and amino acids are more concentrated in female's fingermarks [23]. Unfortunately, the significance of these differences has not been yet fully explored for age estimations of latent fingermarks.

The relevance of substrate-related factors include whether the substrate itself is porous, non-porous, or semi-porous, as well as its texture (roughness), conductivity, waviness, hardness, etc. With porous substrates, the compounds are absorbed into the substrate and less likely to be subject to extrinsic degradation factors, whereas with non-porous substrates, the fingermark material remains on the surface and is much more likely to face degradation while exposed to the elements [22, 36]. Semi-porous items absorb eccrine compounds quickly; however, sebaceous compounds are absorbed much slower due to their less aqueous nature [22]. This topic is discussed more in depth in Chaps. 5 and 6.

Environmental conditions have a significant impact on the chemical composition and topography of latent fingermarks [19, 22, 39, 40]. Factors such as humidity, temperature, exposure to UV radiation, dust, condensation, friction, and air circulation have all been shown to have some degree of effect on the chemical constituents in fingermark residue. Loss of moisture (dehydration and rehydration cycles) and breakdown of chemical components could explain, in part, why there are differences in chemical profiles over time. This topic is discussed more in depth in Chaps. 3, 7, and 8.

Lastly, transfer factors may include pressure exerted during the imprinting process, duration of contact and angle, type of deposition, and time of day [8, 22, 41]. The relative concentration of chemical compounds in sweat can vary throughout the day, depending on levels of perspiration, and whether or not donors have touched areas of their body that produce sebaceous secretions, or the last time the individual washed their hands [23]. Pressure and contact duration have shown to produce stronger coloration when processed with ninhydrin, indicating the more transfer of fingermark residue [22]. Additionally, fingermarks deposited in repeated sequence (succession) show a reduction in contrast when processed with fingerprint powder, indicating that residue is lost in between depositions [41]. This topic is discussed more in depth in Chaps. 5 and 6.

2.3 Redefinition of "Latent Fingermark"

Latent fingermarks were once thought to be an identical copy of friction skin ridges, with an immobile ridge structure (including minutiae) that would not change over time after it was transferred onto a substrate. However, research has demonstrated that latent ridges are complex three-dimensional matrices that, immediately after deposition, are being altered at the chemical, biological, and physical levels. Especially relevant is the latter because of the direct involvement of morphological features in human identifications via minutiae comparisons. In fact, a latent fingermark's residue is a complex biochemical "living" system that is continuously affected by multiple factors as it slowly "evolves" post-deposition. Therefore, latent fingermarks must be understood as "dynamic" structures for both identification and age determination (dating) purposes. As a friction ridge skin impression ages, this biochemical matrix will experience a series of natural decaying phases from an initial "fresh" stage (i.e., when friction skin ridges make contact with a surface) to a complete obliteration. At deposition, the transferred ridge patterns will mostly reproduce the topography of ridge skin, except in instances of severe distortions and smudges caused by inadequate pressure, duration of contact, and angle of contact. Other substrate qualities (e.g., softness, porosity, and roughness/texture) may also affect the visual clarity of the impression. At the "aged" stage, the ridges will experience a loss in integrity and, in some cases, even a physical change in the ridge pattern (possibly) as a result of biochemical degradation caused by interactions among substrate, donor, and the environment. There are numerous factors that will impact the aging process of a fingermark that will affect its "life expectancy." Some of these factors are discussed in Chaps. 5 and 6.

It has been the goal of many scientists for many years to identify which "natural" factors are most relevant or influencing, and which redundant, in latent fingermark aging processes. These can be identified by studying the biochemical and physical (topographical) behavior of the fingermark to certain controlled and monitored factors. By reducing the number of concurrent (influencing) variables to just few that could explain most of the changes detected, researchers could achieve the ultimate goal of developing a mathematical model that could estimate time since deposition.

2.4 Why Is it Important to Determine the Visual Quality of Latent Fingermarks?

The level of visual quality (i.e., assessment of ridge integrity and background noise) and clarity (i.e., the fidelity with which anatomical details are represented) of latent fingermarks become key elements to consider when attempting to identify the source (and age) of the impression. There are a few visual grading scales and computer software programs to aid in determining levels of quality and clarity, and therefore, which fingermarks are best suited for comparison and to be uploaded to an Automated Fingerprint Identification System (AFIS). Chapters 5 and 6 describe grading scales. The idea behind this process is that when a fingermark of poor quality is examined, it would be less reliable to conclude whose it may be. A similar principle applies to age estimation where a minimum level of ridge quality is required to properly conduct the analysis. In addition, the evidence in question must be potentially identifiable, or already identified, before conducting any age estimation analyses.

When a crime suspect presents a (prior) legitimate access to the scene, the time of deposition of the fingermark identified can be exceedingly helpful. Beyond witness testimonies, criminal investigators ideally seek to objectively date the moment a latent fingermark was deposited [3]; however, there is no agreement on the admissibility and weight of testimony regarding opinions on the age of fingermarks [42]. Research and field experience have shown that a visibly high-quality fingermark does not necessarily imply "freshness," and it would even be misleading to make such an assumption. However, a certain level of quality is still necessary for identification as well as for age estimation purposes.

In 1927, Heindl, who was a pioneer in the study of ridge pattern analysis from fingers, stated that older fingermarks were more difficult to enhance [9, 42]. Based on what we know today, this statement could be easily misinterpreted in that all fingermarks which are difficult to enhance are "old" or that fingermarks which are easily enhanced are always "fresh." In addition, some of the chemical enhancement processes used today work better on older impressions, e.g., powder suspension. In 1987, Holyst mentioned that sebaceous-rich fingermarks may have a "life expectancy" 5 times longer than their eccrine-rich counterparts and that in outdoor conditions ridge degradation might be 15 times faster than in indoor conditions [43]. In 1990, Baniuk expanded this knowledge and described the "*narrowing of the fingerprints; loss of stickiness; and loss of continuity* [of ridges]" [44]. Again, these pioneering works must be carefully interpreted as the enhancing techniques, and unreported/uncontrolled endogenous and exogenous factors, may have altered the aging patterns significantly [3]. Holyst finally stated that "*expert opinion* [on fingermark age] *should not be based solely on certain age characteristics, but also* [need to incorporate] *knowledge on the deposition and storage conditions of these traces*." Therefore, a reliable method to date fingermarks must incorporate several meaningful and complementary age indicators (physical and/or biochemical) that are distinctively time-sensitive and that can be useful to monitor and/or correlate endogenous and exogenous factors (pre-, peri-, and post-deposition) in order to reduce possible misinterpretations of age.

2.5 The Analysis of Latent Fingermarks: Identifications

There are several methods and technologies currently available for identification purposes. These may consist of several enhancing and recording techniques, processes and protocols, several AFIS platforms, and complementary software [45–48] such as the Universal Latent Workstation (ULW) developed by the US Federal Bureau of Identification (FBI). ACE-V is the acronym for Analysis, Comparison, Evaluation, and Verification workflow which is widely acknowledged and adopted by examiners and many law enforcement agencies and crime laboratories internationally [14, 20]. Although not perfect and sometimes criticized [49–53], ACE-V (and other similar workflows) is used in fingerprint examinations and can be considered the equivalent to the scientific hypothesis testing (i.e., data collection, testing, conclusion, and peer review) which allows to standardize the process and reduce identification errors such as false positives and false negatives. This workflow, however, is not error-free because it relies on human action and interpretation. Although not common practice yet, the use of a computerized grading system to qualify portions of a fingermark based on ridge clarity can aid in minimizing the level of subjectivity of an examiner [54]. Once the fingermark quality is checked, the AFIS systems use algorithms to search and provide (best) source candidates from a large database, which are then manually compared and verified for source identifications. Indeed, AFIS allows for the automatic search of unknown fingermarks and known fingerprints from persons whose fingerprints have been previously entered in the system. It is up to the examiner to draw conclusions based on whether he/she supports the results provided. Despite the acronym, the system does not "identify" the source of a fingermark, but it helps to establish a hierarchy of candidates in a list of determined size.

2.6 The Analysis of Latent Fingermarks: The Degradation Process and Aging Patterns

In real-life situations, a fingermark will never be immediately recovered from a scene; time goes by, and physical, chemical, and biological influences will begin to alter the initial latent impression at the physical, chemical, and biological levels. Each influencing factor will impact the "life expectancy" of the fingermark in very specific ways, a process that can be monitored, and possibly predicted, by the examination of certain measurable time-sensitive features [37]. Many of these tell-tale features – referred to as "aging parameters or indicators" – as well as the effect of certain influencing factors have been investigated in laboratory settings and have mostly been explored in indoor conditions, with a few exceptions [55]. This common approach is the direct result of two circumstances: 1) In many real scenarios, fingermarks are discovered inside homes or premises which become the most valuable evidence from an investigative perspective. Indeed, a (random) impression outside a burglarized store may be of little probative value or relevance to demonstrate that someone committed the crime. 2) Environmental conditions indoors are more limited, controllable, stable, somewhat predictable, and replicable than outdoors. Attempting to replicate all real-world environmental conditions outdoors is nearly impossible with today's technologies [3]. In indoor conditions, although with many variables at play and some degree of anticipated (estimated) error, the possibility to reliably date a fingermark by using mathematical models becomes more feasible and realistic. The effect of endogenous and exogenous factors on fingermarks is extensively covered in Chaps. 3, 5, 6, and 7.

2.6.1 Physical Degradation of Latent Fingermarks: The Effect of Substrate

The type of substrate a fingermark is deposited on has a major impact on the topographical degradation of latent fingermarks, even more than the environment it is aging in [3, 23, 42]. For most aging studies, non-porous substrates have been used in indoor conditions, including glass, tiles, plastics, metals, etc., and it has been demonstrated that the three types of fingermarks used, eccrine-rich, sebaceous-rich, and "natural," behave differently post-deposition [3, 25, 42]. For example, on a glass substrate, no major visual differences were observed between fresh and aged powdered sebaceous-rich fingermarks up to 5 months after deposition [3]. Overall, fingermark depositions on glass were proven to be more resilient to degradation, and more specifically sebaceous-rich impressions were identifiable for longer periods of time than their eccrine-rich counterparts [3]. On plastic surfaces, "halos" begin to develop for sebaceous-rich fingermarks after only 7 days, a clear sign of aging, although these impressions may still be identifiable [3]. Overall, on plastic, fingermarks were less resilient and had a lower visual quality, especially eccrine-rich depositions enhanced and examined 5–6 months post-deposition.

2.6.2 *Physical Degradation of Latent Fingermarks: The Effect of Environmental Conditions*

Several studies have explored the effect of the environment on fingermark degradation. De Alcaraz-Fossoul et al. (2012) showed that exposure to natural light through a window played a small role in the degradation of fingermarks on both glass and polystyrene plastic for both types of secretions examined, sebaceous-rich and eccrine-rich [3]. Under certain conditions, fingermarks aged in complete darkness showed the fastest ridge degradation, whereas in penumbra/shade, their behavior was more varied. However, it was noted that, in comparison with other environmental conditions, the degree of exposure to natural light did not play a primary role in the aging process and the visual quality of fingermarks.

Fingermarks are affected also by temperature and levels of relative humidity (RH%). Barnett et al. (1977) stored and examined fingermarks at 32%, 69%, 73%, 93%, and 98% relative humidity and aged up to 7 weeks [56]. In this case, RH% of 32%, 69%, and 73% were considered to be low humidity and 93% and 98% considered to be high humidity. The results from this study indicated that the fingermarks were in worse condition after being stored in high humidity for 7 weeks. The authors of the study stated that this was a "surprising" result since it was believed that the dehydration of fingermarks was one of the primary causes of aging. For the fingermarks stored at low humidity, there was no statistically significant difference in quality scores. It is likely that the moisture affected the speed and degree of (chemical) decay, induced topographical alterations due to physical changes, and influenced the survival of microbes feeding on sweat secretions; altogther having an impact on the overall ridge quality.

2.6.3 Chemical Degradation of Latent Fingermarks: The Effect of Substrate and Environment

The endogenous chemical profile of sweat secretions can play a role in the patterns of fingermark degradation and subsequent age estimation. As a latent fingermark degrades, the relative changes of its chemical compounds can be used as a tool in assessing its age [23, 26, 27, 35, 57]. The chemical composition over time is very dependent on environmental factors and varies between individuals. A study performed by Dorakambura et al. (2016) determined that diffusion of endogenous fatty acids from the ridges into the furrows of a fingermark can be used to determine its age. Unfortunately, fatty acids diffuse very quickly and only last about 2 days; therefore, it may be difficult to use this technique for long-term age estimations. In fact, the height of the secretion can be used as an aging parameter as it is affected by the amount of organic and inorganic compounds in the residue and their decay is the result of reacting with one another, the substrate, and the environment [58]. Interactions between chemical composition of the fingermark and the surface of deposition play an important role in the diffusion of molecular compounds from the ridges to the furrows. Higher molecular weight compounds could be used to date fingermarks over longer periods of time, but it is dependent on the properties of the surface of deposition [59].

In regard to temperature, a thermal-degradation experiment was conducted with seven amino acids that are commonly found in fingermark residue, as well as lactic acid and pyruvic acid [25]. These compounds were deposited onto four galvanized steel coupons heated to the desired temperatures (50 °C, 100 °C, and 150 °C), with one specimen at room temperature used as reference. The results demonstrated that all of the amino acids were undetectable after being exposed to 150 °C and the onset of degradation for lactic acid and pyruvic acid was detected at 50 °C. This research implied that the quality of latent fingermarks will be affected (or completely disappear) depending on the temperature they are exposed to. This experiment focused on extreme temperatures, and further research is necessary to examine the effect of this environmental factor on the quality of fingermarks with more conventional temperatures.

2.6.4 Biological Degradation of Latent Fingermarks: The Effect of Skin Microbiomes

Microorganisms (including one's own microbiome) most probably have an effect on the speed, pattern, outcomes, and levels of degradation of latent fingermarks. The human skin microbiome is highly complex and diverse due to skin containing numerous folds and niches that contain fungi, bacteria, viruses, and dead skin cells [30, 60, 61]. While alive, microorganisms on our skin protect us against foreign infections and are adapted to certain areas of the body depending on the physiology of that specific site. For example, some microorganisms tend to colonize drier skin regions, while others will prefer moisture regions of the body [60]. However, there are a handful of factors that influence an individual's microbiome. These factors include the donor's physiology (age, sex, ethnicity), immune responses such as previous exposure to pathogens or skin inflammation, the use of antibiotics, the donor's lifestyle (both occupational and hygienic), and the climate in which the individual lives [61]. There is minimal variation in one's skin microbiological signature over time, making each individual's microbiomes quite unique; therefore, the transfer of a fingermark's microbes on a substrate will be as distinctive. The differences in each person's microbiome profile may influence the aging pattern of a fingermark dependent on the interaction with the substrate and the environment. This topic is explored in more detail in Chap. 11.

Although the impact of microbes on latent fingermark degradation has not been sufficiently studied, it is quite realistic to assume they are present and active figures in the fingermark aging process, as microorganisms live on the human skin and "touch" is a mode of transmission of microbes [29]. Environmental factors, like the degree of exposure to natural light, relative humidity, and temperature, play a role in the life expectancy of microbes and may as well in fingermark degradation post-deposition [62]. Microorganisms, bacteria in particular, grow and thrive in the dark with humid and warm conditions away from damaging sun rays. In fact, research has shown that latent fingermarks stored in complete darkness, of both sebaceous-rich and eccrine-rich origin, degrade to a much higher degree, and especially eccrine-rich showed the most physical degradation [3]. Hence, the topographical degradation of ridges could be in part caused by the action of microorganisms. It could be hypothesized that after deposition, microbes are feeding from fingermark residues but more so from eccrine-rich specimens due to their chemical composition. It is not known how long microorganisms can survive on a fingermark; however, as fingermark components evaporate and degrade, the microorganisms would be expected to reduce or halt their activity and die. Research is necessary to determine the role of chemical and microbial (de)composition in estimating time since deposition and degradation patterns. The level of chemical/ biological decay could then be correlated with (microscopically) visible and measurable physical changes of the fingermark ridges (Fig. 2.1).

2.7 The Analysis of Latent Fingermarks: Visual Indicators of Degradation in 2D

As mentioned previously, the combination of donor, substrate, and environmental factors causes a differentiating effect on the "life expectancy" and the natural fingermark degradation process. At a crime scene, a plethora of factors continuously interact leading to a specific aging pattern of a ridge impression that impacts its physical appearance, as well as its chemical and biological profile. There are several visual

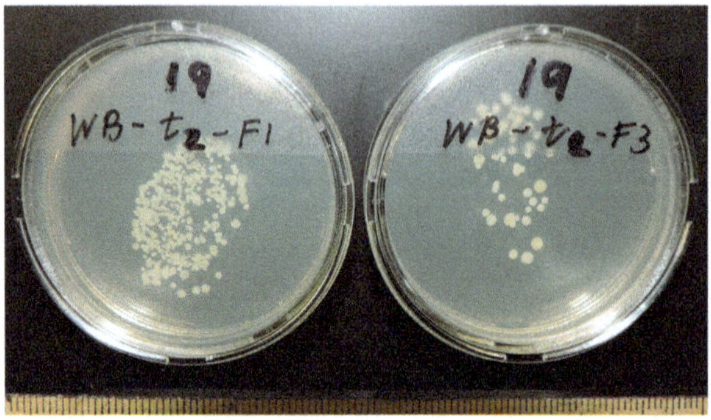

Fig. 2.1 Image depicting microbial colonies from two different fingertip depositions growing on solid agar medium incubated at 37 °C for 6 days. Note the position of the original impression. Fingers were washed and let to recharge 5 min without touching any surfaces before deposition

indicators of degradation that are sensitive to and affected by these factors and are the direct result of their activity. These aging indicators or parameters can be used as age predictors in a (future) mathematical aging model, applicable only to non-porous surfaces indoors. These indicators may include the loss of viable minutiae; a decrease in color contrast between ridges and furrows after enhancement; an increase in the number of discontinuities; and a change in the width of ridges, among others that may become available or discovered, explained in more detail in Chaps. 5 and 6.

In the context of fingermark aging, any morphometrical age indicator could be classified into two main categories [3]:

- Subject-independent parameters (SIPs): such as color contrast and the increasing number of ridge discontinuities
- Subject-non-independent parameters (SNIPs): such as changes in ridge width and number of minutiae

The main fundamental difference between these two categories lies in the information required for the parameter to work. For SNIPs, it would be necessary to know the donor's initial ridge width and the number of minutiae of the area being analyzed for comparison; thus, these indicators are highly dependent on the pre-identification of the donor prior to attempting an age estimation. For SIPs, it is not required to have any additional information from the exemplar or reference fingerprint, as long as some environmental data are known. This provides a good initial approach to a screening process based on age-sensitive visual quotes. In the following subsections, some of the effects of exogenous factors on these parameters are summarized and are a clear example of the differential behavior of the topography of fingermark ridges when exposed to the environment.

2.7.1 *Loss of Viable Minutiae*

In general, powder-enhanced latent fingermarks on glass aged at room temperature have shown to maintain a higher number of minutiae and are visually almost unchangeable over time (at least up to six months) compared to their counterparts on plastic substrates [4]. Of these, sebaceous-rich depositions present a higher number of minutiae and have a slower rate of degradation over time than eccrine-rich fingermarks. Exposure to direct sunlight through a window glass does not always result in a higher degree of physical degradation; however, the number of minutiae of eccrine-rich fingermarks on glass aged in darkness decreases much faster [4].

2.7.2 *Decrease of Color Contrast*

For fingermarks enhanced with a common powder developer, changes in the color contrast between ridges and furrows occurs over time and can become an excellent indicator of degradation [5, 54]. Under certain environmental conditions, the color of ridges (i.e., contrast with background) becomes weaker as the impression ages, either because there is a loss of sensitivity of the powder to the aged ridge compounds or no (less) ridge material is present. Color contrast can be quantified using the histogram feature in some image editing software and corroborates and complements the information revealed by the loss of minutiae on aging patterns. It has been observed that, in some cases, exposure to direct natural light extended the life expectancy of the impression. This aging parameter was able to detect that the type of substrate (glass vs. plastic) and type of secretion (eccrine- vs. sebaceous-rich) were more influencing than the different levels of exposure to natural light. The most "resilient" fingermarks that showed no apparent signs of degradation when examined with this aging parameter were sebaceous-rich deposited on glass substrates.

2.7.3 *Increase in Ridge Discontinuities*

The term discontinuity refers to the naturally occurring interruptions of the ridges that "appear" during the natural aging process of a latent fingermark after enhancement [6]. It can be further defined as a clear and complete interruption on the flow of the ridge at least 1/4 in length of the ridge width along the ridge. This is possibly caused by the chemical degradation of the ridge compounds. In the natural process of degradation, the number of discontinuities will progressively and randomly appear along the ridge and will increase in number compromising ridge integrity and flow. At a certain point in time, the number of discontinuities decrease due to the merging of individual interruptions and leading to a larger portion of the ridge being obliterated. In most cases, discontinuities increase with time; however, for sebaceous-rich fingermarks aged in the dark and for fingermarks aged in direct light, this age indicator may remain constant. Perhaps, this parameter might not be the most suitable nor sensitive to describe aging patterns. As shown by other age indicators, fingermarks on glass substrates were less affected than their counterparts on plastic. Additionally, direct exposure to natural light did not always increase the appearence of discontinuities, and storage in darkness was sometimes detrimental to the quality of the fingermarks [6].

2.7.4 Variations in Ridge Width

The physical modifications in ridge width over time have been explored and can also be reliable indicators of the degradation process [7, 8]. In order to make this parameter most meaningful in real scenarios, the examiners would have to compare any changes in ridge width from aged fingermarks with fresh counterparts or exemplars. Research has shown that plastic substrates may be more suitable for estimating time since deposition because the ridges experience a more noticeable alteration (i.e., narrowing or widening) over time than on glass. This aging parameter was able to detect that eccrine-rich fingermarks degraded much faster than their sebaceous-rich counterparts and, in some cases, the presence of natural light was irrelevant to the aging process. In the absence of light, ridges could experience the most severe levels of physical degradation. When ridge width is used as an indicator of aging, the type of substrate appears to have a higher impact than the secretion type and the degree of exposure to light [7].

2.8 Topography of Latent Fingermark: The Special Case of "Ridge Drift"

Fingermark ridge drift refers to the phenomenon where, under certain environmental conditions, a single ridge can randomly "change" from its original position over time while the adjacent ridges may remain unaltered [8]. As a result, a minutia at a specific location may be altered, for example, from a ridge ending to a bifurcation. This apparent visual modification of a ridge has been reported on powdered fingermarks using different developers (carbon black and titanium dioxide) only on non-porous surfaces, such as polystyrene plastic and glass. Currently, the causes of this phenomenon are not clear. In fact, there are at least three possible mechanisms that could explain the occurrence of ridge drift. The first implies that the ridge microscopically "migrates" over a non-porous surface by means of diffusion of residues. The second is a process named "selective degradation," where the process of aging affects only random specific micro-locations along the ridge. This creates the visual effect of an apparent move, but there is no actual movement of the ridge. The third mechanism could relate to the degree of adherence (sensitivity) of the developer to the degraded residue and residue diffused to the furrows. It is important to note that all these mechanisms are describing a phenomenon observed on non-porous surfaces and visualized with powders (a common method used at crime scenes) that require at least a time-lapse of 3 days after deposition. In addition, the visual quality of the fingermarks must be sufficient for a possible identification which could be critical during the process of attributing the source of the impression [3]. Research has shown that under certain environmental conditions, the prevalence of ridge drift is about 10% of the fingermarks examined. There are currently no other visualization methods that have described this phenomenon, nor has it been proved in 3D analysis [10], although the diffusion of residues on surfaces has been described [58].

2.9 The Analysis of Latent Fingermarks: Visual Indicators of Degradation in 3D

Despite the fact that fingerprint identifications have always been conducted in a two-dimensional plane, fingermarks are indeed three-dimensional "living" objects whose chemical and biological composition degrades and ages in all dimensions. In research, and in the field [63], some aging features that have been overlooked include changes in the height, area, and volume of the ridges post-deposition [10, 64, 65]. These 3D measurable parameters, combined with 2D visual analysis, can advance the understanding of how and why a fingermark degrades. For example, an optical profilometer (OP) is able to measure and record with high precision microscopic changes in the topography of a latent fingermark as it ages [10]. Typically, an OP is utilized in the discipline of material science to detect surface micro-abrasions and surface roughness. This technology is non-destructive and contactless, eliminating the need for fingermark enhancement and minimizing the potential manipulation or destruction of the sample. The analysis of fingermarks in 3D is explored in more detail in Chap. 6.

2.10 Conclusions

Latent fingermarks are complex and dynamic biochemical matrices that experience physical, chemical, and biological modifications over time caused by the continuous exposure to the environment post-deposition and influenced by donor, substrate and deposition conditions. From a topographical perspective, the collection of morphometric data can be a fundamental source of information to study aging processes. If age estimation of latent fingermarks is to become a reality in the near future, there is a need for more scientific research in areas such as designing portable and non-destructive techniques, identifying meaningful aging parameters, and deepening knowledge on the impact that environmental factors have on degradation. The satisfactory achivement of these goals combined should provide the tools and data necessary to formulate a mathematical model that will aid crime investigators in excluding or including suspects at the scene which in turn will strengthen the probative value of evidence and improve the public trust in the judicial system .

Acknowledgments I would like to thank Ms. Marisa San Antonio for her contribution in this manuscript.

References

1. Houck MM, Siegel JA (2015) Chapter 19 friction ridge examination. In: Fundamentals of forensic science, 3rd edn. Academic Press, San Diego, pp 493–518
2. De Alcaraz-Fossoul J, Islam M (2019) Chapter 3 novel technological applications for latent and blood-stained fingermark aging studies. In: Emerging technologies for the analysis of forensic traces. Springer, Cham, pp 33–66
3. De Alcaraz-Fossoul J et al (2012) Determination of latent fingerprint degradation patterns—a real fieldwork study. Int J Legal Med 127(4):857–870. https://doi.org/10.1007/s00414-012-0797-0
4. De Alcaraz-Fossoul J et al (2015) Latent fingermark aging patterns (part I): minutiae count as one Indicator of degradation. J Forensic Sci 61(2):322–333. https://doi.org/10.1111/1556-4029.13007
5. De Alcaraz-Fossoul J et al (2015) Latent fingermark aging patterns (part II): color contrast between ridges and furrows as one Indicator of degradation. J Forensic Sci 61(4):947–958. https://doi.org/10.1111/1556-4029.13099
6. De Alcaraz-Fossoul J et al (2016) Latent fingermark aging patterns (part III): discontinuity index as one Indicator of degradation. J Forensic Sci 62(5):1180–1187. https://doi.org/10.1111/1556-4029.13438
7. De Alcaraz-Fossoul J et al (2019) Latent fingermark aging patterns (part IV): ridge width as one Indicator of degradation. J Forensic Sci 64(4):1057–1066. https://doi.org/10.1111/1556-4029.14018
8. Barros RM et al (2013) Morphometry of latent Palmprints as a function of time. Sci Justice 53(4):402–408. https://doi.org/10.1016/j.scijus.2013.08.002
9. Popa G et al (2010) Method for fingerprints age determination. Rom J Legal Med 18(2):149–154. https://doi.org/10.4323/rjlm.2010.149
10. De Alcaraz-Fossoul J et al (2015) Fingermark ridge drift. Forensic Sci Int 258:26–31. https://doi.org/10.1016/j.forsciint.2015.11.003
11. De Alcaraz-Fossoul J et al (2018) Application of 3D imaging technology to latent fingermark aging studies. J Forensic Sci 64(2):570–576. https://doi.org/10.1111/1556-4029.13891
12. O'Neill KC, Lee YJ (2017) Effect of aging and surface interactions on the diffusion of endogenous compounds in latent fingerprints studied by mass spectrometry imaging. J Forensic Sci 63(3):708–713. https://doi.org/10.1111/1556-4029.13591
13. Champod C et al (2016) Issues related to the exploitation of fingerprints and fingermarks. In: Fingerprints and other ridge skin impressions, 2nd edn. CRC Press, Boca Raton, pp 315–331
14. Champod C et al (2016) Friction ridge skin and prints. In: Fingerprints and other ridge skin impressions, 2nd edn. CRC Press, Boca Raton, pp 1–31

15. Cohen PR (2017) Capecitabine-associated loss of fingerprints: report of Capecitabine-induced Adermatoglyphia in two women with breast Cancer and review of acquired dermatoglyphic absence in oncology patients treated with Capecitabine. Cureus 9(1):e969. https://doi.org/10.7759/cureus.969. PMID: 28191373; PMCID: PMC5298917
16. Bordas L, Bonsutto J (2020) Adermatoglyphia: the loss or lack of fingerprints and its causes. J Forensic Identif 70(2):154–162
17. National Institute of Justice (2014) The fingerprint sourcebook. National Institute of Justice, Washington, DC
18. Galton F (1892) Finger prints. Macmillan, New York
19. Vucetich J (1904) Dactiloscopía Comparada (comparative fingerprinting): the new argentine system
20. Ashbaugh DR (1999) Quantitative-qualitative friction ridge analysis: an introduction to basic and advanced ridgeology. CRC Press, Boca Raton
21. Chaberski M (2008) Level 3 friction ridge research. Biom Technol Today 16(11–12):9–12. https://doi.org/10.1016/s0969-4765(08)70234-0
22. Girod A et al (2012) Composition of fingermark residue: a qualitative and quantitative review. Forensic Sci Int 223(1–3):10–24. https://doi.org/10.1016/j.forsciint.2012.05.018
23. Cadd S et al (2015) Fingerprint composition and aging: a literature review. Sci Justice 55(4):219–238. https://doi.org/10.1016/j.scijus.2015.02.004
24. Kent T (2016) Water content of latent fingerprints – dispelling the myth. Forensic Sci Int 266:134–138. https://doi.org/10.1016/j.forsciint.2016.05.016. Epub 2016 May 24. PMID: 27262684
25. De Paoli G et al (2010) Photo- and thermal-degradation studies of select Eccrine fingerprint constituents. J Forensic Sci 55(4):962–969. https://doi.org/10.1111/j.1556-4029.2010.01420.x
26. Girod A et al (2016) Aging of target lipid parameters in fingermark residue using GC/MS: effects of influence factors and perspectives for dating purposes. Sci Justice 56(3):165–180. https://doi.org/10.1016/j.scijus.2015.12.004
27. Girod A et al (2015) Fingermark initial composition and aging using Fourier transform infrared microscopy (μ-FTIR). Forensic Sci Int 254:185–196. https://doi.org/10.1016/j.forsciint.2015.07.022
28. Kirk PL (1953) Crime investigation: physical evidence and the police laboratory. Interscience Publishers, Inc, New York
29. Hefetz I et al (2019) Touch DNA: the effect of the deposition pressure on the quality of latent fingermarks and STR profiles. Forensic Sci Int Genet 38:105–112. https://doi.org/10.1016/j.fsigen.2018.10.016
30. Noureddine M, Bailey JA (2016) A protocol for the recovery of STR DNA from fingerprints developed on the adhesive side of duct tape. J Forensic Identif 66(6):30
31. Califf K et al (2014) The human microbiome: getting personal. Microbe Mag 9:410–415
32. Carter DO et al (2017) Forensic microbiology. Wiley, Hoboken
33. Metcalf JL et al (2017) Microbiome tools for forensic science. Trends Biotechnol 35(9):814–823. https://doi.org/10.1016/j.tibtech.2017.03.006
34. Meadow JF et al (2014) Mobile phones carry the personal microbiome of their owners. PeerJ 2:e447. https://doi.org/10.7717/peerj.447
35. Oonk S et al (2018) Proteomics as a new tool to study fingermark ageing in forensics. Sci Rep 8(1). https://doi.org/10.1038/s41598-018-34791-z.
36. Yamashita B, French M (2011) Latent print development. In: Holder EH, Robinson LO, Laub JH (eds) The fingerprint sourcebook. National of Justice, US Department of Justice, Washington, DC, pp 7.1–7.67
37. Williams DK, Brown CJ, Bruker J (2011) Characterization of children's latent fingerprint residues by infrared microspectroscopy: forensic implications. Forensic Sci Int 206:161–165. https://doi.org/10.1016/j.forsciint.2010.07.033

38. Lambrechts SAG, van Dam A, de Vos J, van Weert A, Sijen T, Aalders M (2012) On the autofluorescence of fingermarks. Forensic Sci Int 222:89–93. https://doi.org/10.1016/j.forsciint.2012.05.004
39. Bleay SM, Croxton RS, Puit MD (2018) Fingerprint development techniques: theory and application. Wiley, Hoboken
40. Hagan AO, Green S (2018) Crime scene to court: a study on finger-mark aging. Forensic Res Criminol Int J 6(6):491–503. https://doi.org/10.15406/frcij.2018.06.00249
41. Vanderwee J, Porter G, Renshaw A, Bell M (2011) The investigation of a relative contrast index model for fingerprint quantification. Forensic Sci Int 204:74–79. https://doi.org/10.1016/j.forsciint.2010.05.005
42. Girod A et al (2016) Fingermark age determinations: legal considerations, review of the literature and practical propositions. Forensic Sci Int 262:212–226. https://doi.org/10.1016/j.forsciint.2016.03.021
43. Holyst B (1987) Kriminalistische Abschätzung des Spurenalters bei Fingerpapillarlinien [Criminal assessment of the age of fingerprints]. Arch Kriminol 179(3–4):94–103. German. PMID: 3592908
44. Baniuk K (1990) Determination of age of fingerprints. Forensic Sci Int 46(1–2):133–137. https://doi.org/10.1016/0379-0738(90)90151-n
45. Hicklin RA et al (2013) Assessing the clarity of friction ridge impressions. Forensic Sci Int 226(1–3):106–117. https://doi.org/10.1016/j.forsciint.2012.12.015
46. Humphreys JD et al (2008) The quantification of fingerprint quality using a relative contrast index. Forensic Sci Int 178(1):46–53. https://doi.org/10.1016/j.forsciint.2008.02.003
47. Kalka ND, Austin Hicklin R (2014) On relative distortion in fingerprint comparison. Forensic Sci Int 244:78–84. https://doi.org/10.1016/j.forsciint.2014.08.007
48. Pulsifer DP et al (2013) An objective fingerprint quality-grading system. Forensic Sci Int 231(1–3):204–207. https://doi.org/10.1016/j.forsciint.2013.05.003
49. Cole SA (2005) More than zero: accounting for error in latent fingerprint identification. J Crim Law Criminol 95(3):984–1078
50. Cole SA, Roberts AJ (2012) Certainty, individualisation and the subjective nature of expert fingerprint evidence. SSRN Electron J. https://doi.org/10.2139/ssrn.2703676
51. Dror I, Rosenthal R (2008) Meta-analytically quantifying the reliability and biasability of forensic experts. J Forensic Sci 53(4):900–903
52. Dror IE, Charlton D (2006) Why experts make errors. J Forensic Identif 56(4):600–616
53. Mnookin J, Kellman PJ, Dror I, Erlikhman G, Garrigan P, Ghose T, Metler E, Charlton D (2016) Error rates for latent fingerprinting as a function of visual complexity and cognitive difficulty. Report. NIJ Award 2009-DN-BX-K225
54. Matuszewski S, Szafałowicz M (2013) A simple computer-assisted quantification of contrast in a fingerprint. J Forensic Sci 58(5):1310–1313. https://doi.org/10.1111/1556-4029.12224
55. Czech A et al (2019) Changes in fingerprints and the quantity of material forming the print depending on hand cleanliness, gender, and ambient conditions. J Forensic Sci 65:84. https://doi.org/10.1111/1556-4029.14164.
56. Barnett PD, Berger RA (1977) The effects of temperature and humidity on the permanency of latent fingerprints. J Forensic Sci Soc 16:249–254
57. Pleik S et al (2016) Fatty acid structure and degradation analysis in fingerprint residues. J Am Soc Mass Spectrom 27(9):1565–1574. https://doi.org/10.1007/s13361-016-1429-6
58. Dorakumbura BN et al (2016) Nanomechanical mapping of latent fingermarks: a preliminary investigation into the changes in surface interactions and topography over time. Forensic Sci Int 267:16–24. https://doi.org/10.1016/j.forsciint.2016.07.024
59. O'Neill KC, Lee YJ (2017) Effect of aging and surface interactions on the diffusion of endogenous compounds in latent fingerprints studied by mass spectrometry imaging. J Forensic Sci 63(3):708–713. https://doi.org/10.1111/1556-4029.13591
60. Grice EA, Kong HH, Conlan S, Deming CB, Davis J et al (2009) Topographical and temporal diversity of human skin microbiome. Science 324(5931):1190–1192

61. Grice EA, Segre JA (2011) The skin microbiome. Nat Rev Microbiol 9(4):244–253
62. Bumbrah GS, Sharma RM, Jasuja OM (2016) Emerging latent fingerprint technologies: a review. Res Rep Forensic Med Sci 6:39–50
63. O'Hagan A (2018) Crime scene to court: a study on finger-mark aging. Forensic Res Criminol Int J 6(6):491. https://doi.org/10.15406/frcij.2018.06.00249.
64. Liu F et al (2017) Case study of 3D fingerprints applications. PLoS One 12(4):e0175261. https://doi.org/10.1371/journal.pone.0175261
65. Merkel R et al (2012) On non-invasive 2D and 3D chromatic white light image sensors for age determination of latent fingerprints. Forensic Sci Int 222(1–3):52–70

Chapter 3
Estimating the Age of Fingermarks: Relevance, Potential Approaches, and Perspectives

Céline Weyermann and Aline Girod-Frais

Abstract The age estimation of (latent or visible) fingermarks recovered at crime scenes may be particularly useful to be able to discern traces relevant to the investigated event or to help place a particular event in time. The measurement of physical or chemical changes of fingermark characteristics over time has been proposed as a mean to achieve this goal. However, fingermark depositions are complex matrices that are influenced by many factors besides time, such as donor, transfer, and environmental conditions to cite only a few. Thus, the main challenge resides in the identification of aging parameters that evolve over time with as little influence as possible from such factors. The development of reliable aging models that can be implemented in practice to place fingermarks in time requires the acquisition of large amounts of fundamental data on the aging processes of target compounds, if possible, using versatile and easily available analytical techniques. The importance of the data interpretation stage should not be underestimated either. Thus, while age estimation of fingermarks would be particularly useful for crime reconstructions or as evidence in court, much more information is currently needed on the aging of natural fingermarks. A research cycle is proposed to address this issue in a more systematic manner, and several cases in which fingermark dating was either attempted or debated are discussed from a forensic perspective.

Keywords Fingerprint · Latent and visible fingermarks · Composition · Aging · Dating · Time

---

C. Weyermann (✉)
Ecole des Sciences Criminelles, University of Lausanne, Lausanne, Switzerland
e-mail: celine.weyermann@unil.ch

A. Girod-Frais
Criminal Intelligence Service (Bundeskriminalamt), Crime Scene Unit, Vienna, Austria

Department of Criminal Law and Criminology, University of Vienna, Vienna, Austria
e-mail: aline.girod-frais@bmi.gv.at; aline.girod-frais@univie.ac.at

J. De Alcaraz-Fossoul (ed.), *Technologies for Fingermark Age Estimations: A Step Forward*, https://doi.org/10.1007/978-3-030-69337-4_3

Acronyms

AFM	Atomic Force Microscopy
CWL	Chromatic White Light
DNA	Deoxyribonucleic Acid
FTIR	Fourier Transform Infrared Spectroscopy
GC	Gas Chromatography
HIS	Hyperspectral Imaging
LC	Liquid Chromatography
LR	Likelihood Ratios
MALDI	Matrix-Assisted Laser Desorption Ionization
MS	Mass Spectrometry
PA	Peak Area
PCA	Principal Component Analysis
PLSR	Partial Least Squares Regression
QCM	Quartz Crystal Microbalance
RMSE	Root Mean Square Errors
SIMCA	Soft Independent Modeling of Class Analogy
SIMS	Secondary Ion Mass Spectrometry
TLC	Thin-Layer Chromatography
TOF	Time of Flight
UP	Ultra-performance
UV-Vis	Ultraviolet Visible Light

3.1 Introduction

Determining the time of fingermark deposition has long been a major concern for forensic scientists and still remains a complex challenge in practical settings [1–4]. Information on time of placement serves several purposes. On one hand, when the time of the offense is (approximatively) known, it enables the selection of relevant traces[1] to be collected for further examination [5], i.e., only traces transferred during the criminal event, rather than before or after, are collected. Furthermore, it allows the selection of adequate detection techniques to enhance latent traces. For example, while the ideal enhancement techniques for fingermarks should work on all kinds of traces, some are reported to perform better on fresher or older marks, such as iodine vapors or physical developer, respectively. This is especially important when the elapsed time between the offense and the investigation reaches a significant interval

[1]A trace can be defined as a mark, signal, or object. It is an observable sign (not always visible to the naked eye), the vestige of a presence, or an action at the place of the latter.

(e.g., several weeks, months, or even years), thus increasing the risk of degradation of the pertinent enhancement targets, as well as the risk of contamination of the crime scene. On the other hand, when the time of the offense remains unknown, estimating the age of relevant traces may help reconstruct the investigated events in time (e.g., estimation of the time of death through the measurement of body temperature).

While the main use of fingermarks by law enforcement agencies remains the identification of its source (i.e., *who* has left the trace), the answers to *how* and *when* were the marks transferred on crime scene are mostly implicitly inferred from *where* the marks were detected. Indeed, a (latent) fingermark discovered on a glass or bottle of wine will have less significance than one on the murder weapon, particularly when the identified person had a legitimate reason to visit the crime scene before the crime occurred (e.g., family, friends, acquaintances). The location of fingermarks at the scene thus helps reconstruct the activities of the people potentially implicated in the crime (e.g., drinking a glass of wine or handling a knife). However, the question of time has to be explicitly addressed and becomes highly relevant when a person admits to having handled the knife – generally well before the crime occurred – but for legitimate reasons, such as cooking. In such cases, the identity of the person is no longer under question, but the time at which the mark had been transferred [6]. Failing to consider the question of *when* in such cases could lead to potential wrong convictions, which highlights the relevance of estimating the age of fingermarks, as well as other forensic traces.

Focusing principally on fingermarks, three main approaches have been discussed in the literature to situate them in time, considering [1, 6]:

1. Investigative information (e.g., if windows are regularly cleaned, then the maximum interval between transfer and discovery of traces on the glass can be extrapolated)
2. Chronology of deposition (e.g., when two traces overlap, determining which one was deposited first)
3. Physical and chemical changes of fingermark characteristics over time (e.g., degradation of the friction ridge skin detail[2] or chemical components)

This chapter will mainly focus on the third approach, i.e., the measurement of changes, generally aging processes, to estimate time since transfer. The requirements for the development of dating methods will first be detailed. Then, the potential and limitations for implementation in practical settings will be discussed through the presentation of case examples.

[2] Friction ridge details comprise the combination of friction ridge flow, friction ridge characteristics (including levels I, II, and III), and friction ridge structure.

3.2 Requirements for the Development of an Approach to Estimate the Age of Fingermarks

As previously described in the literature, the development of a methodology to estimate the age of fingermark residue is a very complex challenge [4, 6, 7]. While operational methods have been reported for the dating of other forensic traces, such as ink or gunshot residue [8, 9], their implementation in practice still remains to some extent a contentious issue. Girod et al. [6] proposed a research cycle to develop and test such methods for practical application (Fig. 3.1).

3.2.1 Target Characteristics

The key strategy to develop dating methods based on aging processes is to select target characteristics that actually change over time in a reproducible and measurable way (e.g., the decrease of squalene[3] in the residue [10, 11] or the minutiae count in the ridge pattern [12] after fingermark deposition). These characteristics should ideally be present in the residue of all donors in order for the method to work on a broad range of fingermarks found at crime scenes [13–15]. Furthermore, the targets should be easily analyzed using analytical methods available in forensic laboratories. Promising candidates are discussed below (see also Table 3.1; see also Chap. 7 for detailed descriptions of the chemical degradation processes).

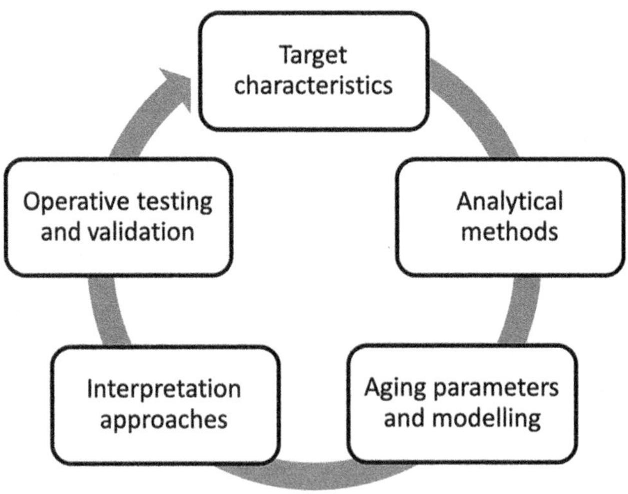

Fig. 3.1 Proposed research cycle to develop dating methods of forensic traces. (Adapted from [6])

[3] Squalene is an organic compound found in human sebum and is an essential precursor in the biosynthesis of cholesterol, steroids, and vitamin D.

Table 3.1 List of possible target characteristics/compounds reported in the literature to describe fingermark aging

Targets	Occurrence	Studied time range	Analysis method	Selected references
Ridge detail	In relevant marks	Up to several years	2D Visual examination 3D Imaging HSI AFM TOF-SIMS	[12, 20–26]
Water	In all fingermarks	A few minutes	QCM	[19, 27]
Chloride	In all fingermarks	ca. 200 days	Silver nitrate	[16, 28, 29]
Tryptophan derivatives	In all fingermarks	A few weeks	Spectroscopy TLC	[30]
Lipids	In all fingermarks	Several weeks	FTIR RAMAN GC-MS LC-MS/MS MALDI-MS	[10, 11, 31–38]
Amino acids	In all fingermarks	ca. 200 days	RAMAN GC-MS	[15, 29, 39]
Proteins	In all fingermarks	A few weeks	Gel electrophoresis LC-MS/MS	[40, 41]

QCM quartz crystal microbalance, *HSI* hyperspectral imaging, *AFM* atomic force spectroscopy, *GC* gas chromatography, *LC* liquid chromatography, *MS* mass spectrometry, *TOF-SIMS* time-of--flight secondary ion MS

In addition to the visual characteristics linked to the ridge detail, many potential target compounds have been identified in fingermark residue and mainly originate from the secretory glands found in the dermis (i.e., the bottom layer of the skin): the apocrine, eccrine, and sebaceous glands [16–18]. Apocrine glands are found only in the genital, mammary, inguinal, and axillary regions, and thus, their secretions do not represent particularly interesting targets for enhancement or dating purposes (i.e., compounds should be found in a majority of fingermarks to be considered relevant targets). Eccrine glands, on the other hand, are found over the whole body with greater masses on the palmar and plantar surfaces. While the initial water content of eccrine secretions can be significant (generally between 20 and 70% of the fingermark's initial weight and up to 90% when hands have been recently washed [19]), several other components have been identified as potential targets to monitor fingermark aging such as proteins, peptides, amino acids, and chlorides. Finally, sebaceous glands are not present on the friction ridge surfaces of the hands and feet. However, many sebum components are still found in fingermark residue because of frequent contact between the hands and body parts such as the face and hair. Lipids secreted by the sebaceous glands have been identified as particularly interesting target compounds to study aging. Natural fingermark residue is generally composed of a mixture of eccrine and sebaceous secretions with external contaminants (e.g., cosmetics, food residue, dust, bacteria, nicotine). The latter are rarely targeted as they are not found in all fingermarks. However, cosmetics often contain compounds that are also found in natural fingermarks, such as fatty acids or wax esters. This can increase the potential of such compounds for enhancement of fingermarks and may influence their aging processes.

It was first proposed to investigate fingermark aging through changes in the quality of the visible two-dimensional ridge details and loss of water [3, 7, 12, 20, 21, 23–25, 32]. While still intuitively used by some investigators, such observations were not deemed reliable for operational purposes yet, and extreme caution should be taken when attempting to correlate pattern quality with time since deposition. Concerning water loss, a recent study indicated that initial water content varied significantly between marks and that evaporation occurred within minutes after deposition [19]. Given the short timeframe of water loss, it cannot be considered as a relevant target component to study aging from a forensic perspective. Since then, several parameters have been suggested to quantitatively describe physical changes in the ridge details over time: minutia count, color contrast, ridge discontinuity, and ridge width and height [12, 23–25]. While promising results were obtained, additional multivariate data is necessary to further investigate fingermark aging using such characteristics.

Lipids are one of the main classes of compounds of interest in aging studies. Free fatty acids, sterols, wax esters, monoglycerides, diglycerides, and triglycerides were previously identified in the fingermarks of many donors [16, 17, 42]. Several studies indicated that fatty acid evolution did not show a clear tendency over time and were thus barely investigated to monitor aging [10, 11]. However, the degradation of two unsaturated fatty acids (Δ6-hexadecenoic acid and Δ8-octadecenoic acid) was recently studied over 14 days, as well as products of the oxidation process of these two target compounds [36]. Squalene, a precursor of steroids, and cholesterol were detected in fingermarks of all donors and decreased relatively quickly in the first days after deposition, following an exponential decay up to ca. 30 days. However, large differences were detected between donors, and normalization of the data was proposed to improve reproducibility [10, 11, 31, 32, 37, 43]. Many wax esters were also identified in fingermarks [44], and some were found to decrease over time, such as isopropyl dodecanoate [32]. Wax esters were, however, detected in relatively small amounts in fingermarks compared to sterols and fatty acids [13], and their decrease over time has not been systematically characterized yet. Finally, intact diglycerides and triglycerides have recently been characterized and shown to decrease rapidly over time following deposition [38, 45–47]. Degradation of the above lipids appears to be largely due to ozonolysis, a process by which ozone reacts with the double bonds of unsaturated compounds, yielding a number of ozonides (1,2,4-trioxolanes) and shorter-chain oxidation products. More systematic studies might yield useful additional data to investigate fingermark aging using such target compounds.

Many studies into fingermark composition actually use "groomed"[4] marks to increase the amount of lipids in the residue. Donors are usually asked to rub their fingers on their forehead or neck before deposition. Thus, most studies focusing on lipids will actually report quantities that may be significantly higher than in natural marks. While some of these studies proved the feasibility of studying the degradation of fingermark lipids under laboratory conditions ("proof of concept"), it would be necessary for operational applications to study the initial lipid composition and subsequent aging of natural fingermarks instead of groomed ones [48].

Eccrine components have been less studied than lipids, and it was only recently suggested that their degradation patterns over time may be of interest for fingermark aging/dating purposes [40]. Amino acids were characterized in natural and groomed fingermarks [15]. However, no data was reported about their aging in ambient conditions, and these compounds were thus not considered yet for estimating the age of fingermarks. While still present in older fingermarks [49], a decrease in amino acids was reported in several publications over longer time periods or due to thermal degradation [29, 39]. Thus, these compounds might present interesting alternatives to study longer time spans if the observed decrease can be confirmed to occur in a reproducible manner. Proteins and peptides were also characterized in fingermarks and recently proposed to monitor aging processes [40, 41]. Among 31 identified proteins, 4 keratins and dermcidin showed significant (increasing or decreasing) trends over 16 days [40]. Possible modifications occurring over time include alkylation, disulfide reduction, ammonia loss, and potassium adducts. As sample preparation can also influence degradation, further tandem mass spectrometry studies over longer time periods would bring a better understanding of these aging processes and potentially help identifying further potential targets.

A decrease in the autofluorescence of fingermarks was also observed and attributed to protein-bound tryptophan derivatives, including indoleacetic acid, (nor)harman, and xanthurenic acid [30]. These compounds were also suggested as potential aging markers, but changes were not studied quantitatively over time. Finally, the diffusion of chloride components was proposed over 40 years ago as an interesting parameter to study the aging of fingermarks on paper substrates [28, 29]. However, the reliability of this approach was not demonstrated, and chlorides were not further evaluated as potential aging markers.

The main difficulty when selecting target compounds to model aging lies in the fact that many intrinsic (related to the fingermark itself) and extrinsic (related to the substrate and environment) factors influence the aging processes of fingermarks [16, 42, 50] (Fig. 3.2). It was previously demonstrated that fingermark composition can vary significantly between different donors physiologically, but also due to activities such as food handling, handwashing, sweating or use of cosmetics ($t < 0$). Then, the initial composition transferred to the substrate will additionally be influenced by the deposition conditions ($t = 0$). Factors such as substrate type,

[4] "Groomed" refers to fingermarks that are artificially charged with, for example, sebaceous secretions by rubbing fingertips to the forehead before deposition.

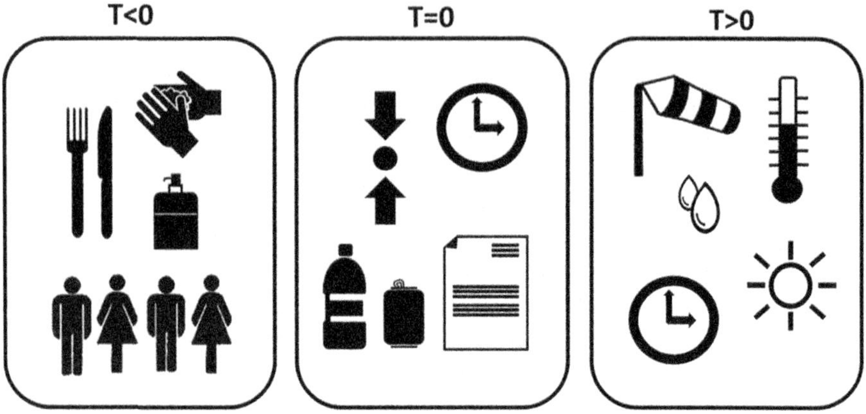

Fig. 3.2 Factors influencing the composition of fingermark residue over time. The deposition occurs at time t = 0

temperature, pressure, and duration of the contact thus have a significant influence. Finally, the conditions and duration of storage will have a major impact on the aged composition of the trace found during crime scene investigation (t > 0). Target compounds should ideally be influenced as little as possible by these factors, or the aging model should be able to account for their impact.

3.2.2 Analytical Methods

The main forensic criteria for adequate analytical methods are cost, rapidity, simplicity, reliability, noninvasiveness, and versatility (see Table 3.2). Thus, selected techniques should be able to carry out analyses efficiently (with minimal resource requirements), in a reproducible way, and detect a wide range of targeted compounds with minimal sample preparation in order to allow subsequent analysis steps. As some approaches may be destructive of the ridge details, a fingermark will generally have to be detected (possibly using enhancement techniques) and recorded before age estimation analysis. Furthermore, if a relevant mark is of insufficient quality to infer who has deposited it, then the sampling and analysis of contact DNA (or touch DNA) will generally be attempted. Thus, analytical techniques and methods of interest for age estimation should be compatible in sequence with both fingermark enhancement techniques and DNA analysis [44, 51].

Several (mainly optical and nondestructive) techniques have been reported to describe physical changes that occur on a fingermark ridge pattern over time. Natural and groomed fingermark aging was studied over 20 h using hyperspectral imaging (HSI) [52]. Atomic force microscopy (AFM) was also suggested to measure changes in the physical characteristics of fingermarks over time [21, 53, 54]. A decrease in the height of the ridges and horizontal migration across (nonporous)

Table 3.2 List of criteria for the techniques able to detect fingermark aging characteristics (physical and chemical) over time

Forensic criteria	Optical examination	Spectroscopic methods	Basic separation methods	Advanced technologies
Examples of techniques	Light source / imaging (3D OP)	HSI, FTIR, RAMAN	GC-MS, LC-MS	UPLC-MS/MS MALDI-TOF-MS
Purchase and maintenance costs	Low	Low to medium	Relatively high	Very high
Analysis time	Quick	A few minutes (+ calibration)	30–60 min (+ calibration)	Variable, longer for imaging (+ calibration)
Simplicity	Basic skills	Basic to advanced skills	Advanced skills	Expert skills
Reliability	Low to medium, mainly qualitative[a]	High, but mainly qualitative	High, quantitative	High, quantitative
Noninvasiveness	Yes	Yes	No	Variable
Versatility and selectivity	Visible pattern	Class of compounds, imaging potential	Volatile and thermally stable lipids, derivatized amino acids	Larger lipids, peptides, and proteins, imaging potential
Availability	All forensic laboratories	Broadly available	Specialized laboratories	Rarely available
Portability	Yes	Yes	Limited	No

[a]Quantitative data can be extracted from high-resolution 2D and 3D imaging (3D optical profilometry); this would potentially increase the reliability of some optical examination techniques

substrate was recorded over 2 months [21]. Physical changes in fingermarks have also been studied over ca. 1 month using electrochemical impedance spectroscopy [55]. Measurement of the contrast modifications between fingermark ridges and background using a chromatic white light (CWL) sensor and/or high-resolution images was also proposed [25, 56]. More recently, very simple optical approaches have also been introduced to study visible changes occurring in the two- and three-dimensional pattern, including minutiae, ridge continuity, contrast, and ridge width and height [12, 20, 23–25]. Optical examination combined with quantitative measurements would make such approaches ideal from a forensic science perspective (see also Chaps. 4, 5, and 6 for further details on the measurement of physical characteristics).

On the chemical composition, while less detailed information is generated using optical methods, these are still generally favored for forensic operational purposes, as they comply with several main criteria such as noninvasiveness as well as simplicity, rapidity (minimal sample preparation), and lower costs, compared to more specific analytical approaches. Fourier transform infrared spectroscopy (FTIR) was one of the main optical techniques used to characterize fingermark composition, specifically the lipid components [57–60]. The amino acid serine was also targeted in one study [58]. While the influence of the donor's age was evaluated using FTIR [61], fingermark aging was also studied over approximately 1 month, taking into account the influence of substrate, light, and moderate temperature [33, 34]. While FTIR is a promising approach to study lipid aging, limitations have also been identified. Indeed, this technique is only semiquantitative and cannot be applied on all type of substrates. It works best on aluminum foil and other nonporous substrates such as glass but cannot be applied on porous substrates such as paper, thus limiting its applicability in forensic practice.

An optical alternative was recently suggested using Raman spectroscopy. While mainly used to identify exogenous compounds such as drugs [62], it has also been applied for the unspecific detection of eccrine and sebaceous secretions [35, 63–65]. Some signals attributed to squalene and fatty acids showed a decrease over 1 month. However, high variability between measurements was also discussed, and results should be carefully interpreted using sufficient replicate analyses of specimens [64].

Many analytical techniques have been proposed to monitor the chemical composition of fingermarks over the years. While generally more invasive than optical methods, they are also more specific, quantitative, and reproducible. Their potential for fundamental research is evident; however, they are also often more expensive, time-consuming, and more complex to apply, and thus less adapted to forensic routine practice. While thin-layer chromatography (TLC) was first employed [66, 67], it was soon replaced by other, more specific techniques such as gas chromatography (GC) or liquid chromatography (LC), nowadays generally coupled to mass spectroscopy (MS). Both techniques allow the study of a wide range of compounds from sebaceous and eccrine secretions. GC-MS generally targets more volatile and thermally stable compounds (e.g., small lipids and amino acids) [10, 11, 31, 32, 36, 68], while ultra-performance (UP)LC-MS/MS is suited to larger molecules such as glycerides, peptides, and proteins [36, 40, 46, 69] as well as to elucidate the structure of the target compounds and aging products [37, 46, 69].

Several alternative mass spectrometric methods have also been used to study the fingermark chemical composition and more specifically chemical imaging of endogenous and exogenous compounds [70–72]. Time-of-flight secondary ion mass spectrometry (ToF-SIMS) imaging was used to study diffusion of saturated fatty acids over time [22]. Some techniques, using lasers and/or electrospray ionization, are minimally invasive at least for less volatile compounds such as lipids and proteins (as most techniques require a vacuum for ionization of the molecules); however, such instrumentation is very expensive, and their use requires a high level of skill and maintenance, making them less adapted to routine application. While optical methods do not suffer from these disadvantages and are generally more easily applicable directly on the crime scene (i.e., when portable devices are used) [73], they are also less selective and reliable to track quantitative changes in several target molecules over time. Thus, no method fulfills all the forensic criteria, and the methods listed in the two central columns of Table 3.2 would represent the best compromise to follow the chemical degradation of fingermarks, with more reliability and selectivity for GC-MS and LC-MS and better availability and less invasiveness for HSI and FTIR.

3.2.2.1 Aging Parameters and Modeling

Once potential target characteristics have been identified, their aging processes must be examined and modeled as a function of time, including the effect of influencing factors to reproduce realistic conditions as encountered in practical settings.

One study focusing on the measurement of the (color) contrast differences between ridges and background proposed an aging parameter based on pixel colors to model fingermark aging over a period of 24 h. It was reported that aging followed a logarithm curve. However, aging could not be modeled over a longer time span (due to large inter- and intra-variability), thus limiting the potential of this method for operational purposes [56]. Several recent publications proposed to follow the evolution of minutiae count, color contrast between ridges and furrows, ridge discontinuities, and ridge width as aging parameters over 6 months using linear regressions for simplicity purposes [12, 23–25]. Their results suggested that the unique combination of environmental factors had a significant distinct impact in ridge degradation. In some instances, no visual degradation could be detected (e.g., sebaceous--rich fingermarks on glass exposed to direct light), while in other cases complete obliteration of the mark was observed (e.g., eccrine-rich fingermarks on plastic stored in darkness) over the exact same time periods. This demonstrated, again, the relevance of the interaction of factors, and the value of each aging parameters, and that they should be considered in the aging models in a multivariate approach to increase reliability.

Concerning chemical target compounds, it was observed that different donors present very variable initial fingermark composition, at least quantitatively. Furthermore, while inter-donor variability is significantly higher than intra-donor variability, variation has also been measured between fingermarks from the same donor [13, 74]. Thus, the calculation of ratios between endogenous compounds has been suggested as an approach to significantly decrease the effects of quantitative variability of individual compounds. The ratio of two main compounds (i.e., the peak area (PA) of squalene was divided by the PA of cholesterol) was proposed to characterize the decrease of these lipid compounds over time [11]. In order to further reduce variability, other preprocessing steps were tested using the PA of compounds analyzed by GC/MS. The best preprocessing technique was determined to be the square root of the PA of a target compound normalized to the sum of PA of other compounds [32]. For the sum of the denominator, seven lipids were selected – isopropyl dodecanoate (IDP), three squalene derivatives (SD1-3), cholesterol (Chol), myristyl palmitoleate (MPO), and myristyl palmitate (MP):

$$\text{Aging parameter MPO} = \sqrt{\frac{PA_{MPO}}{PA_{IDP} + PA_{SD1} + PA_{SD2} + PA_{SD3} + PA_{Chol} + PA_{MPO} + PA_{MP}}} \quad (3.1)$$

Exponential linear regressions were used to model the decrease of the normalized aging parameters. While squalene was reported as a promising target [32], the aging of other compounds can be modeled as shown in Fig. 3.3 for the aging parameter MPO (Eq. 3.1).

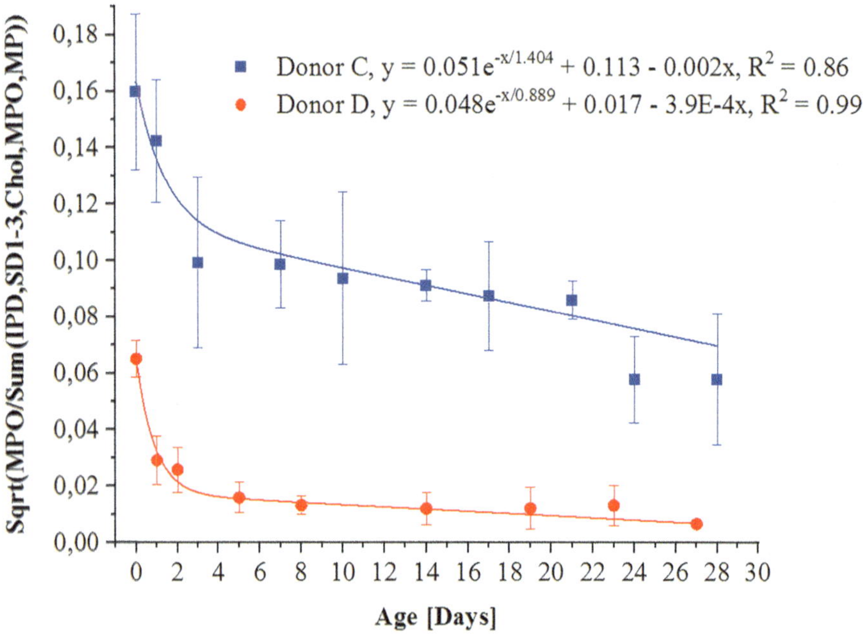

Fig. 3.3 Aging curves for two donors using myristyl palmitoleate (MPO) as the target compound. Fingermarks were deposited on paper and stored in the dark during 28 days before extraction using dichloromethane and analysis using GC/MS [32]. The aging parameter is represented as a function of time and fitted using exponential linear regressions

Ratios were also proposed to follow triglyceride degradation [45]. The decrease of the ratio TG48:1/TG48:0 was followed over 24 days, as well as the increase of the ratios of monoozonide products of TG48:0.

A second approach to follow fingermark aging focused on the use of multivariate models, thus combining the potential of several aging parameters in one model. Partial least squares regression (PLSR) was tested for this purpose, and it was possible to differentiate fresh fingermarks ($t \leq 3$ days) from older ones ($t \geq 10$ days). However, the PLS models showed large spreading due to the variability among the fingermark samples of the same donor (intra-variability) even when using preprocessing techniques. This type of model may thus not be the best choice to infer about the age of fingermarks [32].

As described above, many factors influence fingermark composition and its aging. In an attempt to propose solutions to deal with this issue, factors were classified into two groups [32]. The known factors, such as the donor (if a suspect has been identified), the substrate (on which the fingermark has been detected), and the enhancement techniques (applied to the mark before analysis) can be determined by the forensic investigators, while the time of deposition, pressure, and environmental conditions are more difficult to reconstruct. Known factors can be included in the aging models since it has been demonstrated that substrate and enhancement techniques also have significant influence on the initial quantity recovered and/or the aging processes [32]. The considered time range will depend on the hypotheses formulated during the investigation. Short time spans (e.g., up to 1 month) would be relatively easy to reconstruct during an investigation using the suspect's marks; however, longer time ranges would require additional resources that may not be available even for serious crimes. Thus, research should endeavor to study fingermark aging for longer time periods, for example, up to 1 year. For such longer periods, lipid components will probably be less relevant targets than amino acids or proteins, as their degradation is rather rapid. While pressure has been shown to influence fingermark length [75], it seems to have less influence on chemical degradation than storage conditions [32]. Light was found to accelerate the aging of lipids in particular, while small changes in temperature (15–25 °C) had a more moderate impact [32, 34, 45, 64, 69]. Reconstructing the storage conditions (approximate temperature, humidity, and light exposure) may be possible over short periods of time (using weather reports as done for entomological purposes). Naturally occurring indoor storage conditions will probably have a lower effect on fingermark aging than outdoor environments. However, exposure to light should still be accounted for as it can significantly influence the degradation processes of some lipids, such as squalene and triglycerides [32, 45, 46, 69]. The impact of substrate type, initial composition (e.g., groomed fingermarks), and exposure to light were also found to modify ridge pattern characteristics [12, 23–25].

3.2.3 *Interpretation Processes, Operative Testing, and Validation*

Several approaches to interpret the age of fingermarks can be developed depending on the information available to investigators (e.g., known vs. unknown suspect). From an investigative perspective, particularly if a suspect has not yet been identified, the objective is to situate the traces in time either to evaluate the relevance of the detected traces or to evaluate when the investigated event (or particular actions) occurred [76]. Due to the fact that in such cases the donor would not have been identified yet, the interpretation models would aim to determine an approximate timeframe during which the mark has been left, taking into account the circumstances and including, if possible, the influence factors (e.g., substrate type, storage conditions). Up to date, no interpretation process including validation steps were published using physical characteristics as aging parameters. Concerning chemical characteristics, three different interpretation models were developed based on lipid target compounds and tested for fingermarks left on glass and stored in an office drawer or on an office desk near a window for up to 34 days (see Figs. 3.4, 3.5, and 3.6 adapted from data acquired in [32]).

The first model was based on principal component analysis (PCA) combined with soft independent modeling of class analogy (SIMCA) in order to predict to which predefined age groups a questioned mark belonged. Validation of this

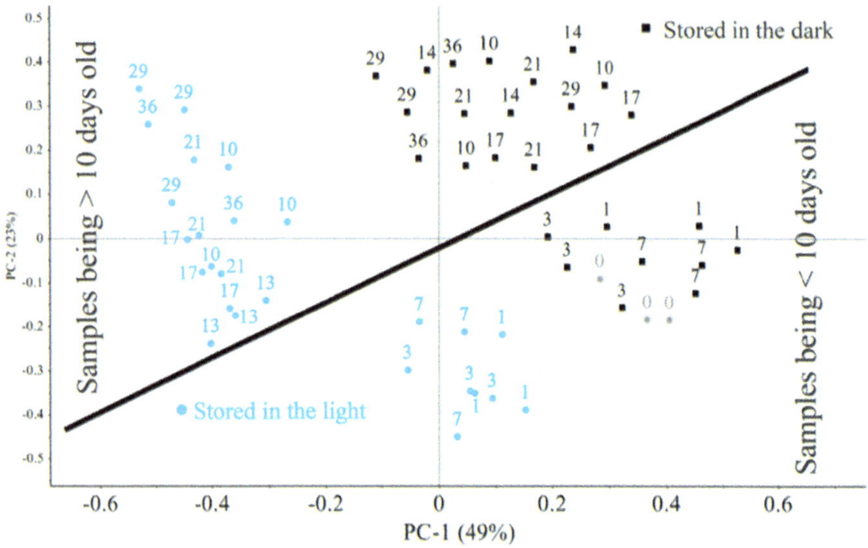

Samples	Real age, storage	SIMCA
FM1	22d, light	N.A.
FM3	1d, light	< 10d
FM5	1d, dark	< 10d
FM7	22d, dark	< 10d

Fig. 3.4 PCA model using six aging parameters and based on latent fingermarks deposed on microfiber filters and left to age in the dark between 0 and 36 days before extraction using dichloromethane and analysis using GC/MS [32]. Based on this model, fingermarks were classified into two groups: < 10 days old and > 10 days old (d = day). One PCA per observed group was built in order to classify blindest samples (e.g., FM1, FM3, FM5, and FM7) in one or the other group using SIMCA. The classifications in green were correct, while the ones in red were wrong

supervised classification method was conducted by an internal cross-validation study and then applied to specimens unknown to the principal investigator through blind testing: substrate and donor were known variables, while age and storage conditions remained unknown to the forensic analyst. Fingermarks were classified as being under or over 10 days old regardless of their storage conditions (dark vs. light). While the model showed very promising results, the storage conditions had a significant impact on the aging of older fingermarks (see Fig. 3.2). Two false results were observed, i.e., two 22- and 34-day-old fingermarks stored in a drawer were classified as being less than 10 days old, showing that aging is much slower when marks are not exposed to light. Moreover, a 34-day-old fingermark exposed to light aged to such an extent in the given timeframe that the model was unable to classify it. Thus, such approaches should take into account storage conditions, and particular care should be taken when evaluating the age of potentially older marks, as some

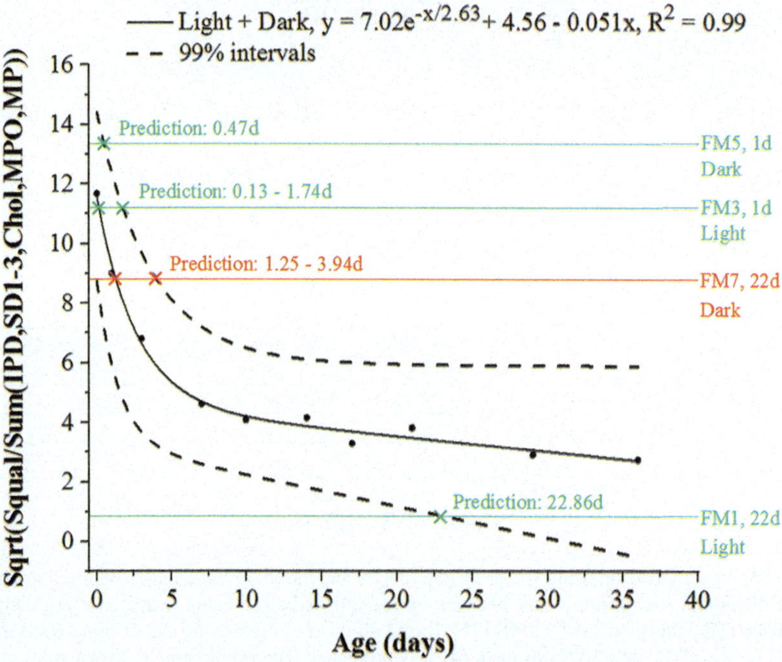

Fig. 3.5 Exponential linear regression using the aging parameter of squalene. Fingermarks deposed on microfiber filters and stored in the dark during 36 days before extraction using dichloromethane and analysis using GC/MS [32]. Based on this regression and using 99% confidence intervals, the ages of blind test samples (e.g., FM1, FM3, FM5, and FM7) measured in days (d) were predicted. The predictions in green were deemed acceptable (actual age correctly predicted within 1 day), while the one in red was incorrect (more than 18 days off)

specific conditions can significantly slow or accelerate aging processes and yield erroneous classification. The practical feasibility of such models should also be further tested using fingermarks from unknown donors and using further apart aging periods, such as very fresh (e.g., less than 3 days) compared to older marks (e.g., more than 2 weeks, 1 month, 1 year). This could potentially increase the feasibility of the model in specific conditions (i.e., predefined timeframes).

The second model was based on univariate exponential linear regressions (see examples in Fig. 3.5). While most studies do not explicitly propose interpretation models, they often present their results using univariate aging curves. The results obtained for eight test marks using a 99% confidence interval yielded five correct and three incorrect estimations [32]. As the model was constructed independently of factors such as light and dark conditions, the errors were quite large, and a correct result could be quite imprecise. For example, one mark was correctly classified as being more than 3 days old but was actually 34 days old. The fresh and old marks' ages were generally correctly determined (even if sometimes imprecisely) using this model. The errors actually occurred for the ages that fell along the center of the regression curve. This may support the hypothesis that it is only possible to

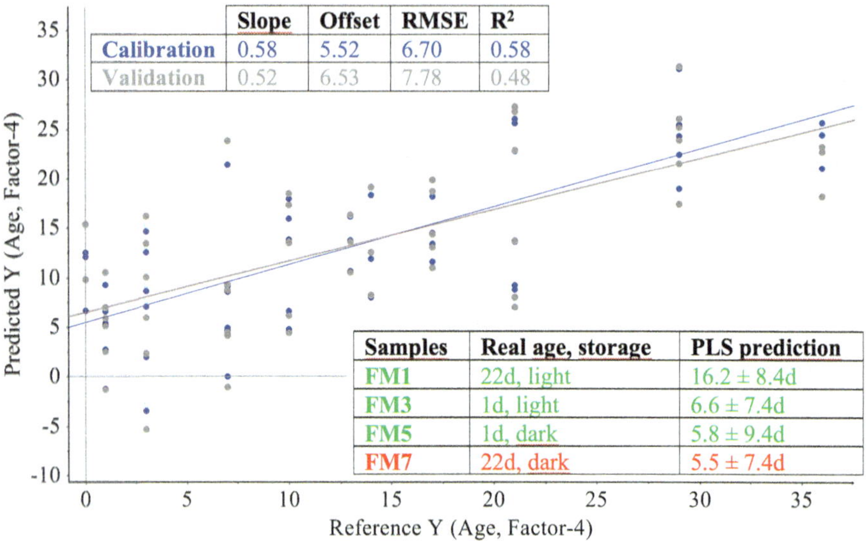

Fig. 3.6 PLS regression using six aging parameters and based on fingermarks deposed on microfiber filters and stored in the dark between up to 36 days before being extracted using dichloromethane and analysis using GC/MS [32]. Based on this regression, the age of blind test samples (FM1, FM3, FM5, and FM7) in days (d) was predicted. The predictions in green were correct (however, with large errors), while the one in red was wrong

differentiate very fresh (less than 3 days old) from older (more than 30 days old) ones, at least based on the rapid decrease of squalene over the first days after deposition. It would be interesting to test other compounds that are known to decrease more slowly for a finer age discrimination and/or estimation over longer time spans.

A multivariate PLS regression was also tested to predict the age of unknown specimens including a prediction error [32]. The robustness of the model was evaluated through the root mean square errors (RMSE) and the coefficients of determination (R^2). Mean RMSE were of ca. 8 days, and obtained R^2 for calibration (0.58) and prediction (0.48) were actually quite low. Thus, predictions were not very precise, and three errors were produced with older fingermarks (22 and 34 days old) as the model could not accurately predict aging in both dark and light exposure conditions. Results varied more significantly over time, and thus prediction of older samples became more difficult if the storage conditions were unknown. These results tend to confirm that storage conditions should be predictable (in addition to substrate and donor) to accurately predict the age of fingermarks over ca. 1 month. Indeed, a PLSR model created specifically for fingermarks deposited on filter paper for one donor and stored in a drawer yielded lower RMSE (ca. 3 days) and higher R^2 (above 0.80). Such a model has also been tested on aluminum and analyzed using FTIR [33]. The regression models showed good linearity for both storage conditions ($R^2 > 0.90$) and errors were relatively low (between 2 and 3 days). Storage conditions had less influence on the results in this study compared to previously obtained results [32]. Unlike GC/MS, FTIR spectra were obtained using continuous measurements from the same fingermarks over time as specimens were not destroyed by the analysis. The influence of initial composition was thus minimized when building the aging model, and expected errors may actually be significantly higher in more realistic conditions.

It is interesting to mention that a fourth type of interpretation model was proposed to allow the comparison of two alternative propositions for the age of a questioned mark, for example, in a court context [32]. For this purpose, likelihood ratios (LR) were calculated using a preliminary Bayesian network. The selected alternative hypotheses for the purpose of illustration were the following: the questioned mark was deposited less than 10 days prior to analysis (i.e., age $\leq$ 10 days old), or it was deposited more than 2 weeks before (i.e., age $\geq$ 14 days old). Two age estimate errors were obtained using the same blind test set (mentioned above), again for older marks of 22 and 34 days stored in drawers. LR values were generally very low (below 4), except for two older marks exposed to light and for which LR values reached ca. 10 and 12. These results highlight, as previously mentioned, that models cannot accurately predict the age of older fingermarks if they do not include the storage conditions (i.e., dark/light exposure). However, when the likelihood ratio approach was tested using separate models for the two storage conditions, it yielded correct estimations for all the tested specimens. Obtained LR values were still very low for dark conditions but reached high values for marks exposed to light. Investigating larger time/age differences would be interesting for further research, particularly for dark conditions under which aging occurred more slowly.

While research continues to gather further fundamental knowledge about fingermark composition and aging, the development and validation of interpretation models remain largely overlooked. In order to develop a reliable fingermark aging method fit for practice, it is essential that additional studies focus on this particular topic. If the validation step fails, the research cycle has to be started anew, focusing either on new target compounds, more reliable analytical methods, or alternative aging and interpretation models. Based on the literature overview, it appears that fingermark dating methods cannot be implemented universally to all encountered cases using the same aging and interpretation models, but should instead be adapted to specific cases considering known substrate, donor, storage conditions, and relevant time span. Even if the feasibility of fingermark dating is limited, acceptable reliability could be guaranteed through adequate testing of the statistical models (including the determination of error rates) and probabilistic approaches. It must be also noted that the majority of research on the composition and aging of fingermarks has been conducted using groomed samples. However, before being able to apply a method in forensic practice, its implementation should also be tested on natural residue, adding another level of complexity to the interpretation. While there are several proposals to control fingermark deposition and environmental conditions, as well as chemical composition for research purposes [43, 77], scientists must be aware that the implementation of dating models in practice needs to be based on results obtained using realistic and uncontrolled specimens and includes blind testing in casework conditions.

3.3 Practical Implementation of Dating Models

A review of 28 court cases between 1961 and 2011, in which age estimation of fingermarks had been attempted, indicated that most cases were burglaries or robberies, attempts included (61%) followed by homicide (32%) [6]. The two remaining cases concerned (sexual) assault and kidnapping. Fingermark age estimation was generally based on the quality of the enhanced friction ridge patterns and experience of the forensic expert, rather than experimentally driven data. While 16 of these judgments were confirmed on appeal based on fingermark evidence and/or other pieces of information, 12 others were reversed on appeal.

It is interesting to note that in the five cases where the defense presented alternative explanations for the presence of the traces, the judgment was reversed [78–81] with the exception of a robbery case in which one suspect was released while the other was convicted [82]. One of these cases concerned a homicide [80]. A 72-year-old man was found dead by his niece in the living room of his house. He had been restrained and shot in the head. The house had been ransacked, and his watch and the money in his wallet were missing. The only evidence which connected the defendant to the murder was a partial fingermark discovered on a small metal box located in another room. The fingermark specialist rightly testified based on his experience that "the length of time a latent print would stay on a metal object like the box found on the desk in the den would vary depending upon conditions, but that a print could stay there for a period of weeks." The niece of the deceased testified that the defendant was unknown to her, that the box was usually kept in a filing cabinet, and that it was only handled by family members. Based on these elements, the defendant was convicted. In the appeal, it was stated that as the fingermark could have been a few weeks old, "during that period the defendant could have either entered the porch unlawfully without the occupant's knowledge or lawfully to transact business with the deceased." This alternative explanation about the presence of the fingermark prompted the judgment to be reversed.

In two burglaries, fingermarks were detected on broken glass fragments [78, 79]. In both appeals, it was concluded that the windows were accessible to the public and the marks could have been deposited for legitimate reasons, such as looking through a shop window. In Townsley v. United States [78], the court also noted that touching a window might be a trivial incident and it would be normal that the defendant had no memory of it when arrested several months later. In Solis v. People [79], the forensic expert testified "that latent prints would remain for about 2 years, unless they were exposed to the elements." As no other evidence was offered by the prosecution, the appeal was accepted in both instances. The fingermark expert's observations regarding the persistence of traces were confirmed by other forensic scientists as reported in the literature [18].

The two last cases dealt with objects that could have been handled by many people in stores before having been purchased and subsequently seized at crime scenes (i.e., a map found in a car and an air freshener found in a bathroom, respectively [81, 82]). In the latter, the court noted the importance of establishing the time of deposition of fingermarks to avoid jeopardizing "the liberty of every person who ever touches anything later found at the scene of a crime [81]." This ruling clearly emphasizes the importance of estimating the age of fingermarks (and other traces) detected on everyday objects found on crime scenes to demonstrate their relevance to the investigated event. While the question "who?" can be answered by dactyloscopic comparison and the location of a fingermark on a crime scene/object largely answer the question "where?", the questions "when?" and to some extent "how?" remain mainly unanswered if the fingerprint residue cannot be situated in time. Indeed, an object could have been handled either in a shop or at the crime scene weeks before the crime occurred, or during the investigated event, and through legitimate (e.g., selecting products during shopping) or illegal (e.g., conducting a search during a burglary) actions (how and for what purpose?).

There are, of course, several other ways to answer the questions mentioned above. Investigative information collected by the police (e.g., testimonies, circumstantial information) should not be ignored, as they might be more straightforward and, at the current stage of knowledge, even more reliable than the age estimation of fingermarks. As discussed in Sect. 3.2, the implementation of dating methods could involve significant costs and resources and may only be justified in serious crimes such as homicides or sexual assaults. Furthermore, fingermark dating would currently mainly be feasible if samples were collected relatively quickly after the occurrence of the criminal event. Based on chemical degradation research, it would be feasible to differentiate fresh fingermarks from significantly older ones (e.g., a few hours/days vs. a few weeks/months) [32]. For example, a fingermark found on a box kept in a closed cabinet (known substrate, storage conditions, and suspect, as in State v. Scott [80]) could probably be classified as having been left a few hours before the crime, rather than several weeks before it occurred. The metallic box was kept in a closed cabinet, and storage conditions could be inferred relatively easily. Using a combination of ridge pattern degradation parameters, a few weeks old fingermarks may be distinguishable from months old traces in certain conditions [25].

During the investigation, it should be noted why a particular object was examined for fingermarks (was the object displaced or used during the burglary?) and how many traces were found (including those of the legitimate users). If only a few fingermarks are recovered on a questioned object, it probably means that this object is cleaned regularly, and the maximum possible age of the collected fingermarks may be estimated. However, as previously mentioned, some objects can often be manipulated by different persons. This is why it is necessary to note the exact placement of the collected fingermark on the object, as its position could give information about the activity conducted when leaving the mark (legitimate or illegal). Such investigative information is very important to be able to estimate fingermark age. Moreover, in order to base these estimations on sound data, research should also focus on the persistence mechanisms of marks on different objects routinely used within our everyday activities. This could help answering questions such as "Is it possible that a fingermark can persist for several weeks on an object that has been bought by other people in a shop?". In the air freshener case example mentioned [81], if the questioned fingermark had been found on the spray button (typical position when using the spray, rather than by touching it in a shop), it would probably not persist for very long, particularly if the air freshener spray was being used regularly. However, experience has also shown that fingermarks of good quality can still be detected several years after deposition on glass windows and on the interior surfaces of apartments as reported in Solis v. People [18, 79], while fingermarks on a drinking glass or a knife would probably not persist after washing the dishes. Knowledge of the persistence mechanisms of fingermarks is thus highly interesting in such contexts.

In summary, situating fingermarks in time could help identify relevant traces (investigation stage) and avoid the conviction of innocent people (court stage). Age estimation can be based on measurable chemical or physical characteristics over time and/or on other relevant information about the deposition and persistence processes. However, further research is needed, as suggested in previous sections following the cycle proposed in Fig. 3.1, in order to develop reliable fingermark dating methodologies. In the meantime, while age estimation is not yet deemed feasible, it is important to consider fingermarks not only at the source level but also at the activity level. This is why crime scene investigation is crucial. Precise information about the placement of marks should always be recorded to enable sound interpretation of the evidence, particularly when a suspect had legitimate reasons to visit the crime scene (e.g., he/she was living in the apartment several months before the crime occurred) or to handle objects later found at the crime scene (e.g., air freshener or metallic box). Thus, situating fingermarks in time should be carried out in a global investigative framework endeavoring to answer all relevant questions surrounding the criminal event (who, what, where, when, how, why, how often, etc.).

One also has to consider that in cases in which a partial mark of poor quality is encountered, the risk of erroneous identification is increased, and age estimation cannot always provide a solution [18]. In the Brandon Mayfield case, for example, the fingermark found at the crime scene in Madrid was actually relevant to the investigation as it led to the identification and arrest of a suspect, but it was not left by Brandon Mayfield as initially thought by several fingermark experts [83]. Thus, particularly when a fingermark is partial and/or of poor quality, extreme care should be taken in the interpretation of the data, and probabilistic approaches should be preferred to categorical identification to avoid erroneous conclusions and decisions [84, 85].

3.4 Conclusion and Perspectives

Beyond the basic question "who?", answers to other relevant queries such as "when?", "where?", or "how" could help minimizing erroneous judicial decisions in criminal inquiries. This requires close collaboration and communication between police investigators, crime scene technicians, and forensic scientists.

In this context, situating fingermarks in time would be particularly useful to select relevant traces at crime scenes and, subsequently, if the persons at the source of those traces are identified, to infer their presence (or actions) at the time of the investigated event. While forensic scientists have often attempted to estimate the "age" of fingermarks (i.e., their time of deposition on objects), a reliable dating method has yet to be reported. It is widely accepted that fingermark patterns, at least on porous surfaces, can persist for a very long time when not exposed to extreme conditions (e.g., direct exposure to water, light, or higher temperatures). Thus, visual observation of the quality of a fingermark conducted with the naked eye cannot be easily correlated to its age and more data is still needed to build and test reliable aging models based on the objective measurements of physical characteristics.

Physical and chemical changes occurring over time have frequently been proposed to estimate the age of fingermark residue. For example, lipid components are known to decrease in quantity over time through diverse degradation and diffusion processes. Squalene, a steroid precursor generally present in relatively high quantities in the initial fingermark residue, has received particular interest, and its decrease after deposition was studied over several weeks. While several potential aging and interpretation models have been proposed (for squalene and other target compounds), many factors play a role in the aging processes and should be further studied to evaluate the practical feasibility of such approaches. It has been suggested that the person at the source of the questioned trace must be identified to reduce the high variability of the initial composition of inter-donor fingermarks. The ideal model should also strive to include substrate type and storage conditions to minimize the risk of making erroneous age estimations. Dating methodologies based on measuring changes in the chemical composition of fingermarks, while interesting and promising, are complex and may never be suitable to determine precise ages. Preliminary results indicate, however, that such approaches could be able to discriminate between fresh (i.e., a few days old) and older fingermarks (i.e., a few weeks old). Blind tests should be systematically carried out at the end of the validation process to evaluate the methods' feasibility and reliability. Furthermore, as the development and application of dating approaches for practical implementation demand large amounts of resources, age estimation should primarily be applied in the investigation of serious crimes such as homicides, even if it would be useful for any type of crime.

Academic research on the composition and aging of fingermark residue also plays an important role in the development of dating methodologies. Future research should focus on the systematic acquisition of fundamental data following the cycle proposed in this chapter. At the current stage of knowledge, it is advised to promote in-depth studies of selected aging parameters and processes (including 2D and 3D changes), rather than preliminary proof-of-concept studies. In fact, new technologies might be less crucial in resolving the dating challenges than data processing and interpretation. With this objective in mind, much more information is currently needed on the aging of more *realistic* specimens (i.e., natural fingermarks stored under different conditions) in order to build reliable aging and interpretation models.

Acknowledgments The authors wish to thank the Swiss National Science Foundation (SNF no. PP00P1_123358/1 and 205121_169677) which enabled the preparation of this collaborative chapter. A great thank also goes to Dr. Amanda Frick and Dr. Ana Moraleda from the School of Criminal Science of the University of Lausanne for their suggestions for improving this chapter.

References

1. Weyermann C, Ribaux O (2012) Situating forensic traces in time. Sci Justice 52(2):68–75
2. Margot P (2000) A question of time. Sci Justice 40(2):64–71
3. Wertheim K (2003) Fingerprint age determination: is there any Hope? JFI 53(1):42–49
4. Midkiff C (1993) Lifetime of a latent print. How long? Can you tell? JFI 43(4):386–392
5. Margot P (2011) Commentary on the need for a research culture in the forensic sciences. UCLA Law Rev 58:795–801
6. Girod A et al (2016) Fingermark age determinations: legal considerations, review of the literature and practical propositions. Forensic Sci Int 262:212–226
7. De Alcaraz-Fossoul F, Barrot Feixat C, Roberts KA (2015) The paradigm of fingerprint age determination. Jacobs J Forensic Sci 1(1):006
8. Aginsky VN (1996) Dating and characterizing writing, stamp, pad and jet printer inks by gas chromatography / mass spectrometry. Int J Forensic Doc Examiners 2(2):103–116
9. Andersson C, Andrasko J (1999) A novel application of time since the latest discharge of a shotgun in a suspect murder. J Forensic Sci 44(1):211–213
10. Archer NE et al (2005) Changes in the lipid composition of latent fingerprint residue with time after deposition on a surface. Forensic Sci Int 154:224–239
11. Weyermann C, Roux C, Champod C (2011) Initial results on the composition of fingerprints and its evolution as a function of time by GC/MS analysis. J Forensic Sci 56(1):102–108
12. De Alcaraz-Fossoul J et al (2016) Latent fingermark aging patterns (part I): minutiae count as one Indicator of degradation. J Forensic Sci 61(2):322–333
13. Girod A, Weyermann C (2014) Lipid composition of fingermark residue and donor classification using GC/MS. Forensic Sci Int 238:68–82
14. Frick AA et al (2015) Investigations into the initial composition of latent fingermark lipids by gas chromatography–mass spectrometry. Forensic Sci Int 254:133–147
15. Croxton RS et al (2010) Variation in amino acid and lipid composition of latent fingerprints. Forensic Sci Int 199(1):93–102
16. Girod A, Ramotowski R, Weyermann C (2012) Composition of fingerprint residue: a qualitative and quantitative review. Forensic Sci Int 223(1–3):10–24
17. Frick AA, Fritz P, Lewis SW (2013) Chemistry of print residue. In: Siegel JA, Saukko PJ, Houck MM (eds) Encyclopedia of forensic sciences, 2nd edn. Academic, Waltham, pp 92–97
18. Champod C et al (2016) Fingerprints and other ridge skin impressions, 2nd edn. CRC Press, Boca Raton
19. Keisar O et al (2019) Measuring the water content in freshly-deposited fingermarks. Forensic Sci Int 294:204–210
20. De Alcaraz-Fossoul J et al (2019) Application of 3D imaging technology to latent fingermark aging studies. J Forensic Sci 64(2):570–576
21. Popov KT, Sears VG, Jones BJ (2017) Migration of latent fingermarks on non-porous surfaces: observation technique and nanoscale variations. Forensic Sci Int 275:44–56
22. Muramoto S, Sisco E (2015) Strategies for potential age dating of fingerprints through the diffusion of sebum molecules on a nonporous surface analyzed using time-of-flight secondary ion mass spectrometry. Anal Chem 87(16):8035–8038
23. De Alcaraz-Fossoul J et al (2019) Latent fingermark aging patterns (part IV): ridge width as one indicator of degradation. J Forensic Sci 64(4):1057–1066
24. De Alcaraz-Fossoul J et al (2017) Latent fingermark aging patterns (part III): discontinuity index as one indicator of degradation. J Forensic Sci 62(5):1180–1187
25. De Alcaraz-Fossoul J et al (2016) Latent fingermark aging patterns (part II): color contrast between ridges and furrows as one indicator of degradation. J Forensic Sci 61(4):947–958
26. Merkel R, Dittmann J, Vielhauer C (2011) Approximation of a mathematical aging function for latent fingerprint traces based on first experiments using a Chromatic White Light (CWL) sensor and the binary pixel aging feature. In: International federation for information processing (IFIP). CMS 2011, LNCS 7025, pp 59–71

27. Kent T (2016) Water content of latent fingerprints – dispelling the myth. Forensic Sci Int 266:134–138
28. Angst E (1962) Procédés pour la détermination de l'âge d'empreintes dactyloscopiques sur le papier. Rev Int Criminol Pol Tech 16:134–146
29. Cuthbertson F (1972) The chemistry of fingerprints. Atomic energy authority, Atomic Weapons Research Establishment (AWRE Report No. 013/69). Aldermastone, United Kingdom
30. van Dam A et al (2016) On the autofluorescence of aged fingermarks. Forensic Sci Int 258:19–25
31. Frick AA et al (2016) Monitoring compositional changes of the lipid fraction of fingermark residues deposited on paper during storage. Forensic Chem 2:29–36
32. Girod A et al (2016) Aging of target lipid parameters in fingermark residue using GC/MS: effects of influence factors and perspectives for dating purposes. Sci Justice 56(3):165–180
33. Girod A et al (2015) Fingermark initial composition and aging using Fourier transform infrared microscopy (μ-FTIR). Forensic Sci Int 254:185–196
34. Johnston A, Rogers K (2017) The effect of moderate temperatures on latent fingerprint chemistry. Appl Spectrosc 71(9):2102–2110
35. Yang J-H, Yoh JJ (2018) Reconstruction of chemical fingerprints from an individual's time-delayed, overlapped fingerprints via laser-induced breakdown spectrometry (LIBS) and Raman spectroscopy. Microchem J 139:386–393
36. Pleik S et al (2016) Fatty acid structure and degradation analysis in fingerprint residues. J Am Soc Mass Spectrom 27(9):1565–1574
37. Dorakumbura BN, Busetti F, Lewis SW (2020) Analysis of squalene and its transformation by-products in latent fingermarks by ultrahigh-performance liquid chromatography-high resolution accurate mass Orbitrap™ mass spectrometry. Forensic Chem 17:100193
38. Hinners P, Thomas M, Lee YJ (2020) Determining fingerprint age with mass spectrometry imaging via ozonolysis of triacylglycerols. Anal Chem 92:3125–3132
39. De Paoli G et al (2010) Photo- and thermal-degradation studies of select eccrine fingerprint constituents. J Forensic Sci 55(4):962–969
40. Oonk S et al (2018) Proteomics as a new tool to study fingermark ageing in forensics. Sci Rep 8(1):16425
41. Drapel V et al (2009) Identification of promising antigenic components in latent fingermark residues. Forensic Sci Int 184(1):47–53
42. Cadd S et al (2015) Fingerprint composition and aging: a literature review. Sci Justice 55(4):219–238
43. Dorakumbura BN, Busetti F, Lewis SW (2019) Investigations into sampling approaches for chemical analysis of latent fingermark residue. Forensic Chem 14:100166
44. Koenig A, Girod A, Weyermann C (2011) Identification of wax esters in fingermark residues by GC/MS and their potential use as aging parameters. JFI 61(6):606–630
45. Pleik S et al (2018) Ambient-air ozonolysis of triglycerides in aged fingerprint residues. Analyst 143(5):1197–1209
46. Frick AA et al (2020) Changes in latent fingermark glyceride composition as a function of sample age using UPLC-IMS-QToF-MSE. In preparation
47. Frick AA, Weyermann C (2019) An untargeted lipidomic approach for qualitative determination of latent fingermark glycerides using UPLC-IMS-QToF-MSE. Analyst 144(11):3590–3600
48. Almog J et al (2014) Guidelines for the assessment of fingermark detection techniques International Fingerprint Research Group (IFRG). JFI 64:174–197
49. Hansen DB, Jouille M (2005) The development of novel ninhydrin analogues. Chem Soc Rev:408–417
50. Bleay SM, de Puit M (2018) Ageing of fingermarks. In: Bleay SM, Croxton RS, de Puit M (eds) Fingerprint development techniques. Wiley, pp 69–97
51. Templeton JEL, Linacre A (2014) DNA profiles from fingermarks. BioTechniques 57(5):259–266

52. Merkel R (2015) Latent fingerprint aging from a hyperspectral perspective: first qualitative degradation studies using UV/VIS spectroscopy. In: 2015 ninth international conference on IT Security Incident Management & IT Forensics
53. DelRio FW, Cook RF (2017) Quantitative scanning probe microscopy for nanomechanical forensics. Exp Mech 57(7):1045–1055
54. Dorakumbura BN, Becker T, Lewis SW (2016) Nanomechanical mapping of latent fingermarks: a preliminary investigation into the changes in surface interactions and topography over time. Forensic Sci Int 267:16–24
55. Rosa R et al (2017) Electrochemical impedance spectroscopy: a deeper and quantitative insight into the fingermarks physical modifications over time. Forensic Sci Int 273:144–152
56. Merkel R (2014) New solutions for an old challenge: chances and limitations of optical, non-invasive acquisition and digital processing techniques for the age estimation of latent fingerprints. PhD Thesis, Otto-von-Guericke University, Magdeburg, Germany
57. Fritz P et al (2013) Infrared microscopy studies of the chemical composition of latent fingermark residues. Microchem J 111:40–46
58. Johnston A, Rogers K (2018) A study of the intermolecular interactions of lipid components from analogue fingerprint residues. Sci Justice 58(2):121–127
59. Tahtouh M et al (2005) The detection and enhancement of latent fingermarks using infrared chemical imaging. J Forensic Sci 50(1):64–72
60. Williams DK, Schwartz RL, Bartick EG (2004) Analysis of latent fingerprint deposits by infrared microspectroscopy. Appl Spectrosc 58(3):313–316
61. Hemmila A, McGill J, Ritter D (2008) Fourier transform infrared reflectance spectra of latent fingerprints: a biometric gauge for the age of an individual*. J Forensic Sci 53(2):369–376
62. Day JS et al (2004) The detection of drugs of abuse in fingerprints using Raman spectroscopy I: latent fingerprints. Spectrochim Acta A Mol Biomol Spectrosc 60(3):563–568
63. Dorakumbura BN et al (2018) Revealing the spatial distribution of chemical species within latent fingermarks using vibrational spectroscopy. Analyst 143(17):4027–4039
64. Andersson PO et al (2017) Towards fingermark dating: a Raman spectroscopy proof-of-concept study. ChemistryOpen 6(6):706–709
65. Figueroa B et al (2017) Label-free chemical imaging of latent fingerprints with stimulated Raman scattering microscopy. Anal Chem 89(8):4468–4473
66. Dikshitulu YS et al (1986) Aging studies on fingerprint residues using thin-layer and high performance liquid chromatography. Forensic Sci Int 31:4261–4266
67. Duff JM, Menzel ER (1978) Laser-assisted thin-layer chromatography and luminescence of fingerprints: an approach to fingerprint age determination. J Forensic Sci 23(1):129–134
68. Szabóová Ž et al (2017) GC–MS/MS method for age determination of fingerprints. Monatshefte für Chemie – Chem Monthly 148(9):1673–1678
69. Mountfort KA et al (2007) Identification of oxidation products of squalene in solution and in latent fingerprints by ESI-MS and LC/APCI-MS. Anal Chem 79:2650–2657
70. Huynh C, Halámek J (2016) Trends in fingerprint analysis. TrAC Trends Anal Chem 82:328–336
71. Francese S et al (2013) Beyond the ridge pattern: multi-informative analysis of latent fingermarks by MALDI mass spectrometry. Analyst 138(15):4215–4228
72. Wolstenholme R et al (2009) Study of latent fingermarks by matrix-assisted laser desorption/ionisation mass spectrometry imaging of endogenous lipids. Rapid Commun Mass Spectrom 23(19):3031–3039
73. Edelman GJ et al (2012) Hyperspectral imaging for non-contact analysis of forensic traces. Forensic Sci Int 223(1):28–39
74. Gorka M et al (2019) Molecular composition of fingermarks: assessment of the intra- and inter-variability in a small group of donors using MALDI-MSI. Forensic Chem 12:99–106
75. Fieldhouse SJ (2015) An investigation into the effects of force applied during deposition on latent fingermarks and inked fingerprints using a variable force fingerprint sampler. J Forensic Sci 60(2):422–427

76. Kind S (1987) Chapter 5: time and sequence. In: The scientific investigation of crime. Forensic Science Services, Harrogate, pp 106–389
77. Steiner R, Roux C, Moret S (2019) Controlling fingermark variability for research purposes: a review. WIREs Forensic Sci 1(4):e1338
78. Townsley v. United States, 236 A.2d 63. 1967
79. Solis v. People, 485 P.2d 903. 1971
80. State v. Scott, 251 S.E.2d 414; 296 N.C. 519. 1979
81. Matter of J.M.C., Jr., 502 A.2d 472. 1985
82. United States v. Collon (United States v. Garside), 426 F.2d 939. 1970
83. U. S. Department of Justice, O.o.t.I.G (2006) A review of the FBI's handling of the Brandon Mavfield Case, O.a.R.D. Office of the Inspector General, ed. Washington, DC
84. Langeburg G (2012) A critical analysis and study of the ACE-V Process. PhD thesis, in School of Criminal Sciences, University of Lausanne
85. Dror IE et al (2011) Cognitive issues in fingerprint analysis: inter- and intra-expert consistency and the effect of a 'target' comparison. Forensic Sci Int 208(1):10–17

Chapter 4
Non-contact Technologies and Digital Approaches to (Latent) Fingermark Aging Studies

Ronny Merkel

Abstract Significant developments in the field of noncontact, noninvasive surface capturing techniques have unlocked new and exciting possibilities for studying the degradation behavior of latent fingermarks. These novel technologies enable the iterative capture of specimens and thus the design and digital processing of image-based degradation time series. This chapter provides an overview of such noninvasive devices and explains the potential of three of these options: two different high-end surface measurement devices and off-the-shelf flatbed scanners. Going through the four main phases of fingermark capture, preprocessing, feature extraction, and age estimation, the benefits of these digital approaches are highlighted. They include the introduction of novel aging features based on the statistical distribution of image pixels as well as the digital computation of known aging features, such as ridge or pore size. The computation of quantitative age estimation measures provides first approximations of the quality of such approaches. As a result, fingermarks can be referred to as being fresher or older than a selected time threshold with a certain accuracy. In addition, dust deposition is shown to constitute a surprisingly promising aging feature. Although the studies presented herein can, in their current form, not be considered as being practically usable for casework or court, they highlight the potential of the noninvasive capture and digital processing for fingermark age estimations. They furthermore outline possibilities of future inquiry into this topic, by pointing out a freely available data collection as well as showing the general feasibility of off-the-shelf flatbed scanners as economical and widely available alternatives to expensive high-end capturing devices.

Keywords Latent fingermark aging · Fingermark age determination · Contactless fingermark acquisition · Noninvasive fingermark capture · Print dating · Long-term aging · Short-term aging · Aging features · Machine learning · Digitized forensics · Computer forensics · Crime scene investigation · Fingermark forensics · Age estimation · Time of a crime · Optical capturing device · High-resolution fingermark

---

R. Merkel (✉)
Otto-von-Guericke-University Magdeburg, Magdeburg, Germany

J. De Alcaraz-Fossoul (ed.), *Technologies for Fingermark Age Estimations: A Step Forward*, https://doi.org/10.1007/978-3-030-69337-4_4

Acronyms

CWL	Chromatic white light
FBS	Flatbed scanner
CLSM	Confocal laser scanning microscope
UVS	Ultraviolet spectroscopy
UV	Ultraviolet
VIS	Visible light
UVC	Ultraviolet camera
CCD	Charge-coupled device
IR	Infrared
FTIR	Fourier transform infrared
IRC	Infrared camera
MPLSM	Multiphoton laser scanning microscope
SERS	Surface-enhanced Raman spectroscopy
IFM	Interferometry
AFM	Atomic force microscopy
SLI	Structured light illumination
EFM	Electric field microscopy
GM	Gloss measurement
DA	Data acquisition
PP	Preprocessing
FE	Feature extraction
AE	Age estimation
tt	Time threshold

4.1 Introduction

When investigating crime scenes, there is often a competitive situation between different techniques for the analysis of fingermark evidence. While forensic investigators may develop a latent fingermark for the identification of a suspect, they may also be interested in performing touch DNA analysis. Sometimes fingermarks are overlapped, requiring a post-visualization separation of ridge patterns and detection of the sequence of deposition. Additionally, the age of a fingermark may have to be evaluated, since it is not always sufficient to link a latent mark to a donor or suspect. This becomes especially relevant for crimes committed in public spaces, such as restaurants, jewelry stores, and banks, when a suspect can easily claim to have been at the scene prior to or after the crime, requiring the investigator to establish an additional connection between the fingermark and the time of the crime.

When using traditional fingermark development and lifting techniques, it is often the case that not all desired tests can be performed. For example, a mark subject to ninhydrin-bathing or cyanoacrylate fuming may no longer be useful for age estimation, since its chemical composition has been altered during development. If a mark is scraped off a surface (i.e., by swabbing) and residues transferred to a test tube for DNA analysis, it can no longer be used for visual identification. This competitive situation between different fingermark analyses requires the use of noninvasive techniques for visualization, lifting, and processing. In this situation, a mark is captured from a surface without chemical or physical development, allowing for its noninvasive age estimation and leaving it unaltered for DNA analysis. Noncontact technologies can assist forensic investigators in this challenge.

In recent decades, many high-resolution capturing devices have been developed or significantly improved, allowing the capture of very small ridge details without direct contact. Most of them are based on electromagnetic radiation being emitted by a radiation source in narrow (e.g., lasers) or wider (e.g., white light) wavebands onto a surface while capturing the radiation reflected from or emitted by the sample; very few approaches also utilize other properties, such as electromagnetic charge or interaction between the tip of a cantilever and the sample (see Sect. 4.2). Therefore, noninvasive capturing devices might allow for suspect identification based on image-guided techniques while leaving the mark unchanged for additional analytical methods. Furthermore (and so far only in a laboratory setting and for experimental purposes), nondestructive techniques can capture fingermarks several times in regular intervals, creating a time series of its degradation behavior and thus enabling age estimation studies. Without such noninvasive devices, accurate aging studies would hardly be possible.

Apart from being noninvasive, noncontact capturing devices often offer very high resolutions, allowing for the analysis of very tiny structures, such as pores, minutiae, or certain residue particles (e.g., sweat secretion droplets consisting of water or fat as well as grains of dust or dirt), which can accumulate on a fingermark [1]. Such a high level of detail might provide additional insights into certain characteristics of a mark, for example, local changes in size or form of residue particles as well as global changes in the statistical properties of pixels from digitized mark images, potentially benefiting forensic investigations. In addition, mobile scanners might be used to automatically scan certain areas of crime scenes for latent marks, which is often a challenge to forensic investigators in case of extensive crime scenes, large number of potentially relevant (or not relevant) marks, or inconspicuous latents. For this purpose, the devices may apply very low capturing resolutions only, identifying latent marks or other traces of interest with a comparatively high speed by detecting physical or chemical differences to the background substrate of a crime scene [2].

Closely linked to the noncontact capture of latent marks is the digital representation and processing of the resulting images. While chemically developed fingermarks can also be digitized, e.g., by photographing or scanning, noncontact acquired marks are directly available in digital form. Great benefits arise from the digital processing of a mark, such as digital distortion removal, contrast enhancement, or residue particle segmentation. Digital processing furthermore enables the design of new features for the study of latent mark aging, such as features based on mathematical measurements of image pixels, topographic height information, or dust segmentation. Also, several studies have explored digital image-guided methods for the so-far unsolved challenge of separating and sequencing overlapped marks (e.g., [3, 4] or [5]).

Noncontact capturing techniques may also have other practical applications, such as the data protection-sensitive (anonymous) analysis of public crime scenes [6]. In such a scenario, marks are only coarsely scanned, where the capturing resolution is not sufficient for identification. Alternatively, only small areas of the mark might be scanned in detail, exhibiting insufficient amounts of minutiae for identification. If the age of a captured "anonymous" mark could successfully be determined, a preselection might be performed with respect to a mark's age. A subsequent detailed scan of the mark would only be required if its age falls within the time period of the crime while being discarded otherwise. Using such procedure, fingermarks not related to the crime should not have to be further examined, and the respective subjects would not have to be identified, potentially minimizing psychological harm or damage of reputation in case of being falsely accused or even investigated. Similarly, anonymous marks might even be scanned in a preventive context for age estimation. Only if a determined age falls within a suspicious period (e.g., a forbidden time period in a restricted area), the mark will be captured in detail. Such scenarios of anonymously determining a mark's age using noncontact devices may allow for a more data privacy-sensitive mark evaluation in the midterm future.

The plethora of advantages of noncontact devices for the capture of latent fingermarks also comes with certain (limited) disadvantages. Often, the devices are expensive and bulky, requiring sufficient funds for their acquisition as well as additional logistic efforts, such as transporting them to the crime scene or moving the respective objects from the crime scene to the laboratory. Furthermore, scanning large or nonplanar objects may pose additional challenges. However, these disadvantages might be overcome with sufficient resources.

A much more relevant challenge to noncontact devices is posed by its limitations in terms of capturing latent marks from different types of substrates. While smooth and well-reflecting surfaces are often ideal for the devices and can lead to very good results, porous or structured substrates are of a much greater challenge [7]. Porous substrates absorb the fingermark material, reflecting the electromagnetic radiation at the surface and therefore preventing the utilization of a mark's reflective properties. Structured surfaces reflect electromagnetic radiation diffusely, preventing large portions from reaching the detector. Of course, these effects may vary significantly between different devices (i.e., the capturing technique or utilized waveband [8]). However, some surfaces are particularly challenging for many devices alike, such as the porous nature of paper.

Traditional latent fingermark development techniques have attempted to overcome the challenge of a mark's visibility depending on the type of substrate by the application of different developers to different substrates, such as powdering for nonporous surfaces and ninhydrin-bathing for porous ones, which has shown to be quite effective. A similar approach to differentiate between certain substrate categories might be applied to noncontact capturing devices in the future, e.g., by mounting different sensors onto a single device, utilizing different capturing techniques and electromagnetic wavebands as appropriate to a specific substrate. However, certain development techniques such as ninhydrin-bathing, where ninhydrin penetrates the surface of the paper, cannot easily be reproduced by noncontact devices. Overall, fingermark acquisition will remain a substrate-specific discipline. Nonetheless, the advantages of noncontact capturing devices appear to be overwhelming if they can successfully be applied to a wide variety of substrates.

Numerous studies on latent mark development, capture, and analysis have been performed in the literature. A very extensive collection of scientific articles is provided in [9], fingermark composition and aging studies are summarized in [10], and invasive and noninvasive emerging capturing techniques are introduced in [11–13]. Overall, noninvasive technologies represent only a small part of latent mark capturing techniques; however, they are very much worth investigating. Section 4.2 briefly summarizes current examples of these noncontact and noninvasive capturing devices. Section 4.3 introduces practical age estimation approaches based on three kinds of these devices, namely, a chromatic white light (CWL) sensor, a confocal laser scanning microscope (CLSM), and flatbed scanners (FBS). Section 4.4 draws conclusions from the findings of these studies and points out potential future work in this area.

4.2 Noncontact, Noninvasive Latent Fingermark Capturing Techniques

For the development of successful age estimation strategies, the degradation behavior of latent fingermarks needs to be studied and translated into an adequate aging model. For this purpose, the design of time series is required, where specific fingermark characteristics are captured in regular time intervals to identify changes occurring over the course of time. Due to the significant variation in the aging behavior as a result of varying internal and external conditions [14], a single time series should ideally be captured from the same fingermark sample only, avoiding the pooling of marks for time series which is sometimes reported in the literature, e.g., laying 100 fingermarks at the beginning of an experiment and developing 10 different marks at each point in time, therefore introducing potential inter-mark differences into the time series.

Most noninvasive devices applied to the capture of latent fingermarks utilize, in one way or another, the different reflection, refraction, or absorption properties of the substrate and the residue particles (like sweat droplets or dust particles) when interacting with electromagnetic waves. These different properties often result in varying quantities of radiation being reflected by each material, leading to a certain contrast between substrate and fingermark material at the detector. This section gives a brief overview over selected noninvasive techniques studied in the literature for capturing latent marks. It is by no means considered to be a comprehensive review, but rather to provide a few examples to illustrate the wide range of possibilities when applying noninvasive devices to capture latent fingermarks and for age estimation.

4.2.1 Latent Fingermark Capturing Devices Utilizing the Ultraviolet Waveband of Electromagnetic Radiation

Ultraviolet (UV) radiation covers electromagnetic wavelengths in the range of 10 nm–400 nm. Radiation with lower wavelengths, such as X-ray or gamma ray, is very high in energy and, if used, would most likely result in significant molecular alterations of a latent mark, e.g., an increased degradation. Therefore, these types of rays are considered very invasive and not feasible for fingermark aging studies. Furthermore, ultraviolet radiation of wavelengths lower than 200 nm is also considered as not being suitable, since it is absorbed by ambient air, thus requiring vacuum spectroscopy, which would potentially alter the latent fingermark. Higher wavelengths of the ultraviolet spectrum could be considered; however, these may increase a mark's degradation to a certain extent due to photodegradation, polymerization of organic material, or DNA degradation. Ultraviolet techniques for examining latent fingermarks are therefore not ideal due to their potentially destructive nature. However, this also depends heavily on the wavelength and intensity of the emitted radiation as well as the time of exposure.

Ultraviolet spectroscopy (UVS) utilizes the ultraviolet spectrum and is sometimes coupled with visible light (VIS) spectroscopy to a UV/VIS spectrometer. In [8], such a device is studied using a Fries Research & Technology (FRT) 330 UV/VIS spectroscope [15], by capturing latent marks from different substrates in a spectral range of 163 nm–826 nm (2048 different spectra). Images were rastered using a movable measuring table, which shifts the sample laterally underneath a fixed sensor, creating a two-dimensional fingermark image with a pixel size of 100 μm. The device was also used to study mark degradation in [16], showing that aging trends seem more uniform in the visible part of the investigated waveband and vary in the ultraviolet band. It was further demonstrated that spectroscopy in the UV band can, to a certain extent, differentiate between the aging behavior of eccrine and sebaceous components of fingermarks. However, the differentiation between such substance groups is far less accurate than in the infrared waveband, i.e., using Fourier transform infrared spectroscopy.

Ultraviolet cameras (UVC) can be considered as an off-the-shelf alternative to UV spectroscopes. Charge-coupled device (CCD) area detectors can be used to capture mark images, which need to be illuminated by a separate radiation source operating in the UV range. These CCD detectors require certain band-pass filters since they are receptive to a wide spectrum of electromagnetic radiation. In [17], such a technique has been successfully utilized to visualize different kinds of forensic traces, including latent fingermarks. Although CCD area detectors can achieve smaller pixel sizes than the 100 μm used for the earlier introduced UV/VIS spectroscope, they cannot differentiate between different spectra for a single pixel and would require a fixed installation of camera and light source, with an ambient illumination of UV radiation instead of a focused radiation beam. Therefore, UV spectroscopes might be more accurate than UV cameras, potentially achieving a higher accuracy for degradation studies.

4.2.2 *Latent Fingermark Capturing Devices Utilizing the Visible Waveband of Electromagnetic Radiation*

The visible (VIS) spectrum of electromagnetic radiation comprises wavelengths in the range of 400 nm–700 nm. As such, it might be considered as being largely non-invasive, at least for short exposure times, and therefore well-suited for the design of latent mark time series and the analysis of degradation patterns. Since this waveband is visible to the human eye, a wide range of common photographic techniques can be applied.

Off-the-shelf *cameras* using CCD area detectors are an example of such techniques. However, the contrast between fingermark residues and substrate is often very low, requiring additional features, such as the utilization of specular reflection and polarization under specific illumination angles, as demonstrated in [18]. Considering the much more precise alternatives in the VIS spectrum, this seems to be a rather coarse approach.

Gloss measurement (GM) can be considered as another capturing technique in the VIS waveband which has been demonstrated in [19]. In this case, a diffractive glossmeter is used which incorporates a diffractive optical element to project a light beam onto the sample. Gloss is measured as the property of how well the light beam is reflected in a specular direction by the sample. The study shows that fingermark residues can be distinguished from the substrate using gloss properties. As light source, a helium-neon laser is applied, with a beam diameter of 10 μm. Little more is known about this technique at the current time, making further studies worthwhile exploring.

Chromatic white light (CWL) sensors are high-resolution capturing devices from the field of surface measurement. They are based on the chromatic aberration of white light at lenses, which is the phenomenon of splitting the beam of white light into differently colored beams due to the different degrees of refraction at the lens. Such devices have been explored to capture fingermarks from different surfaces [7, 20]. The studies conducted include the use of a Fries Research & Technology MicroProf200 CWL600 device [15], which can raster images using a movable measuring table with a pixel size of 2 μm. In addition to capturing the intensity of reflected visible light, the sensor can also capture height information with a longitudinal resolution of 20 nm by making use of the chromatic aberration. Because of its comparatively high precision and lateral resolution, the device seems well-suited for the capture of fingermark time series and has been investigated in several aging studies, which are summarized in [1].

Confocal laser scanning microscopes (CLSM) can achieve even higher resolutions than CWL sensors, utilizing white or red laser light and excluding non-focused light using a pinhole. The Keyence VK-X110 confocal laser scanning microscope [21] has been used in various studies to capture time series for the investigation of fingermark aging behavior, which are summarized in [1]. The device can be equipped with different magnification lenses to achieve very high lateral resolutions, e.g., a pixel size of 65 nm for a 100× lens. It provides rastered images of latent marks in different forms, such as intensity images based on the reflected laser light, topographic height information based on different focus levels, as well as color information captured with a CCD detector. The experiments conducted indicate that 10× magnification lenses (using a pixel size of 1.3 μm) may be sufficient for latent fingermark aging studies, since higher magnifications will, due to the very high resolution, be able to capture only a few ridge details per scan. Overall, the device seems very adequate for aging studies. A research collection (library) of captured digital mark aging series of both CWL and CLSM devices has been published as open source for further studies in [22].

Flatbed scanners (FBS) may also be utilized for the noninvasive capture of latent marks, if a sample can be adequately fixed onto the measuring surface without touching it. In [23, 24], an Epson Perfection 1660 Photo flatbed scanner (using a pixel size of 8 μm) [25] as well as a Canon CanoScan 9000F Mark II (using a pixel size of 5.3 μm) [26] was studied with respect to latent mark aging. The movable white light source in combination with a CCD sensor cannot be considered as being as precise as a CWL or CLSM device. However, if sufficiently accurate aging models could be established, flatbed scanners might provide a resourceful alternative to these high-end devices.

4.2.3 Latent Fingermark Capturing Devices Utilizing the Infrared Waveband of Electromagnetic Radiation

Infrared (IR) electromagnetic radiation covers a range of 700 nm–1 mm and might be considered as noninvasive to fingermark residues in appropriate doses, leading to the vibration of residue molecules rather than their disintegration. Although the spectrum is comparatively wide, studies in this area are often focused on *Fourier transform infrared (FTIR)* spectroscopy, such as in [27, 28] or [29]. Such technique can be used to determine qualitative and quantitative amounts of chemical compound groups within a latent mark, where specific frequencies of the emitted mid-infrared radiation are absorbed. In contrast to other capturing techniques, this one allows for the analysis of the degradation of specific chemical compounds. In comparison, UV/VIS spectroscopy might only be able to differentiate between very broad groups of components, such as eccrine vs. sebaceous particles. Due to its potential for image-based analysis of time series of fingermark-specific chemical compounds, FTIR spectroscopy appears to be a very promising technique for latent mark age estimation.

Infrared cameras (IRC) provide an off-the-shelf alternative to IR spectroscopes; however, these perform much worse in terms of precision. A separate infrared radiation source is required, and single spectral frequencies cannot be differentiated. Although different types of forensic evidence have been visualized using infrared cameras in [30], fingermarks are not among them. In general, infrared cameras seem to be much less suitable for aging studies than IR spectroscopes.

4.2.4 Latent Fingermark Capturing Devices Utilizing Luminescence and Raman Spectroscopy

Luminescence as a form of spontaneous emission of light may be studied in respect to its capabilities for visualizing latent marks. It can be induced by electromagnetic radiation at different wavelengths. First approaches have investigated this technique by studying fluorescence of a mark after excitation with lasers or lamps, such as in [31–34]. However, it is not clear whether they can be considered as being noninvasive, because the ultraviolet radiation used for exciting the mark might alter its chemical components due to its high energy. Such hypothesis also depends heavily on the time and intensity of exposure. A solution may lie on the application of a multiphoton laser scanning microscope (MPLSM) as suggested in [31], for multiphoton excitation at wavelengths in the range of 780 nm–900 nm. However, for a reliable estimate of the technique's feasibility, further studies are required.

Raman spectroscopy is a technique based on the inelastic scattering of IR, UV, or VIS monochromatic light at molecules. Surface-enhanced Raman spectroscopy (SERS) has been used in [35] for capturing latent fingermarks; visual differences in fresh and aged marks have been shown to exist in [36]. Although this technique might be suitable for fingermark age estimation, practical performance estimates have not been provided yet.

4.2.5 Latent Fingermark Capturing Devices Utilizing Height Information

The utilization of height information of ridges may be a valuable asset when studying mark degradation. Such studies are also referred to as profilometry. Several techniques have been proposed for the noncontact acquisition of topographic images from latent marks. They vary in their capturing technique, but all have in common that they can obtain local height information of a specimen, creating topographic images by means of lateral rastering.

Chromatic white light (CWL) sensors introduced in Sect. 4.2.2 utilize the different focal points of certain wavebands (colors) appearing when white light is refracted at a lens. In few words, the waveband which focal point best matches the distance between sensor and the measured object is reflected with the most intensity, allowing to determine the height of the object. The technique has been investigated in several studies summarized in [1]. Although the technique seems promising for latent mark age estimation, it is also subject to a certain amount of noise, resulting from radiation being refracted or diffusely reflected at residue particles, such as sweat secretion droplets or dust grains.

Confocal laser scanning microscopes (CLSM) can, apart from intensity images (see Sect. 4.2.2), also produce fingermark height information. Due to the exclusion of non-focused light using a pinhole, the red or white laser light can be used to trigger different longitudinal focus levels, scanning an object plane-wise and, therefore, capturing height information. The technique has been studied for latent mark capture and aging in [1], and a research collection including CWL as well as CLSM topographic time series is available in [22]. The CLSM sensor faces similar challenges than its CWL counterpart with respect to noise and other distortions. Nonetheless, the technique seems promising.

Interferometry (IFM) is a technique able to provide 3D volume representations of fingermarks, utilizing different optical paths of emitted electromagnetic radiation and producing interference upon recombination. From the interference fringes, differences in optical path lengths are derived, and height information can be computed. Such technique is considered as noninvasive, since it is usually used in conjunction with a near-infrared light source. *Swept-source optical coherence tomography* has been applied to capture fingermarks in [37] and was able to detect mark residue underneath a dust layer. *Coherence scanning interferometry* has been used in [38] to study the degradation of ridge height information over the course of 1 year. This promising technique is introduced in more details in Chap. 6.

Atomic force microscopy (AFM) can be utilized to capture surface properties of residue particles at nanometer to micrometer scale. In AFM, a cantilever spring with a very sharp tip mounted onto a piezoelectric scanner is interacting with a sample. Height information can be captured by measuring the deflections of the cantilever that result from the forces acting between the tip and the specimen analyzed. In [13], PeakForce quantitative nanomechanical mapping atomic force microscopy has been used for the investigation of different characteristics of residue particles over time, such as sweat secretion droplet adhesive properties to dust powder particles glued to the microscope cantilever. Capturing the topography (or other properties) of fingermark particles might be feasible to identify characteristic changes during aging; however, it requires further studies in the future.

Structured light illumination (SLI) is a technique for capturing the shape of an object by projecting a striped light pattern onto it. Deformations of the stripes are caused by the object, which may be used by a camera system to construct a topographic image. The technique has been mentioned in [2] as potential use for topographic representations of fingermarks; however, it seems to currently not be of sufficient resolution for this field of application.

4.2.6 Latent Fingermark Capturing Devices Utilizing Electric Field Microscopy

In a proof-of-concept study [39], *electric field microscopy* (EFM) has been successfully used to reconstruct a latent mark by measuring electric charges induced in insulating surfaces during mark deposition. Its discharge over time was analyzed to provide a timeline of degradation. The technique achieves a pixel size of 100 μm and is considered to be noninvasive, since the measurement process does not alter the surface charges. Apart from the limitations of the technique being applicable only to insulating surfaces and only to aging periods of a few weeks, because of the comparatively quick discharge of the surface, this technique may be worth further investigations.

4.3 Approaches to Digital Image-Guided Latent Fingermark Age Estimation

Although various noncontact and noninvasive techniques for age estimation of latent fingermarks have been proposed in the literature, only very few have been investigated beyond a proof-of-concept stage. The age estimation approaches using the CWL, CLSM, and FBS capturing devices from [1, 22–24] are introduced in this section, which are, at this point, the only approaches providing specific aging models and formal (quantitative) age estimation performance measurements. They are all image-based, utilizing the visible waveband of electromagnetic radiation, and include measured intensity and topographical information. They demonstrate the potential and limitations of using time series based on noncontact techniques for age estimations and could be complemented in the future by additional techniques, including fusion approaches combining time series of different capturing devices.

The investigation of latent mark aging using the digital processing of image-based, noninvasively captured time series can be modeled using a general processing pipeline, which consists of four major steps: acquisition, preprocessing, feature extraction, and age estimation (see Fig. 4.1). Each step of the pipeline faces particular challenges, for example, the removal of environmental and sensor-specific artifacts, the selection of adequate parameters and methods for segmentation and feature extraction, the design of well-suited aging models taking into account the variety of different influences (environmental, sweat composition, type of substrate, etc.), as well as choosing precise formal quality measures for the evaluation of age estimation results. A detailed study of the different aspects of such pipeline can be found in [1]. The technique of transferring the complete processing chain into the digital domain is also referred to as computational forensics [40] or digitized forensics [41].

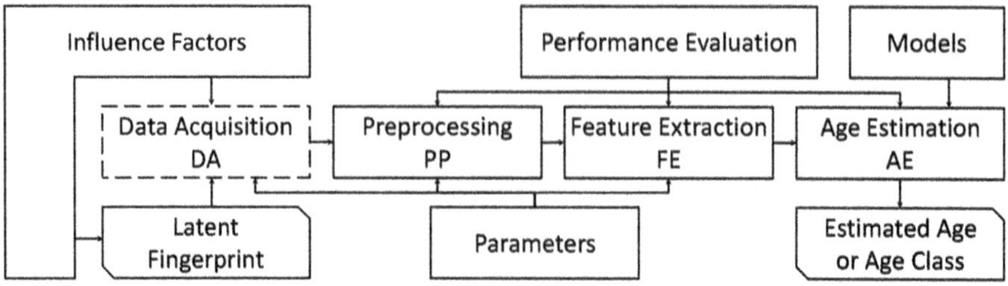

Fig. 4.1 Digital processing pipeline for the study of latent fingermark time series, extracted from [1]

The following sections give a brief insight into each of the four steps of the processing pipeline, outlining the performance and challenges of each step with regard to the CWL, CLSM, and FBS capturing devices.

4.3.1 Noninvasive Data Acquisition (DA) of Latent Fingermark Time Series

Creating a digital and image-based time series of a certain fingermark area is the main aim of this step. Different parameters are of interest, such as the specific capturing technique, image resolution, and time between two consecutive scans. CWL and CLSM devices are subject to the most extensive studies so far in this field [1, 22], leading to a variety of different time series, including intensity- and topographic-based image series. A comprehensive collection of these time series has been published in [22], which is freely available to the scientific community. Furthermore, few studies have aimed at reproducing findings from these devices for off-the-shelf FBS [23, 24]. The measurement setup for these devices is illustrated in Fig. 4.2. Latent fingermarks were lifted from hard disk platters (round disks from hard drives), which reflect radiation very well and are considered as ideal substrates for proof-of-concept studies; other nonabsorbing surfaces, such as glass, plastic, or metal, could be studied in a similar way, as was demonstrated in [42]. Acquisition parameters include pixel sizes of 20 μm to 0.141 μm and capture periods of 24 h (short-term aging, captured hourly) as well as periods of up to 3 years (long-term aging, captured weekly). Examples of images captured with these devices are depicted in Figs. 4.3, 4.4, and 4.5.

Fig. 4.2 Measurement setup for the noninvasive capture of latent fingermark time series using a CWL (left upper image, extracted from [1]), CLSM (right upper image, extracted from [1]), and FBS device (lower images, extracted from [24])

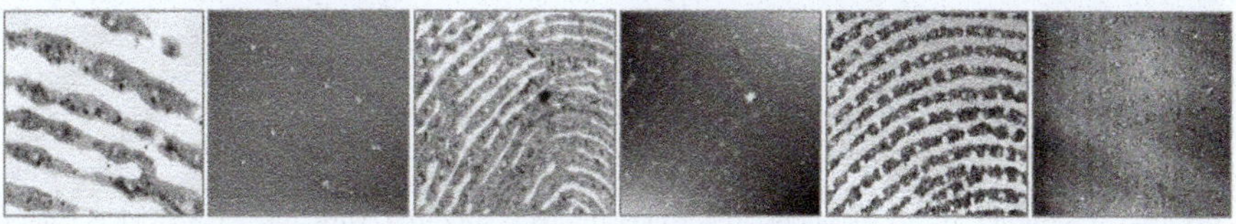

Fig. 4.3 Examples of fresh latent fingermark images captured with a CWL device (from [22]). From left to right (in pairs of intensity and corresponding topography image, providing pixel size, image size in mm, and image size in pixels): [2 μm, 2 × 2 mm, 1000 × 1000 pixels], [10 μm, 5 × 5 mm, 500 × 500 pixels], [20 μm, 4 × 4 mm, 200 × 200 pixels]

4.3.2 Time Series Preprocessing (PP)

Captured time series are a product not only of the intrinsic changes occurring during fingermark degradation but also of many other influencing factors from the environment (e.g., temperature, humidity, dust), the fingermark deposition process (e.g., smeared marks, varying pressures), the capturing device (e.g., capturing technique, resolution, specific sensor properties), the sweat composition (e.g., eccrine vs. sebaceous compounds, contaminations with exogenous substances, diet, drugs, medications), as well as the surface material (e.g., well-reflecting, roughness, porosity). While some of these effects cannot easily be removed and therefore require addressing in the scope of the aging model, such as differences in the particular chemical

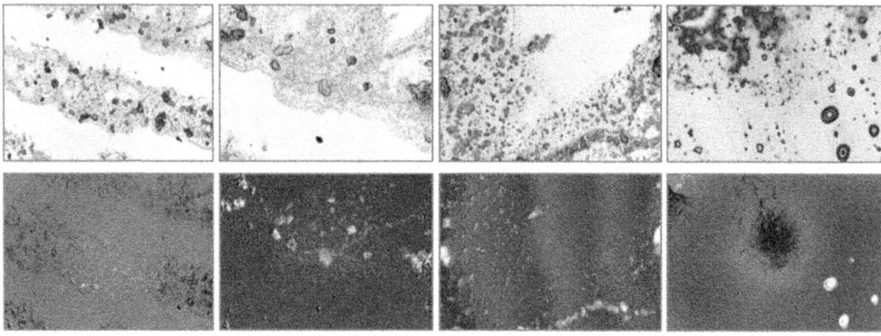

Fig. 4.4 Examples of fresh latent fingermark images captured with a CLSM device (from [22]). Upper row from left to right: intensity images (providing magnification, pixel size, image size in mm, and image size in pixels): [10×, 1.3 μm, 1.2 × 0.9 mm, 924 × 668 pixels], [20×, 0.69 μm, 0.64 × 0.46 mm, 924 × 668 pixels], [50×, 0.273 μm, 0.225 × 0.155 mm, 824 × 568 pixels], [100×, 0.141 μm, 0.116 × 0.080 mm, 824 × 568 pixels]. Lower row: corresponding topography images

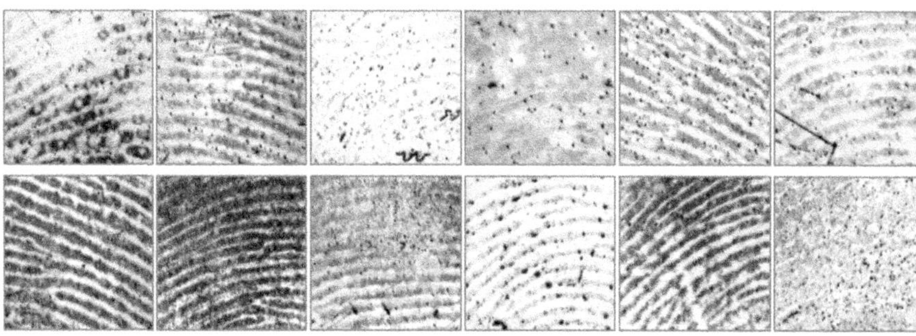

Fig. 4.5 Examples of fresh latent fingermark images captured with the FBS device Canon CanoScan 9000F Mark II (from [24]). Images are captured with a pixel size of 5.3 μm and an overall size of 5 × 5 mm (945 × 945 pixels) and are of eccrine (upper row) and sebaceous (lower row) nature

composition of a fingermark, others could be reduced or removed by adequate preprocessing methods, including certain distortions from the capturing process.

Preprocessing methods may comprise spatial alignment of consecutive fingermarks if required, planarization of topographic images, compensation of fluctuations in the overall image brightness due to variations in the light source intensity, artifact or noise removal, as well as segmentation of residue particles (e.g., secretion droplets or dust particles), ridges, or pores. In this scope, a wide range of digital image processing algorithms could be applied. However, it is important to not alter the characteristic changes induced by mark degradation, which may sometimes be challenging. The resulting time series of a certain device are in a comparable form and can be used as input for the development of aging models if a successful preprocessing has been performed. Examples of time series after preprocessing are depicted in Figs. 4.6 and 4.7 for the CWL and CLSM devices. Unfortunately, no images are available for the time series of FBS devices from the studies of [23, 24].

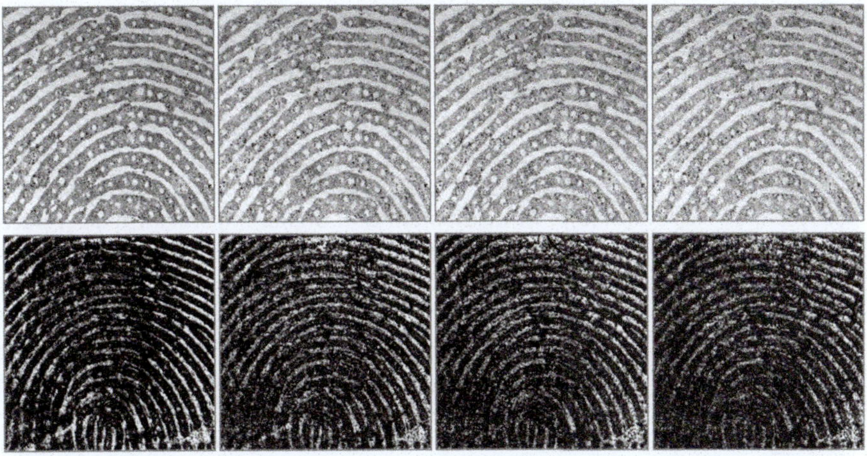

Fig. 4.6 Examples of latent fingermark time series captured with a CWL device and preprocessed for distortion removal (from [1]). Upper row: intensity-based gray scale time series captured at points in time $t_1 = 0.5$ h, $t_2 = 8$ h, $t_3 = 16$ h, and $t_4 = 24$ h (short-term aging) with a pixel size of 20 μm and an overall size of 4 × 4 mm (200 × 200 pixels). Lower row: topography-based, binarized time series captured at points in time $t_1 = 1$ day, $t_2 = 35$ weeks, $t_3 = 79$ weeks, and $t_4 = 120$ weeks (long-term aging) with a pixel size of 10 μm and an overall size of 8 × 8 mm (800 × 800 pixels)

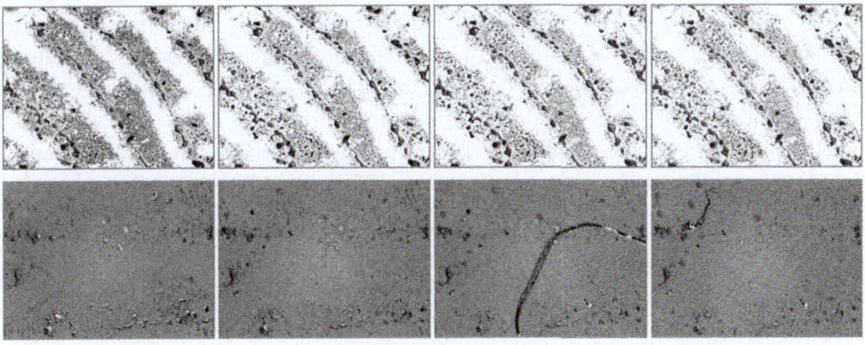

Fig. 4.7 Examples of latent fingermark time series captured with a CLSM device and preprocessed for distortion removal (from [1]). Images are captured with a pixel size of 1.3 μm and an overall size of 1.3 × 1 mm (1024 × 768 pixels). Upper row: intensity-based, binarized time series captured at points in time $t_1 = 0.5$ h, $t_2 = 9$ h, $t_3 = 17$ h, and $t_4 = 24$ h (short-term aging). Lower row: topography-based gray scale time series captured at points in time $t_1 = 1$ day, $t_2 = 10$ weeks, $t_3 = 20$ weeks, and $t_4 = 30$ weeks (long-term aging)

4.3.3 Aging Feature Extraction (FE)

Classical aging features from the literature include morphological features like ridge thickness or pore size. While these were also studied using digital aging time series, they were amended by newer features such as dust settling on the substrate over time. Furthermore, morphological features were complemented by new statistical features based on the pixel distribution of a digital fingermark image. Such novel image-based features include the relative frequency of substrate (non-fingermark) pixels after binarization (becoming more during aging), the global or local mean of pixel values, standard deviation, variance, gradients, roughness, coherence, contrast, or even Benford's law digit distribution. These features are able to measure the overall changes within a mark image, which are assumed to be the result of the evaporation of water and short-chained fatty acids as well as other possible degradation processes. These changes show themselves in the form of a loss in contrast of the fingermark pixels throughout the complete image, which can be hard to detect by the human eye and cannot be captured with classical morphological features. Therefore, digital image-based processing techniques enable a new feature class, which provides very interesting additional insights into the degradation behavior of latent marks.

The morphological features of mean ridge thickness, number of pores, and mean pore size are observed in [1, 24] as not exhibiting a characteristic change over time (study period: 1.5 and 2.3 years), which is in contrast to earlier findings of analogue investigation techniques, such as [43]. Examples of images visualizing the changes in pores and ridges over time are depicted in Figs. 4.8 and 4.9.

Concluding from these findings, latent fingermark degradation characterizes itself by showing a continuous loss of image contrast, where ridge pixels become brighter or disappear altogether during aging. These changes appear throughout the

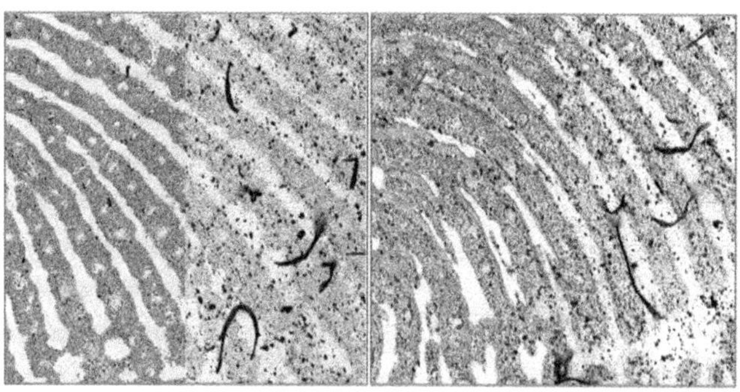

Fig. 4.8 Exemplary intensity-based gray scale images of latent marks captured with a CWL sensor (from [1]) at times $t = 1$ day (left half of each image) and $t = 2.3$ years (right half of each image). Images are captured with a pixel size of 10 μm and an overall size of 5.6 × 5.6 mm (560 × 560 pixels). A visual comparison of the ridge transition between the fresh and aged part of the mark shows a loss of ridge contrast as a result of mark aging rather than a decrease in ridge thickness

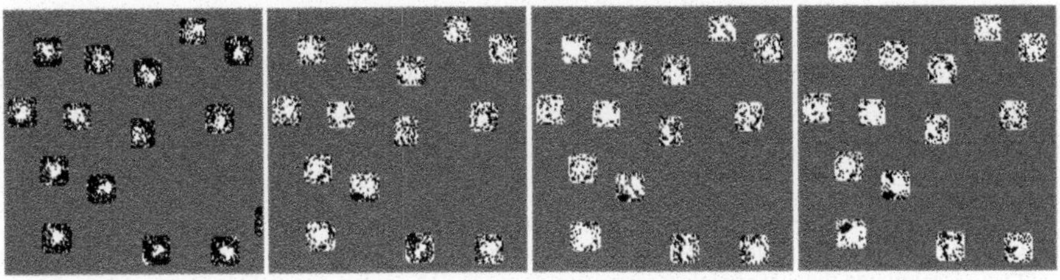

Fig. 4.9 Exemplary intensity-based, preprocessed, and binarized image of latent mark pores captured with a CWL sensor and masking non-pore areas in gray (from [1]) at times t_1 = 1 day, t_2 = 35 weeks, t_3 = 79 weeks, and t_4 = 120 weeks). Only parts are depicted from images captured with a pixel size of 10 μm and an overall size of 5.6 × 5.6 mm (560 × 560 pixels). It can be seen that during mark aging, an increasing number of pixels from all over the segmented area change its value from black to white, indicating a general loss of the overall image contrast rather than an increase in pore size

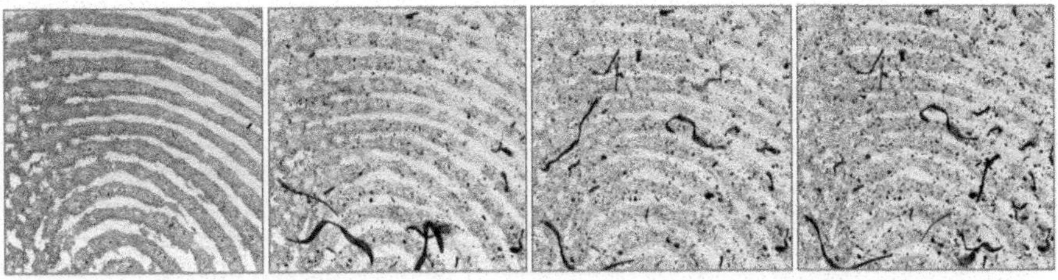

Fig. 4.10 Exemplary intensity-based, preprocessed latent fingermark time series captured with a CWL device (from [22]) at times t_1 = 0 years, t_2 = 1 year, t_3 = 2 years, and t_4 = 3 years. Images are captured with a pixel size of 10 μm and an overall size of 5 × 5 mm (500 × 500 pixels). A continuous settlement of dust on the sample can be observed with time

complete image rather than at certain areas like the edges of ridges or pores. Therefore, statistical image-based features may be better suited to study fingermark aging than morphological ones. Fluctuations in ridge thickness or pore size observed in the literature, e.g., [43–45], could be the result of less sweat material being available at the edges of a ridge. When the quantity of sweat material becomes lower over time, the original width of the ridge cannot be properly segmented anymore, giving the impression that the ridge becomes smaller. This applies to visual inspection, visualization with adhesive powder, and segmentation using digital algorithms in a similar way. However, there is no indication that the observed reduction in ridge width or pore size is an actual shrinking. It rather seems to be the illusion of shrinkage, notwithstanding the fact that the rate of sweat material degradation appears to be similar throughout all areas of a ridge. However, other studies suggest that certain areas of the same latent can degrade differently [38, 46, 47] leaving this topic open for further debate.

A very promising morphological feature may be seen in the amount of dust settling on a sample over time, which is visualized in Figs. 4.10 and 4.11. Since the amount of settled dust seems to constantly increase with the age of the mark, it

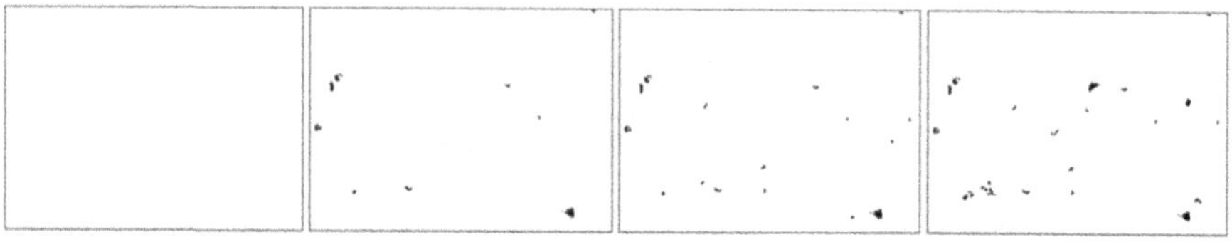

Fig. 4.11 Exemplary intensity-based, preprocessed, and binarized time series of dust particles with a size greater than 50 pixels captured with a CLSM device (from [1]) at times t_1 = 1 day, t_2 = 11 weeks, t_3 = 20 weeks, and t_4 = 30 weeks. Images are captured with a pixel size of 1.3 μm and an overall size of 1.3 × 1 mm (1024 × 768 pixels). For an increased accuracy, dust particles are only recorded from substrate areas not being covered by mark residue. It can be observed that the amount of dust particles with the respective minimum size is constantly increasing with time

might only be a matter of extracting characteristic dust particles from the image to achieve a reliable aging feature.

4.3.4 Age Estimation (AE) Results

Based on the selected aging period, e.g., short term (hours to days) or long term (weeks to years), aging features (e.g., morphological or statistical), and age estimation technique (e.g., formula-based or machine learning-based), different accuracies of estimated ages might be achieved. The following subsections introduce a formula-based (Sect. 4.3.4.1) and a machine learning-based (Sect. 4.3.4.2) approach to fingermark age estimation. These models focus on statistical (image-based) features, since these have been identified as much more promising than morphological ones. They can capture changes that occur throughout the fingermark rather than only at the edges of ridges or pores (Sect. 4.3.3). Furthermore, theyare suitable for the analysis of short-term aging periods of up to 24 h, which are hard to be examined using morphological features, often requiring much longer time periods for changes to become noticeable. Dust deposition, however, is a morphological feature with great potential and therefore included in the here presented age estimation approaches. The section will conclude with some suggestions on the application of neural networks and deep learning to latent fingermark age estimation (Sect. 4.3.4.3).

4.3.4.1 Formula-Based Age Estimation for Latent Fingermarks

In a first approach, a formula-based age estimation has been studied for short-term ages of up to 24 h[1]. A time series containing a total of 770 fingermarks was captured with a CWL sensor in an interval of 1 h, leading to a total of 15,400 samples. For these samples, ages were computed using the mathematic function best fitting the observed aging curves, and the differences between such computed ages and the corresponding real ages were determined. However, a significant challenge arose

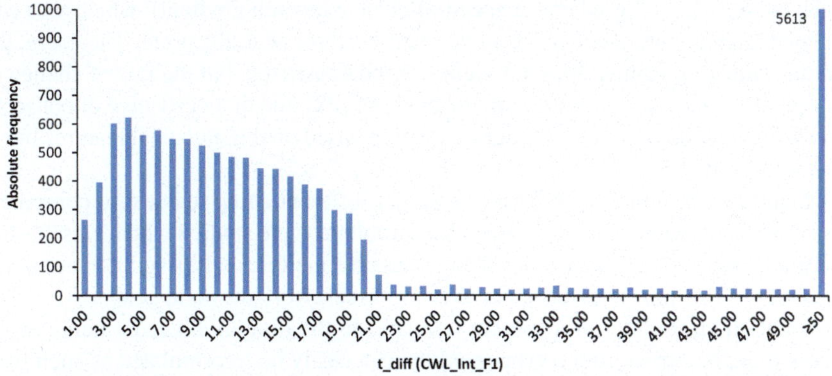

Fig. 4.12 Distribution of the time difference t_diff (in hours) between computed and real mark ages from the study of [1] using a CWL sensor and a time series of 770 fingermarks. Marks are captured every hour for a total time period of about 24 h, leading to approximately 24 samples per time series (15,400 samples in total). One feature is used (referred to as CWL_Int_F1 in the image), measuring the relative frequency of substrate (non-fingermark) pixels after binarization. Image with permission from [1]

when using such approach, based on the logarithmic nature of the aging curves resulting from many of the features proposed in Sect. 4.3.3. Logarithmic trends have the property of producing extreme y-values for x-values close to zero. Therefore, small changes in computed feature values may lead to significant changes in the corresponding ages calculated and, therefore, to unreliable results. Furthermore, it has been observed in [1] that for approximately 40% of the samples, the fingermark's age could not even be computed, since the logarithmic function was not defined by such values. For the remaining samples, median differences between computed and real ages were in the range of 3.4 –6.8 h. The distribution of the time difference between the computed and real mark ages is shown in Fig. 4.12.

Taking into account the high number of non-computable ages and that the marks were aged to a maximum of 24 h, the results can be considered interesting, yet far from being reliable. The potential influences from the environment, sweat composition, specific way of fingermark deposition, substrate, and capturing device lead to a substantial variation in the computed aging curves and therefore the predicted ages.

4.3.4.2 Machine Learning-Based Age Estimation for Latent Fingermarks

In a second approach, automated model training and age classification using machine learning were investigated by studying a time series containing 550 fingermarks from a CWL and 639 from a CLSM device [22]. Images were captured every hour over a time period of 24 h. Features were computed from the digital image of a mark at each capturing point and combined to an overall aging curve, representing the changes of a feature value occurring during aging. However, the best age estimation results were achieved using the first or second derivative of a specific point on such curve (its local gradient or speed of change of such gradient), which was calculated from a window of consecutive feature values near this point. Therefore, the actual value of a feature does not seem very characteristic, but the rate of change of consecutive feature values does. In the study of [22], the first derivative of consecutive feature values was used for age estimation based on the statistical features introduced in Sect. 4.3.3.

Machine learning allows for the fusion of different aging features to compute a result with increased accuracy. Two-class problems were studied which classify fingermarks as younger as or older than a certain time threshold (tt). Such approach seems to be the most realistic, since in criminal casework the main goal is to confirm or refute the testimony of a suspect in regard to the time of allegedly being at the scene (e.g., before or after a crime), which can easily be reformulated as such two-class problem. The age estimation performance achieved by the study is depicted in Fig. 4.13, showing the classification performance as kappa value[1] in respect to different time thresholds (tt), representing different two-class problems of the following form: marks younger as or older than 1 h, younger as or older than 2 h, etc.

The results depicted in Fig. 4.13 show that the age classification performance is higher when combining several of the features introduced in Sect. 4.3.3 (kappa improvement of about +0.1). This approach performs best for time thresholds between tt = 4 hand tt = 8 h, with a certain fluctuation for lower thresholds and a continuously decreasing performance for higher thresholds. For time thresholds of tt < 4 h, the exact reasons for the decreased performance are unknown and might be attributed to a lower number of available samples for that class and, therefore, an insufficient training of the model. The lower performance for time thresholds of

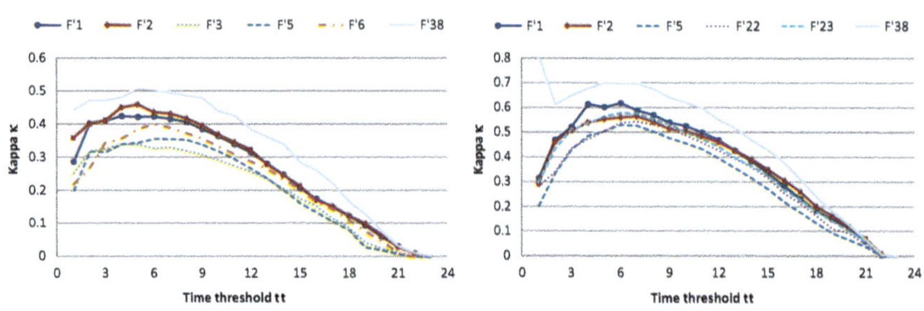

Fig. 4.13 Age classification performance from the study of [22] using a CWL sensor (left) in a 550 sample time series and a CLSM device (right) in a 639 sample series. Marks are captured every hour for a total time period of about 24 h, leading to approximately 24 samples per time series. Results are depicted in respect to two-class problems for different time thresholds $tt = \{1,2,\ldots,23\text{ h}\}$. Performances are presented in the form of kappa values. The five best performing features for each device have been depicted in the graph, as well as a combined approach fusing all investigated features (F'38). Images from [22]

[1] Kappa is a classification metric commonly used in the field of machine learning, which removes the influence of chance on the metric. A kappa value of zero represents pure guessing, whereas a kappa value of one represents error-free classification. More information about the kappa measure can be found in [48].

tt > 8 h implies that many changes are occurring during the first hours after deposition, while marks are decaying much slower and therefore become less distinguishable during the later stages of their decay. The highest performance in age classification is observed for both devices for time thresholds of tt = 5 h, exhibiting kappa values of $\kappa = 0.51$ (CWL) and $\kappa = 0.70$ (CLSM). Such kappa values correspond to accuracies of 75.5% (CWL) and 85% (CLSM) in case of equal class sizes for model training.

Overall, the results indicate that while aged fingermarks usually exhibit a low rate of change, fresh marks can exhibit either a high or a low rate, probably depending on their particular sweat composition. In turn, this means that latent fingermarks exhibiting a high rate of change in their feature values can be classified as fresh marks with a high degree of accuracy. However, marks exhibiting a lower rate of change might be either fresh or aged – in this case, not much information can be extracted from them. As a consequence, fingermarks with a high rate of change are much more valuable to forensic investigations, since they exhibit a high probability of being recently placed, whereas not much information can be gathered from other marks.

Apart from the comparatively short aging periods of 24 h investigated in [22], the study also examined a very promising feature based on dust accumulating on the substrate over time periods of up to 29 months, using time thresholds tt = {1,2,…,29 months}. Such feature achieves, without being combined with other features, very good age classification results for time thresholds of $5 \leq tt \leq 20$ months ($\kappa > 0.7$, 85% accuracy for equally sized training classes; best performance for tt = 11 m; $\kappa = 0.85$, 92.5% accuracy for equally sized training classes). However, the investigation is based on the samples of 20 fingermark time series only, which may be of insufficient number for statistically significant results.

The dust feature is of particular importance to future studies of fingermark aging because it does not measure any property of the fingermark itself, but rather the amount of particles accumulating on it. This feature is not subject to the fingermark-inherent influence factor of sweat composition, rendering it much more robust. While outdoor crime scenes are exposed to wind, rain, and other environmental influences, potentially rendering the dust feature unfeasible for age estimation, it could be of great value for determining a mark's age in indoor scenes.

In [23, 24], the aging studies using CWL and CLSM sensors have been reproduced using FBS devices to investigate whether similar accuracies can be achieved. The study of [24] performs practical age estimation and provides a comparison between the CWL and CLSM results, which is depicted in Fig. 4.14. However, it is to note that such comparison is not very realistic, due to the significantly different test setups, leading to a potentially worse performance for the CWL and CLSM devices. The comparison should therefore only be used to highlight the potential of using a FBS device for age estimation and not as a realistic benchmark against its CWL and CLSM counterparts.

The comparison of Fig. 4.14 shows that using only two classes for age estimation leads to a better performance than having to separate marks into three different classes. Furthermore, there appears to be no significant loss in accuracy when using

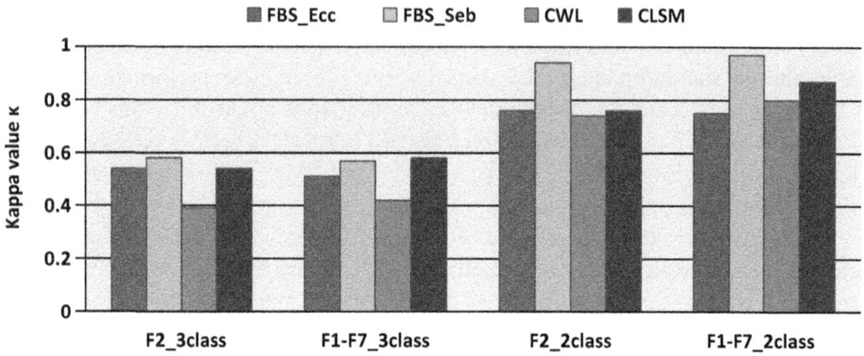

Fig. 4.14 Age estimation performance comparison between the Canon CanoScan 9000F Mark II FBS [24] and the high-resolution CWL and CLSM devices from [1] (image is from [24]). Age estimation performance is provided using the kappa value κ. Results are depicted in respect to a three- and two-class classification (three-class ages, [0,2 h] [8–12 h] [22–26 h]; two-class ages: [0,2 h] [8–26 h]). Furthermore, results are depicted in respect to eccrine (FBS_Ecc) and sebaceous (FBS_Seb) marks as well as in respect to a single feature (the best performing feature F2) and the fusion of seven different features (F1–F7)

only the best performing feature instead of a fusion of seven different features, indicating a high interdependency between the features. Also, sebaceous marks captured with an FBS device lead to better results than capturing eccrine marks. This might be attributed to a better visibility of the sebaceous marks, making it easier to detect changes during aging.

Most importantly, the results show that FBS-based age estimation performs similarly well, and in some cases even better, than CWL- and CLSM-based approaches. However, the results may, to a certain degree, be distorted, due to the differing test setups, potentially introducing disadvantages for the CWL and CLSM devices. In particular, fresh marks examined with the FBS were usually captured only a few minutes after deposition, whereas CWL and CLSM scans were used from fingermarks up to 2 h in age. Although there is the possibility that the comparison is not completely analogous, it nonetheless shows that FBS devices have the potential to achieve similarly good age estimation results than the expensive, high-resolution CWL and CLSM devices. Therefore, FBS devices could render this field of research accessible to a wide array of researchers, by removing the need for the acquisition of expensive high-end capturing devices.

4.3.4.3 Neural Network and Deep Learning-Based Latent Fingermark Age Estimation

The use of neural networks and deep learning techniques has shown very promising results for all sorts of classification challenges. Therefore, it is very likely that these techniques would also yield performance improvements for fingermark age estimation. They could be applied to samples captured with CWL, CLSM, or FBS devices utilizing contrast, dust, or height information, but also to samples from other devices, such as FTIR sensors, capturing images of certain fingermark compounds. A combination of morphological, statistical, and chemical fingermark features from differing devices may also be a possibility with great potential.

At this point in time, no study is known to have been applying neural networks or deep learning to the fingermark age estimation challenge, even if the respective algorithms are readily available. The main deterrent for the application of these techniques may stem from the absence of large sample databases needed for model training. Therefore, the research collection of [22] can be considered as a starting point for evaluating the applicability and potential of these artificial intelligence techniques for the age estimation of latent fingermarks. However, an even greater number and variety of fingermarks and time periods will be required in the future.

4.4 Conclusion

While there is a plethora of digital capturing devices available that can be applied to noninvasively capture latent fingermarks, only few have been studied in detail at this point. This may be attributed to the often high cost of such devices and does not necessarily pose a limitation in terms of feasibility. However, the most promising and best-studied approaches to date appear to utilize the visible range of the electromagnetic spectrum as well as the infrared range (using FTIR-based approaches). The use of high-resolution devices, such as CWL and CLSM sensors, has led to the first measurements of the accuracy of age estimation, which is a significant step forward in the long-lasting struggle for robust and reliable fingermark age estimation models.

Since latent fingermarks exhibit a wide spectrum in their degradation behavior – due to their differing composition, ambient conditions, mode of deposition, and underlying substrate – the accuracies achieved are far below of what is required for a practical use by crime laboratories and in court, leaving much room for improvement. The combination of information obtained from different capturing devices (e.g., image-based features in the visible range with chemical traits captured by FTIR spectroscopes) and the application of alternative models (e.g., based on neural networks and deep learning algorithms) are potential ways forward that could provide further improvements. However, since very high accuracies are required for age estimates to be admissible in court, the challenge might not easily be overcome.

One very promising approach is the use of alternative indicators for the time of fingermark deposition. As discussed earlier, dust settlement has shown to be much more constant over time than the intrinsic process of fingermark degradation. In addition, it allows for the development of specific crime scene "dust models" even after a crime has been committed. Therefore, information on the quantity of dust particles present on a fingermark could be regarded as a valuable addition to the mere study of the fingermarks themselves, yet being limited to long-term aging periods.

In order to ease and widen the range of studies in fingermark aging, the availability of samples plays an important role. Especially the cost of many high-resolution acquisition devices may hinder researchers in obtaining a sufficiently large number of samples. The publication of datasets from previous studies is, therefore, of vital importance for the reproducibility and the application of novel processing and classification techniques. The research collection of [22] makes a first step in that direction, containing most of the previously investigated CWL and CLSM time series. It is available to the scientific community and presents a free of charge possibility to study CWL and CLSM time series of latent fingermarks without actually having to obtain the respective devices. Furthermore, it provides a formal benchmark to compare against, and assess, the performance of different and new approaches.

FBS devices can be considered a low-cost alternative for studying fingermark degradation behavior. Although the FBS results presented only indicate their potential and cannot be reliably compared to the high-resolution CWL and CLSM devices of earlier studies, they can be easily acquired off the shelf and used for reproducing and extending the CWL and CLSM conclusions.

For a practical application in crime scene investigations, it is likely that, at first, only fingermarks exhibiting a (very) fast rate of degradation will be of use, because this type of marks have a high probability of actually being fresh, whereas in all other cases it is more difficult to obtain reliable age information. It is unlikely that an absolute guarantee could be given concerning the reliability of the age of latent fingermarks for all possible scenarios, since there will be cases of abnormal degradation rates due to certain influencing factors, such as mark contamination with external substances. Due to the large variety and complex interrelation of different effects on residue degradation, an estimate of age should be considered as an additional indicator complementing other types of evidence rather than a standalone proof of time of deposition. However, significant progress is required in latent fingermark aging research to achieve the necessary accuracy, even for its application as an indicator.

References

1. Merkel R (2014) New solutions for an old challenge: chances and limitations of optical, non-invasive acquisition and digital processing techniques for the age estimation of latent fingerprints. Logos, Berlin. Also available at: http://nbn-resolving.de/urn:nbn:de:gbv:ma9:1-4875. Last accessed 24 Aug 2019
2. Kiltz S, Hildebrandt M, Dittmann J, Vielhauer C (2012) Challenges in contact-less latent fingerprint processing in crime scenes: review of sensors and image processing investigations. In: 20th European signal processing conference, Bucharest, Romania
3. Schott M, Merkel R, Dittmann J (2012) Sequence detection of overlapping latent fingerprints using a short-term aging feature. In: International workshop on information forensics and security, Tenerife, Spain
4. Qian K, Schott M, Zheng W, Dittmann J (2014) Context-based approach of separating contact-less captured high-resolution overlapped latent fingerprints. IET Biometrics 3(2):101–112

5. Stojanović B, Nešković A, Marques O (2017) A novel neural network based approach to latent overlapped fingerprints separation. Multimed Tools Appl 76(10):12775–12799
6. Merkel R, Pocs M, Dittmann J, Vielhauer C (2012) Proposal of non-invasive fingerprint age determination to improve data privacy management in police work from a legal perspective using the example of Germany. In: 7th international workshop on data privacy management, Pisa, Italy
7. Hildebrandt M, Merkel R, Leich M, Kiltz S, Dittmann J, Vielhauer C (2011) Benchmarking contact-less surface measurement devices for fingerprint acquisition in forensic investigations: results for a differential scan approach with a chromatic white light sensors. In: 17th IEEE international conference on Digital Signal Processing (DSP), Corfu, Greece
8. Hildebrandt M, Makrushin A, Qian K, Dittmann J (2013) Visibility assessment of latent fingerprints on challenging substrates in spectroscopic scans. In: IFIP CMS 2013, LNCS 8099, Magdeburg, Germany
9. Egli N, Moret S, Bécue A, Champod C (2013) Fingermarks and other impressions – a review (August 2010 – June 2013). In: 17th interpol forensic science symposium, Lyon, France
10. Cadd S, Islam M, Manson P, Bleay S (2015) Fingerprint composition and aging: a literature review. Sci Justice 55(4):219–238
11. Bumbrah G, Sharma R, Jasuja O (2016) Emerging latent fingerprint technologies: a review. Res Report Forensic Med Sci 6:39–50
12. O'Hagan A, Green S (2018) Crime scene to court: a study on finger-mark aging. Forensic Res Criminol Int J 6(6):491–503
13. Dorakumbura B (2017) Fundamental studies into the chemical and physical properties of latent fingermarks. Dissertation, Curtin University
14. Merkel R, Gruhn S, Dittmann J, Vielhauer C, Braeutigam A (2012) On non-invasive 2D and 3D chromatic white light image sensors for age determination of latent fingerprints. Forensic Sci Int 222(1):52–70
15. FRT GmbH (2019) Fries research & technology (FRT) [Online]. Available: www.frt-gmbh.com/en/, www.frt-gmbh.com/en/microprof-200.aspx. Accessed 25 July 2019
16. Merkel R (2015) Latent fingerprint aging from a hyperspectral perspective: first qualitative degradation studies using UV/VIS spectroscopy. In: Proceedings of the 9th international conference on IT security incident management and IT forensics (IMF2015), Magdeburg, Germany
17. Richards A, Partner OP (2010) Reflected ultraviolet imaging for forensics applications [Online]. Available http://www.uvcorder.com/pdf/Reflected_UV_Imaging_for_Forensics_V2.pdf. Accessed 24 Feb 2014
18. Lin SS, Yemelyanov KM, Pugh Jr EN, Engheta N (2006) Polarization- and specular-reflection-based, non-contact latent fingerprint imaging and lifting. J Opt Soc Am A 23(9):2137–2153
19. Kuivalainen K, Oksman A, Peiponen K (2010) Definition and measurement of statistical gloss parameters from curved objects. Appl Opt 49(27):5081–5086
20. Leich M, Kiltz S, Dittmann J, Vielhauer C (2011) Non-destructive forensic latent fingerprint acquisition with chromatic white light sensors. In: Proceedings of SPIE 7880, media watermarking, security, and forensics III, 78800S, San Francisco, USA
21. Keyence Corporation (2019) Keyence color 3D laser scanning microscope[Online]. Available: www.keyence.com, www.keyence.com/products/microscope/laser-microscope/vk-x100_x200/models/vk-x110/index.jsp. Accessed 25 July 2019
22. Merkel R, Dittmann J, Vielhauer C (2017) A first public research collection of high-resolution latent fingerprint time series for short-and long-term print age estimation. IEEE Trans Information Foren Secur 12(10):2276–2291
23. Merkel R, Vielhauer C (2012) On using flat bed scanners for the age determination of latent fingerprints: first results for the binary pixel feature. In: Proceedings of the 14th ACM workshop on multimedia and security (MM&Sec), Coventry, UK
24. Merkel R, Dittmann J, Vielhauer C (2016) Resource-efficient latent fingerprint age estimation for Adhoc crime scene forensics: quality assessment of flat bed scans and statistical features.

In: The sixth international conference on image processing theory, tools and applications, Oulu, Finland
25. Seiko Epson Corporation (2019) Epson perfection 1660 Photo [Online]. Available: https://www.epson.de/en/products/scanners/consumer-scanners/epson-perfection-1660-photo. Accessed 25 July 2019
26. Canon (2016) Canon CanoScan 9000F Mark II [Online]. Available http://www.canon.com. Accessed 2016
27. Crane NC, Bartick EG, Schwartz Perlman R, Huffman S (2007) Infrared spectroscopic imaging for noninvasive detection of latent fingerprints. J Foren Sci 52(1):48–53
28. Antoine KM, Mortazavi S, Miller AD, Miller LM (2010) Chemical differences are observed in children's versus adults' latent fingerprints as a function of time. J Foren Sci 55(2):513–518
29. Williams DK, Brown CJ, Bruker J (2011) Characterization of children's latent fingerprint residues by infrared microspectroscopy: forensic implications. Forensic Sci Int 206(1–3):161–165
30. Lin AC, Hsieh HM, Tsai LC, Linacre A, Lee JCI (2007) Forensic applications of infrared imaging for the detection and recording of latent evidence. J Forensic Sci 52(5):1148–1150
31. Barcikowski S, Bunte J, Ostendorf A, Aehnlich J, Herrmann R (2004) Contribution to the age determination of fingerprint constituents using laser fluorescence spectroscopy and confocal laser scanning microscopy. In Proceedings of SPIE 5621, Optical Materials in Defence Systems Technology, London
32. Saitoh N, Akiba N (2005) Ultraviolet fluorescence spectra of fingerprints. Sci World J 5:355–366
33. Dam A, Schwarz J, Vos J, Siebes M, Sijen T, Leeuwen G, Aalders C, Lambrechts S (2014) Oxidation monitoring by fluorescence spectroscopy reveals the age of fingermarks. Angew Chem Int Ed 53(24):6272–6275
34. Akiba N, Kuroki K, Kurosawa K, Tsuchiya K (2018) Visualization of aged fingerprints with an ultraviolet laser. J Forensic Sci 63(2):556–562
35. Connaster RM, Prokes SM, Glembocki OJ, Schuler RL, Gardner CW, Lewis SA, Lewis LA (2010) Toward surface-enhanced Raman imaging of latent fingerprints. J Foren Sci 55(6):1462–1470
36. Andersson P, Lejon C, Mikaelsson T, Landström L (2017) Towards fingermark dating: a Raman spectroscopy proof-of-concept study. ChemistryOpen 6(6):706–709
37. Dubey SK, Mehta DS, Anand A, Shakher C (2008) Simultaneous topography and tomography of latent fingerprints using full-field swept-source optical coherence tomography. J Opt A Pure Appl Opt 10(1):015307–0153015
38. De Alcaraz-Fossoul J, Mancenido M, Soignard E, Silverman N (2019) Application of 3D imaging technology to latent fingermark aging studies. J Forensic Sci 64(2):570–576
39. Watson P, Prance RJ, Beardsmore-Rust ST, Prance H (2011) Imaging electrostatic fingerprints with implications for a forensic timeline. Forensic Sci Int 209(1):41–45
40. Srihari SN (2010) Computing the scene of a crime. IEEE Spectr 47(12):38–43
41. Hildebrandt M, Kiltz S, Grossmann I, Vielhauer C (2011) Convergence of digital and traditional forensic disciplines: a first exemplary study for digital dactyloscopy. In: Proceedings of the 13th ACM workshop on multimedia and security (MM&Sec), New York, USA
42. Merkel R, Braeutigam A, Kraetzer C, Dittmann J, Vielhauer C (2011) Evaluation of binary pixel aging curves of latent fingerprint traces for different surfaces using a chromatic white light (CWL) sensor. In: Proceedings of the 13th ACM workshop on multimedia and security (MM&Sec), New York, USA
43. Popa G, Potorac R, Preda N (2010) Method for fingerprints age determination. Roman J Legal Med 18(2):149–154
44. Baniuk K (1990) Determination of age of fingerprints. Forensic Sci Int 46(1):133–137
45. Barros R, Faria B, Kuckelhaus S (2013) Morphometry of lantent palmprints as a function of time. Sci Justice 53:402–408

46. De Alcaraz-Fossoul J, Mestres Patris C, Balaciart Muntaner A, Barrot Feixat C, Gené Badia M (2013) Determination of latent fingerprint degradation patterns – a real fieldwork study. Int J Legal Med (4):857–870
47. De Alcaraz-Fossoul J, Roberts KA, Barrot-Feixat C, Hogrebe G, Gené Badia M (2016) Fingermark ridge drift. Forensic Sci Int 258:26–31
48. Carletta J (1996) Assessing agreement on classification tasks: the kappa statistic. Comput Linguist 22(2):249–254

Chapter 5
Latent Fingermark Aging in 2D: Qualitative and Quantitative Analytical Approaches

Josep De Alcaraz-Fossoul and Jennifer Chisholm

Abstract (Latent) Fingermarks have long been an instrumental part of criminal investigations. A recurring point of contention in court is often summarized in the following two questions: *How long has the fingermark been on the item in question?; Does the trace line up with the time frame of the crime?* With the emergence and adaptation of novel techniques and methodologies adopted and/or adapted for fingermark age estimations, scientists can begin to answer these questions. To date, there have been two main approaches to fingermark age research: chemical (i.e., analyses of compounds) and physical/optical (i.e., measurements of the topography of ridges). The latter can be further sub-classified based on the type of fingermark features to be analyzed: 2D- and 3D-based. This chapter will specifically focus on methodologies that describe morphometric features in 2D and have been made available through research although not yet ready for deployment in the field. A series of 2D imaging approaches as well as advantages and disadvantages of each of these techniques will be discussed. While scientific research on fingermark age estimation has come a long way since the 1950s, there are still a number of challenges and limitations, including elucidating the effect of environmental factors to ridge degradation and the potential use of mathematical models to crime scene scenarios, among others. It is anticipated that in the near future, there will exist a computer-based, automatic and portable "measuring tool" which will be able to objectively estimate time since deposition with a known degree of accuracy.

Keywords 2D · Morphometrics · Age estimation · Ridge width · Color contrast · Minutiae count · Ridge discontinuities · Powders · Titanium dioxide · Carbon black · Degradation · Fingermark · Fingerprinting · Time since deposition

---

J. De Alcaraz-Fossoul (✉) · J. Chisholm
Henry C. Lee College of Criminal Justice and Forensic Science,
Forensic Science Department, University of New Haven,
West Haven, CT, USA
e-mail: jfossoul@newhaven.edu

J. De Alcaraz-Fossoul (ed.), *Technologies for Fingermark Age Estimations: A Step Forward*, https://doi.org/10.1007/978-3-030-69337-4_5

Acronyms

2D	Two dimensional
3D	Three dimensional
ACE-V	Analysis, Comparison, Evaluation, and Verification
ULW	Universal Latent Workstation
CAST	Center for Applied Science and Technology
SIP	Subject-Independent Parameter
SNIP	Subject-Non-Independent Parameter
BMI	Body Mass Index
DI	Discontinuity Index
PMI	Post-Mortem Interval
RH%	Relative Humidity

5.1 Introduction

Latent fingermarks are neither static nor immutable friction ridge skin impressions. Rather, under a specific combination of donor, environmental, and substrate conditions, fingermarks (ridges) can become quite dynamic and distinctively change over time, physically, chemically, and biologically. In fact, ridge impressions start degrading from the very moment they are created [1–6]. There is a wide array of topographical/physical modifications that latent fingermarks can experience, the most characteristic being fluctuations in ridge width, changes in the level of (color) contrast with the background, loss of integrity of ridges, evaporation of moisture, and loss of mass, among others [1, 4–10]. Once measured, all these modifications combined can provide meaningful information about the nature and aging behavior of a fingermark beyond its intrinsic potential – and original purpose – of identitying its source. Indeed, there is much more to fingermarks than just ridge patterns and minutiae recognition. On these grounds, one of the main challenges in the forensic subdiscipline of friction ridge analysis has been to develop robust methods to estimate time since deposition, also referred to as the *age* of the fingermark. To date, scientific research has mostly focused on developing approaches and characterizing the chemical and physical changes of ridges in laboratory settings with little priority to advancing methods for applicability to real crime scene scenarios.

Age determination of latent fingermarks remains an unsolved issue despite its relevance and the numerous attempts researched and published since the 1950s [11, 12]. In fact, the development of robust methods and techniques for age estimation in 2D is still in its infancy. Perhaps, a broad pathway to achieve this ultimate goal should be laid out. First, more research needs to be conducted to comprehend the relationship between the decay of chemical constituents and the changes in physical parameters (or indicators) of degradation (i.e., time-senstive morphometrics), including any possible effects of the enhancement processes (i.e.,visualization techniques) aplied during the analysis. Then, a thorough understanding of the impact of the most influencing environmental factors in aging is necessary. Once this is accomplished, mathematical correlations between environmental factors and levels of chemical/physical/biological decay could be formulated before a tentative design of any aging model is made available. Lastly, it is hoped that in the near future mathematical models, computer software, and portable hardware could be developed to assist in automating fingermark age estimation with the click of a button.

In the forensic community, the ability to provide a reliable estimate of the age of latent impressions would be an incredibly useful tool for crime scene investigations. For instance, crime scene investigators could exclude or include potential suspects by assessing when a fingermark was left on a surface. This could be extremely helpful for evidence that may have been handled by different people, including the actual perpetrator, or when suspects have had legitimate access to the scene prior to the crime. If the time of deposition does not correspond with the time of the crime, then the donor could be reasonably excluded. This not only helps to focus the investigation process but litigation as well. Once in court, the time of deposition of a fingermark may be easily contested by one of the legal parties as a defense strategy, but a robust estimate of the age of evidence would create a convincing argument for juries to consider. In instances of cold cases, *age* could be an important and key piece of information in discerning relevant from non-relevant evidence irrespective of the potential identification of the questioned fingermark. For example, a case where age estimation would have been useful was experienced by the Los Angeles Police Department (LAPD) in August 1983, where a church was burglarized on two occasions. A fingermark was developed at the time of the first burglary. A year later, the same fingermark was recovered and identified when the church was burglarized a second time [1]. If this fingermark had not been discovered in the original investigation of the first burglary, and only found at the time of the second burglary, the LAPD would not have known that this fingermark was indeed irrelevant to their current case. With the development of fingermark age estimation methodologies, this potential scenario could be avoided.

Throughout this chapter, we will explore what constitutes a latent fingermark, an overview of the major variables that affect its chemical composition, and what physically may occur during its degradation process. We will then review the scientific research conducted on the application of 2D age estimation methodologies and discuss the advantages and limitations of each aging parameter and technique.

5.2 Relevance of Age Estimation of Latent Fingermarks: How Can Age Help?

A recurrent question asked in courts of law, particularly by defense attorneys attempting to cast doubt for the jury, is the inevitable: "*Is it possible to scientifically determine the time the fingermark in question was left behind by my client?*" Forensic science should be able to answer this frequently asked question. The solution relies on linkage analysis where an identified fingermark relates a person to the evidence and the evidence to the crime scene –and, in the near future, to the time of the event. Altogether, this evidence-related intelligence would provide a solid probative value to a fingerprint examiner's conclusions and expert testimony. Not only will it be possible to link a person to the scene and the time of a crime, but suspects may be excluded from the criminal event: if the time of deposition does not coincide with a resonable timeframe of the crime the individual could be discarded as a prime suspect. This is particularly important for locations or items that are publicly accessible. Indeed, courts have reversed judgments based on this argument alone, claiming that the questioned fingermark could have been deposited at any time prior to the crime [1]. Girod et al. (2016) sum it up perfectly: "*Failure to establish an accurate deposition age (or age range) for the fingermark could ultimately jeopardize the liberty of every person who has ever touched anything later found at the scene of a crime*" [1].

Age determination of ridge pattern impressions could not only help differentiate prime suspects from unrelated persons but assist with appropriate enhancement/visualization methods. The (microscopic) topographical dynamism of the degradation process is the product of various chemical compounds evaporating and oxidizing as well as the result of physical interactions with the substrate, i.e., diffusion and absorption, and the effect of the environment. Then, a better understanding of the mode of action of (fingermark) enhancing reagents can be achieved by targeting these compounds. In fact, numerous studies have developed best practices for processing aged latent fingermarks [1, 2, 4, 15–17]. For instance, a commonly used reagent for degraded latent fingermarks is physical developer (PD). While PD is used last in a sequential process of enhancement due to its "damaging properties" to fingermark residues, age estimation can help to discard techniques that are known not to work well on aged impressions, such as powder-based methods or Oil Red O [15, 17]. Understanding the process of physical/chemical degradation can be invaluable in establishing novel processing/sequential techniques by targeting chemical compounds that are most stable over time.

Other uses of age estimations include assisting with cold cases, crime scene reconstruction, and unusual cases where a previously developed latent from an already processed crime is redeveloped (or discovered for the first time) when the location becomes the scene of a second crime. Examples exist in the literature where police redeveloped fingermarks that were discovered from a previous crime or developed fingermarks of first responders at a scene long before [1]. Regarding cold cases, one of the challenges is being able to link a newly developed fingermark to the time of the crime. Usually at this stage in the investigation (months, years, or decades after the incident), few evidence items could provide additional information on the circumstances of the event. If a piece of evidence is reexamined and a latent fingermark is developed, it would be extremely beneficial to the case to have a time since deposition estimation. In addition, age estimations can assist with crime scene reconstruction efforts through the best application of sequential enhancement of fingermarks [18, 19]. In this particular case, it is most useful on overlapping impressions. A sequence of events can be hypothesized by being able to discern which fingermark was deposited first. With age estimation, fingermarks could be quickly dismissed if determined that they are not relevant to the current scene in question. This would save time and effort for the investigators and latent fingerprint examiners, as well as minimize errors in misattributions and, therefore, increase the probative value of evidence in court [5].

5.3 The Process of Estimating the Age of Latent Fingermarks in 2D

What is our recommended pathway for developing a model for age estimations based on measuring topographical (and visible) features, commonly known as morphometrics?

First, the fingermarks must meet a certain baseline set of criteria. It is important to note that, at present, age estimation by morphometrics is feasible only to non-porous substrates located indoors and to fingermarks enhanced by traditional (powdering) methods. These criteria include that the questioned fingermark:

1. Must have clear ridges, i.e., contain sufficient ridge detail for comparisons and identifications, ideally the minimum standard of ridge clarity at the level 2 or 3 ridge detail;
2. Must contain sufficient surface area of good visual quality, i.e., contain enough clear ridges as for a traditional identification;
3. Must be associated with recorded basic environmental conditions from the scene; as these have a distinctive effect on degradation patterns (and "life expectancy") of latent impressions and should be known and regularly documented on scene [2, 14, 15], including type of substrate, degree of exposure to natural light, temperature, and relative humidity (RH%)[1]
4. Must have been transferred with medium pressure, approximately two to seven kilograms, i.e., depositions with too little or too much pressure can exhibit lower visual quality and be distorted [13, 20–22].

Despite its immense potential as an investigative tool, the age determination of fingermarks does come with certain limitations. While the first three criteria could be met, unfortunately, the last criteria will never be known with precision, but it could be inferred based on certain visual indicators (discussed further in Sect. 5.9.1). On sufficient quality, the area of analysis of the fingermark must meet a minimum level of clarity, either defined by the FBI software Universal Latent Workstation (ULW), or by the fingermark grading scale recommended by the Center

[1]These variables are, for the most part, similar to those collected when estimating post-mortem intervals (PMI) with deceased persons. In this instance, environmental factors and body type, among other factors, cause different outcomes and rates in decomposition. However, as long as these are considered when attempting to estimate time since death, a time frame can be set, barring limitations. Similarly, it is recommended age estimation of latent fingermarks be handled in an analogous way as the estimation of PMI. Time frames, such as "within 72 hours" or "greater than 72 hours," can be provided as bracketed estimates.

Table 5.1 CAST grading scale [25, 26], a scale of quality level commonly used in the literature to determine the value of an enhanced latent fingermark

0	No evidence of fingermark
1	Weak development; evidence of contact, but no ridge detail
2	Limited development, about 1/3 of ridge details are present, but probably cannot be used for identification purposes
3	Strong development, between 1/3 and 2/3 of ridge details; identifiable fingermark
4	Very strong development; full ridge details; identifiable fingermark

for Applied Science and Technology (CAST) [23–25], as shown in Table 5.1. These quality ratings help examiners to determine the visual quality of an impression. Using the "image clarity" map in the ULW software, the fingermark is "color-coded" based on areas marked as "debatable ridge flow, debatable minutiae, and definitive minutiae" [23]. This software automatically calculates the levels of quality within the fingermark and helps to determine how clear the fingermark is. The ideal or recommended ULW-based levels of quality for age estimations still needs to be resolved as no research has been conducted using ULW on fingermark degradation patterns.

Second, it is necessary to select for testing – and future practice in the field – a specific method/technology for the detection and collection of time-sensitive ridge alterations, e.g., topographical changes over time detected by the application of carbon black powder and captured with digital photography. The choice will be contingent on the type of substrate being analyzed, financial constraints, manpower, expertise, portability of instrumentation, and relevance of the case, among others.

Third, the ridge features to be considered and examined must correspond to relevant visual parameters/indicators that are time-sensitive and that are measurable, e.g., variations in ridge width. This is necessary to develop a robust dataset as foundation for modeling degradation patterns under controlled or monitored conditions (explained in more detail in Sects. 5.6 and 5.7); eventually, to apply these models to blind studies and uncontrolled situations and ultimately determine the age of a questioned latent impression.

Fourth, regardless of the technology and aging parameters selected, it is recommended that the practice of age estimation by optical (visual) means be a two-step process. More details on this process are provided in Chap. 6. Briefly, after determining the questioned impression is of sufficient quality (clarity) and with recorded environmental conditions, the first step is a *primary (binary) screening*. This helps investigators to discern whether or not the fingermark is either "fresh" or "aged". In this case, "fresh" refers to latents that are "younger" than 3–7 days; "aged" are those older. This step does not provide a specific age estimate of the fingermark, but "class" characterizes the evidence into two main categories. For example, if the crime occurred and was discovered within 72 h, and the binary screening indicated that the fingermark was not "fresh," this fingermark could be initially excluded as prime evidence, reducing the number of items initially processed. This screening is based on the examination of certain morphometric parameters of an enhanced fingermark, such as color contrast between ridges and furrows and presence of ridge discontinuities. These age estimation parameters are described as Subject-Independent Parameters (SIPs) because they do not necessarily require any biometric information from the donor's fingerprint exemplars, as long as certain environmental conditions are known [10]. The primary information collected from the questioned impression can then be compared with that from an exemplar fingerprint (inked or scanned). If the fingermark lines up with the time frame of the crime and it is individualized to a person, the *secondary screening* step is then conducted. This is where more detailed information about time of deposition can be obtained. This step would additionally include measurements and comparisons of the ridge width and the number of viable minutiae between questioned and exemplars [9, 10]. These two are described as Subject-Non-Independent Parameters (SNIPs) because they require certain morphometrical information from the donor's exemplars [10]. In this step, a more accurate age estimate with an error rate is possible to be obtained. It is also important to ensure that the primary and secondary screenings are consistent with one another. Depending on the type of variables observed, the morphometric parameters measured, the visualization technique applied, and the mathematical model used, further consideration should go into analyzing error. This could be achieved by concurrently examining the topographical and chemical changes of ridges over time, understanding the effects of variables in aging, and modeling degradation patterns, as long as certain environmental conditions are known and/or inferred from the crime scene.

5.3.1 Where May Age Estimation Fit in the ACE-V Process?

Age determination would be included at two parts in the standard ACE-V process (Analysis, Comparison, Evaluation, and Verification) (Fig. 5.1). The first step could be viewed as a *primary screening*. This is the binary "fresh" vs. "aged" classification mentioned earlier using aging parameters such as ridge/furrow color contrast and presence of ridge discontinuities. This would occur prior to the ACE-V process entirely. The goal of this step is to include or exclude a fingermark as an item of interest in an investigation. The *secondary screening*, thought of as a confirmation screening, is placed between the comparison and the evaluation step of ACE-V. This step occurs necessarily after a possible identification when more information about the donor can be collected and used to attempt a more accurate age estimate of the questioned impression. This may include comparing the known fingerprint's ridge widths or minutiae count from the ridge areas of interest. Additionally, for serious crimes or high-profile cases, a known impression from a suspect could be further subjected to specific experimental conditions by having the individual deposit fingermarks and let them degrade under similar conditions as the questioned impression, such as substrate type, level of exposure to light, relative humidity, and temperature.

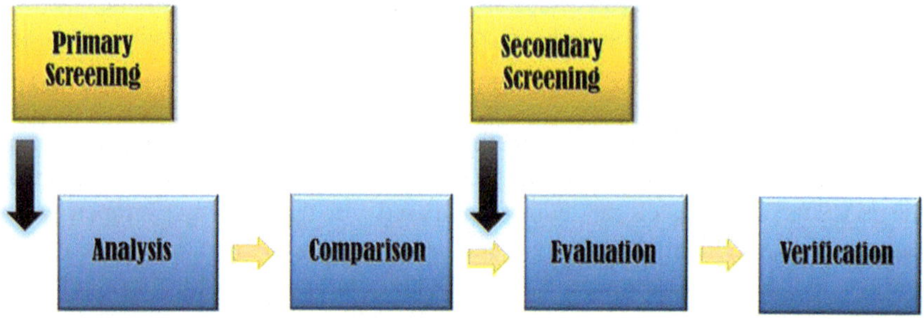

Fig. 5.1 Graphical representation of the ACE-V process with primary and secondary screening steps where age estimates would be recommended to be performed

By incorporating primary and secondary screening steps in the ACE-V process, the examiner would be able to evaluate the relevance of the questioned latent fingermark and either include or exclude it as primary evidence. This two-step screening method strengthens the evidentiary value of the impression by decreasing the probability of misidentifications due to time inconsistencies. If the results of these two screenings disagree with each other, it could be either a sign of a weak model or a poor quality of the impression, or that additional aging parameters and analytical techniques should be considered.

5.4 Factors That Affect Latent Fingermark Degradation

Before any age estimation model or protocol can become a standard, one must have a solid foundation of the compositional makeup and degradation behavior of latent fingermarks. To date, an abundance of research has been published on the composition and, to some extent, degradation of sweat components; however, specific aging patterns in relation to the effect of the environment are more difficult to find. Latent fingermarks contain a convoluted mixture of endegenous compounds secreted by the human skin which may be highly variable between and within individuals [1, 14, 15, 27, 28] based on a number of factors including diet, environment, age, or biological sex [1, 14, 15, 27, 29]. These compounds are in the sweat but it can also contain exogenous substances such as drugs, cosmetics, dirt, and others that may end up on a person's fingertips [4, 14].

The level (intensity) and speed (rate) of degradation of latent impressions are affected by many factors, resulting in distinctive chemical and physical (topographical) aging patterns. Research into the aging patterns and age estimation models must consider the substrate, donor (including deposition conditions), environment, and interaction of all these factors over time. It is important to highlight that the degradation process is a dynamic continuum and very fluid, making it impossible to include all influencing variables in a single mathematical model of aging;

nevertheless, an ideal and realistic approach could encompass just the most relevant factors that could be easily measured or inferred from the crime scene. Therefore, it becomes key in research to identify these most influencing variables.

The best way to describe the aging process, from deposition to total obliteration of the fingermark, and how the variables affect it, is by classifying a discrete timeline as pre-deposition, peri-deposition (or just deposition), and post-deposition. At each stage in this process, the effects of the influencing factors are distinctive and influence the fingermark at different degrees. Several variables interact with each other which result in a varied array of outcomes. The pre-deposition stage includes the circumstances that lead up to depositing the fingermark, regardless of substrate. The major source of variability during this stage includes donor (in terms of physical, chemical, and biological factors) and to a lesser extent the environment. Peri-deposition, or "at deposition," encompasses all the conditions that affect the transfer process itself, such as donor, substrate, mode of deposition (i.e., length of contact, pressure, angle, etc.), and the environment. Post-deposition involves the effect of variables after the fingermark has been deposited where all influencing factors work in conjunction and that result in the inevitable gradual and natural degradation of the impression, unless a mechanical force disrupts this process (Fig. 5.2).

5.4.1 *The Effect of Donor*

The donors themselves are the source of a large number of variables that may influence the speed of degradation of the latent impression. Fingermarks' topography vary depending on the size of the donor's fingers, ridge density, scars, skin conditions, region of skin from where they originated, biological sex, and ethnicity, among others [29–36]. Several studies have consistently shown that females tend to have a higher ridge density than males [30, 33, 34, 36], so much so that it has been proposed that ridge density could be a potential forensic anthropological tool for sex identification [34]. Additionally, biological age can affect several aspects of a donor's fingermark. As a child gets older, their hands will increase in size, thus affecting the dimension of the fingers and ridges [30]. For elderly people, ridges can flatten, making the person's fingermarks less prominent [30]. Diet and body mass index (BMI) have also been explored, showing that BMI can affect the quality of the impression, possibly due to how much sweat the donor produces [35]. The donor may also introduce contaminants into the fingermark's chemical composition [14] which can include cosmetics, dirt, dust, and grease, to name a few.

Donor-dependent deposition conditions: The concentration of compounds within latent fingermarks can vary depending on how and for how long the transfer onto the surface was. Pressure, duration of contact, angle of deposition, and the type of residue on the donor's fingers can affect enhancement and detection as well as degradation patterns [15]. It has been observed that higher pressure at deposition and longer times of contact with the surface can produce a much darker coloration of the fingermark when processed with ninhydrin [14]. This would be expected as more

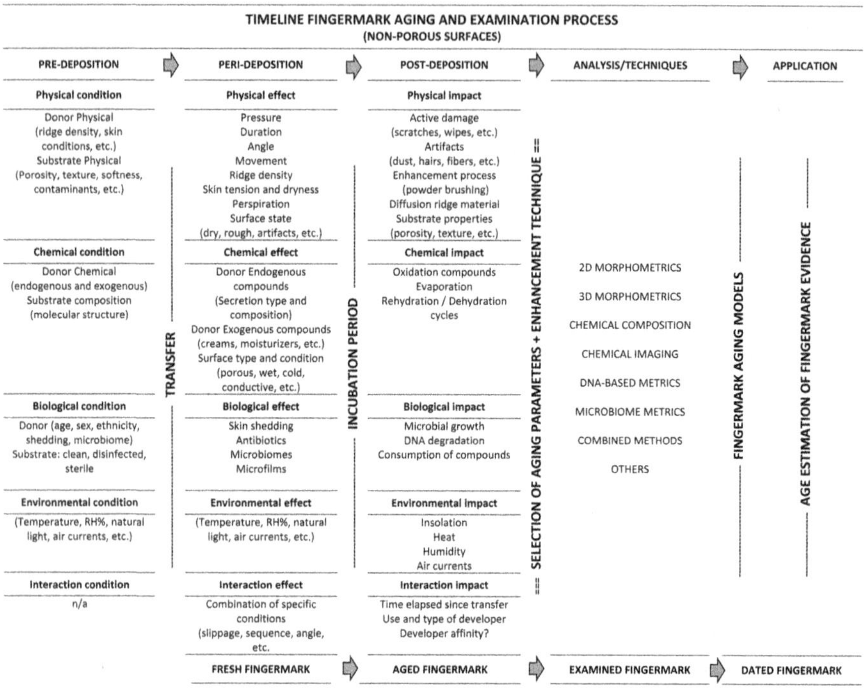

Fig. 5.2 Timeline of latent fingermark aging and examination process for non-porous surfaces. Three stages are noted (pre-deposition, peri-deposition, and post-deposition) where each one is characterized by a number of conditions, effects, interactions, and impacts based on certain variables. The fate of a fresh fingermark is affected prior to being deposited on a surface by a series of physical, chemical, biological, and environmental factors. The initial deposition is then subject to the interaction of all these factors, including the enhancement/detection technique before examination. The aging parameters must be established according to the technique applied before developing mathematical models of aging which is the last step before providing an age estimate

residue is transferred to the surface [14]. Concentration of lipids may also affect degradation. One study examined changes in color contrast of fingermarks with different lipid contents and observed that those with high lipid content had the longest period of no visible change in contrast over 60 days [40]. Fingermarks with low and moderate lipid content showed initially a period of no change; however, it was followed by a decrease in color contrast against the background. Specimens of low lipid content had a shorter period of no change and a more significant decrease in contrast when compared to impressions of moderate lipid content [40].

5.4.2 The Effect of Substrate

One of the most influencing variables on aging is substrate type, i.e., the surface where the fingermark is deposited. For forensic purposes, substrates are generally classified by their porosity (absorbance ability). This classification includes porous, non-porous, and semi-porous materials [2, 14, 29]. As for aging studies, the type of substrate is considered the most relevant variable in fingermark processing (visualization) since the enhancement method of choice is highly dependent on the porosity and other physical properties of the receiving surface. For porous types, the fingermark residues will be mostly absorbed. The higher the porosity, the more residue is absorbed [14]. For example, on the degradation of squalene, it was observed that it degraded within 1–7 days on non-porous substrates as compared to 30 days for the porous substrate [14]. This supported the idea that for porous substrates, the fingermark is being absorbed and protected from environmental influences, making the residues more resilient to decomposition. Some components, with time, can eventually diffuse within the substrate of porous materials such as urea and chlorides which, over time or with high humidity, can display a diffused effect [2]. On non-porous substrates, the fingermark compounds will lay on the surface fully exposed to the environment which will cause it to dehydrate and rehydrate in cycles [2, 8, 9, 16, 35, 38, 39].

To date, most fingermark aging studies have involved non-porous substrates (i.e., glass, metals, plastics, etc.) since the effect of degradation, especially when analyzed by morphometrics, can be observed and measured easily. Although the fingermark is more exposed to the elements, these materials are mostly independent from rates of absorbance of its residues into the substrate, an additional and not desirable new variable difficult to measure that should otherwise be quantified in porous surfaces. Among non-porous types, there are significant physical and chemical differences that will also affect the properties of the substrate and its interaction with the fingermark (see Sect. 5.4.4). For example, aging patterns on glass and polystyrene (plastic) have proven to be remarkably different when visualized with powders, regardless of other variables such as temperature, degree of exposure to natural light, and donor, demonstrating the large impact that the type of substrate alone has over others.

5.4.3 The Effect of the Environment

The environment plays a central role on degradation after the substrate. Some of the more important variables include UV radiation, temperature, relative humidity, and air circulation. Studies on UV radiation have shown conflicting results. Some state that direct sunlight can have severe degradation effects which include the increased rate of degradation of squalene [14]. Squalene disappears faster when exposed to sunlight as opposed to dark conditions, while saturated fatty acids are not affected by light conditions, lending more evidence to other research that observed the little impact of light on degradation. Interestingly, some other studies have concluded that natural light through a window glass does not negatively influence the topographical (physical) degradation of latent fingermarks. In fact, fingermarks stored in the dark may degrade faster than those exposed to natural light [6, 9] as examined by morphometrics. According to the literature, it is then possible, and probable, that degradation at the molecular level (chemical/biological) may not directly translate to a topographical and measurable change on the quality of ridges.

Temperature and humidity are environmental factors that may closely relate to the degree of exposure to natural light. It is known that heat and relative humidity determine how quickly liquids evaporate from a surface [2]. As a latent fingermark dries out, ridges may appear shrunken upon visual inspection. With the loss of water, adhesion of physical enhancers worsens (i.e., powder formulations) which makes the enhancement process less effective [2]. High levels of humidity can have a deleterious effect, particularly on fingermarks deposited onto porous surfaces where chemical constituents are more affected resulting in their breakdown [2, 14]. Bleay and de Puit (2018) explain it perfectly: "*high humidity may play a more important role in promoting the migration of the water-soluble constituents in the fingermark, in particular those deposited on porous surfaces. As moisture is absorbed into a substrate, components diffuse and migrate*" [2]. With thermal degradation, eccrine sweat produces constituents that can fluoresce on porous substrates [2]. Additionally, fingermarks processed with silver nitrate in high humidity conditions appear to produce worse results than those processed in lower humidity [14]. However, in high humidity scenarios, eccrine-rich impressions show better enhancement when processed with cyanoacrylate, indicating that eccrine secretions may be more affected by humidity than sebaceous-rich secretions [14]. This information is relevant when attempting to draw patterns of degradation over time under certain (known) environmental conditions.

Air circulation also plays a role in the drying of latent fingermarks and more than likely works in conjunction with temperature and humidity. Wind can have a drying effect and can also displace droplets or components within the impression [2]. Studies have shown that fingermarks left outdoors degrade faster than those left indoors, indicating a possible air circulation effect. Other environmental variables that impact degradation may include dust, rain, condensation, friction, and atmospheric contaminants [14]. However, most of these factors are of little relevance to indoor scenarios.

5.4.4 Interactions

In practical scenarios, all the aforementioned variables do not act isolated but in conjunction. This is best demonstrated by the triangle of interaction (Fig. 5.3). Once a latent fingermark is deposited on a surface, its "life expectancy" is impacted by the interrelation between its composition (donor-related), the surface, and the environment. Each one of these elements influences each other in a unique but relatively

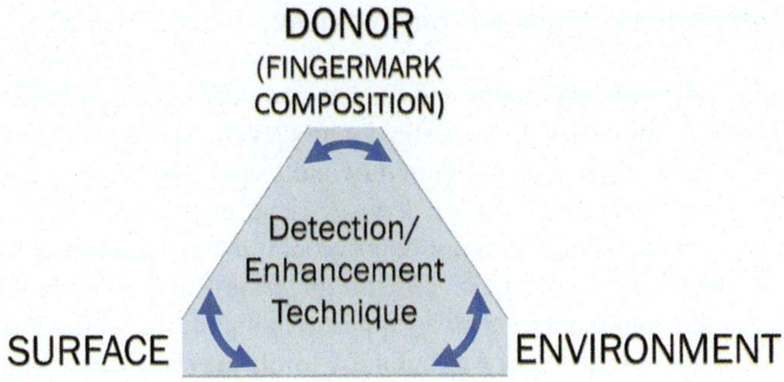

Fig. 5.3 The "triangle of interaction" is a simplified representation of how the donor (i.e., the chemical composition of the fingermark), surface, and environment interact with each other during the deposition and aging of the latent fingermark. The successful detection of the fingermark will depend on the technique/s applied for visualization. (Figure adapted from Refs. [2, 15, 25])

predictable way resulting in several but limited number of aging outcomes [2, 15]. For example, an eccrine-rich fingermark left on a piece of glass and next to a fire is going to have a much different behavior than its sebaceous-rich counterpart left on a piece of paper in a room with average temperature and humidity. The fire (high temperature) may affect the fingermark by dehydrating it and covering it with soot. Since it is an eccrine secretion, with more soluble contents [4, 14], it is most likely going to evaporate very quickly. It is one of the goals of fingermark age estimation research to discover commonalities between different variables, aging patterns, final outcomes and the driving influence of each variable to be able to develop mathematical models that can, to a certain degree, explain and predict the most probable outcomes for questioned fingermarks.

5.4.4.1 Interactions: Substrate and Donor

The chemical components of a latent fingermark will interact with the substrate from the moment it is deposited, and vice versa, which may in turn have an effect on its visual appearance. For example, certain types of metals are corroded by fingermark residues [2, 39]. As described earlier, the impression will behave differently depending on the nature of the substrate (porous, non-porous, and semi-porous). On porous substrates, eccrine compounds are absorbed quicker than sebaceous compounds [14]. This leaves the latter to be more affected by environmental factors over time, while eccrine compounds are more preserved within the substrate [15]. In semi-porous substrates, eccrine compounds, like porous, are absorbed faster than sebaceous secretions; however, both eccrine and sebaceous secretions are subject to an even slower absorption rate compared to porous substrates [14]. With semi-porous surfaces, both of these secretions are more affected by environmental factors [15]. Non-porous substrates do not absorb either of the secretions, leaving the fingermark exposed until degraded, naturally or by a mechanical action.

5.4.4.2 Interactions: Donor and Environment

The interaction between the donor and the environment affects the life expectancy of a fingermark significantly. Indeed, the donor's level of perspiration is influenced by environmental temperature and humidity and can affect the chemical composition of the fingermark before it is even deposited (pre-deposition) or during the time of deposition (peri-deposition). At pre-deposition, if the temperature is high enough to cause the donor to sweat intensively, the fingermark will contain much higher levels of eccrine gland components, as opposed to situations with colder temperatures with no active sweating. Lower humidity levels may also cause higher rates of skin shedding and change the viability of microbiomes which might be involved in degradation processes. The climate may also indirectly cause the donor to behave in certain ways, by using cosmetics, such as moisturizers, thus adding contaminants on the skin. After deposition, the environment, now by itself, continues to degrade the fingermark. Increased temperature and dehydration/rehydration cycles may induce a faster degradation [2].

5.4.4.3 Interactions: Substrate and Environment

Temperature and humidity may not only affect the fingermark itself but also the substrate. It can cause the substrate to melt, burn, oxidize, or degrade chemically [2]. The corrosion of a substrate is influenced usually by temperature, humidity, natural light, and air circulation [15]. The substrate can also dry out or more readily absorb water depending on humidity [2]. Condensation can build up on non-porous surfaces and influence the rate of degradation of the fingermark, possibly by diluting it in repeated cycles of dehydration and rehydration [35]. This was observed in a study that examined variations of the width of ridges over time. A decrease in the width was reported initially; however, when the room temperature increased, a 3% increase in the width of ridges was detected [35]. This exemplifies once more the importance of monitoring environmental factors in research and in the field.

5.5 What Happens to a Latent Fingermark as It Ages?

Several scientific studies have investigated the degradation patterns and outcomes of latent fingermarks by observing the aforementioned variables and interactions [1–10, 15].

5.5.1 Patterns of Latent Fingermark Degradation on Non-porous Surfaces

To properly comprehend the natural process of degradation, it has to be understood as the result of the concomitant occurrence of biological, chemical, and physical (topographical) changes over time. Yet, no study to date has explored these three aspects altogether. One of the main reason is the difficulty of defining meaningful aging indicators for each category, especially when the technique used for analysis is destructive to the specimen under examination, making the experiment design extremely complex and costly. Despite the absence of studies analyzing all these changes simultaneously, some meaningful discoveries have been reported that can help to develop models for estimating time since deposition (Table 5.2).

It has been described that the chemical profile of a latent fingermark changes significantly post-deposition and that these variations are dependent on the environment and substrate [4, 14, 41, 45]. In fact, Chaps. 3 and 7 of this book discuss this topic extensively. Regardless of the reported initial content of water, studies have shown that most will evaporate immediately after deposition, with the fingermark losing up to 98% of its mass within 72 h, leaving behind only organic compounds and inorganic salts [4, 14]. One example is the behavior of squalene [4, 14, 44, 48] which may remain present for years but its quantity decreases over time [44]. It has been observed that, when stored at 90 °C, free fatty acids, cholesterol, and squalene all degrade in this particular order [44]. Squalene derivatives, and cholesterol, may

Table 5.2 Summary of typical changes that are observed during latent fingermark aging from biological, chemical, and physical perspectives

Physical[a]	Biological[b]	Chemical[c]
Ridge width changes (Decrease or increase) [4, 9, 14, 38]	Decrease in human DNA quantity (Touch DNA) [4, 14, 39]	Decrease in free fatty acids, cholesterol, squalene, and squalene derivatives [1, 4, 14, 15, 41, 44]
Changes in color contrast (Affinity to developers) [4, 7, 14, 38, 40]	Microbial growth (Degradation of DNA and chemical compounds) [14]	Evaporation of water content [14, 45]
Ridge breakages (Discontinuities) [4, 8, 14, 38]		Fluorescence [4, 11, 46–48]
Decrease of viable minutiae [4, 10, 14, 38]		
Ridge drift [41–43]		

[a]Further discussed in Chaps. 4, 6, and 9
[b]Further discussed in Chaps. 10 and 11
[c]Further discussed in Chaps. 3, 7, and 8

all be useful in estimating the age of a fingermark, but it has been recognized that these variables are highly affected by substrate, donor, and environmental factors as well as enhancement methods [48].

Microbiological changes also occur within the fingermark over time. This may include degradation of touch DNA and other chemical compounds due to microbial activities. Microbiomes may also have an effect on the life expectancy of a latent fingermark. Indeed, squalene is affected by the action of microbes and further degrades due to their presence [14]. The impact of the biological effect on fingermark degradation is further discussed in Chaps. 10 and 11.

Physically, a latent fingermark experiences numerous changes from the instant it is deposited as a result of the continuous chemical and biological alterations; however, not all alterations will translate into a noticeable and measurable topographical modification; in fact, many chemical/biological changes occur unnoticed by optical means. The first example of this is the loss of mass. As mentioned earlier, a fingermark may lose most of its mass within a few hours, which is attributed to the evaporation of water content [4, 14]. The loss of moisture leads to a gradual loss of affinity of some developers, such as physical methods by powdering [14, 38, 41]. It has been demonstrated that, once deposited, fingermarks are not the static object that they were thought to be. Rather, they are quite dynamic at the molecular and morphological levels. Indeed, ridge compounds are degraded and ridges themselves even "move" over time [41–43]. In one instance, it was observed, using atomic force microscopy, that the height of fingermarks decreased and the ridges migrated horizontally on the surface after a 2-month period [41]. Another study examined the phenomenon coined as "ridge drift" (Fig. 5.4) where a single ridge may randomly change its original position, despite other neighboring ridges remaining unaltered [42]. These dynamic behaviors are probably the result of the residue diffusion into the material of non-porous surfaces, the loss of affinity of the developers to degraded ridge residues, or the adherence of developer to diffused material to the furrows.

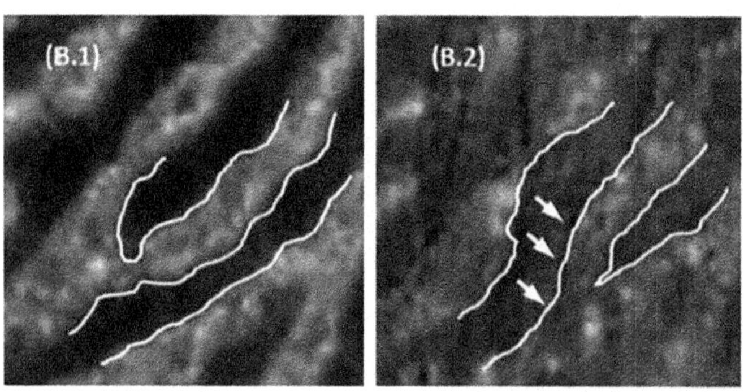

Fig. 5.4 Example of ridge drift. The marked ridge in white, enhanced with titanium dioxide powder, apparently changes from a leftward to a rightward position, modifying its original location. (B.1) Corresponds to a fresh sebaceous-rich deposition on plastic; (B.2) corresponds to a 49-days-old sebaceous-rich deposition aged in darkness on plastic. (Figure reprinted with permission from Ref. [42]. Copyright 2016, Elsevier Science & Technology)

Other examples of morphological modifications include the width of ridges. It has been reported that the average ridge width size of a fingermark varies with time, decreasing or increasing depending on conditions [1, 9, 16, 35, 38, 39]. Another visual and quantitative illustration of degradation is observed in the color contrast between ridges and furrows after enhancement with powders. Over time, the contrast is reduced [7, 40] on non-porous surfaces. The increase in ridge discontinuities (breakages) also occurs, particularly for eccrine-rich fingermarks [8, 38]. In many cases, this leads to a lower number of identifiable minutiae [10, 39]. The shapes and distance between pores also change; while some disappear altogether, others may appear to have formed a "chain" by decreasing their distance [39]. These morphometric aging parameters are discussed in more detail in Sect. 5.7. In addition to these 2D morphological features that appear to be time-sensitive, fingermarks also experience change in the third dimension (see Chap. 6 for details). Research has demonstrated that latent impressions lose height in their ridges, despite showing unaltered from a 2D perspective [49].

5.5.2 *Visual Outcomes of Latent Fingermark Degradation on Non-porous Surfaces*

Based on morphometric alterations in 2D, a number of distinctive outcomes have been reported in the literature. One particular study [10] delineates and names them all and describes the circumstances where they most often occur based on donor + substrate + environmental conditions. The outcomes emerge after powdering the fingermark and may indicate different stages of aging, or different pathways to aging, until the moment the fingermark is discovered or until it is destroyed by natural or mechanical means: "no degradation," "fragmentation," "constellation," "diffusion," and "obliteration" [7, 10], as shown in Fig. 5.5 and summarized in Table 5.3. "No degradation" defines a fingermark that has aged for a significant period of time but shows no visible signs of degradation. This pattern displays fine ridge edges and detail, textures, pores, and clear minutiae, and it most often occurs in sebaceous-rich depositions on glass substrates when exposed to natural light over a period of, at

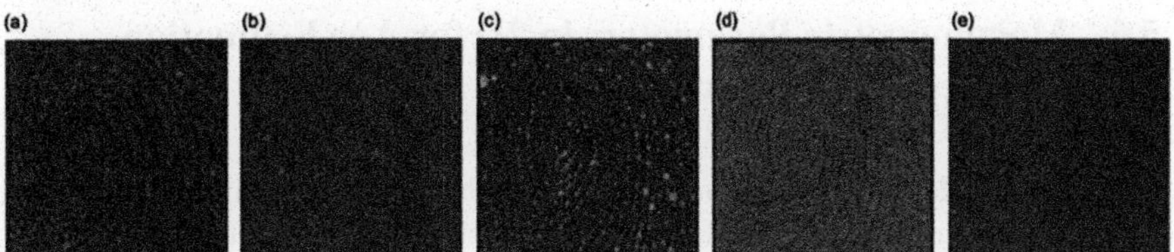

Fig. 5.5 Examples of five different outcomes of latent fingermark degradation visualized with titanium dioxide powder (ridges in white). (**a**) No degradation. (**b**) Fragmentation. (**c**) Constellation. (**d**) Diffusion. (**e**) Obliteration. (Figure reprinted with permission from Ref. [10]. Copyright 2016, Jon Wiley & Sons)

Table 5.3 Summary of typical conditions observed for each degradation type on latent fingermarks powdered with titanium dioxide over a period of 170 days. These conditions are not exclusively associated with each degradation type, nor are the degradation types mutually exclusive. Fingermarks can experience multiple types of degradation over time depending on aging conditions

Degradation type	Secretion	Substrate	Environment	Onset time
No degradation	Sebaceous [10]	Glass [10]	Direct light [10]	n/a[a] [7, 10]
Fragmentation	Eccrine [8]	Plastic [10]	Penumbra and darkness [8] [10]	7 days [10]
Constellation	Eccrine [10]	Plastic [10]	Direct light [7]	7 days [10]
Diffusion	Sebaceous [10]	Plastic [10]	Darkness [7] and high temp [50]	7 days [10]
Obliteration	Eccrine [10]	Plastic [10]	Darkness [6]	3 months [10]

[a]After 6 months with no visible degradation

least, up to 6 months. "Fragmentation" designates fingermarks that display breakages in ridge flow, a reduction in ridge width, and a loss in the number of viable minutiae. "Fragmentation" is characterized by the aging indicator named as ridge discontinuity and occurs to eccrine- and sebaceous-rich secretions, primarily on plastic, starting at 7 days after transfer. "Constellation" patterns exhibit dotted ridges with multiple interruptions and are observed in eccrine-rich fingermarks on plastic as soon as 7 days post-deposition. "Diffusion" occurs when the ridges appear to widen into the furrows increasing in width. The presence of halos around the perimeter of the deposition and an overall faded appearance in ridge detail are also characteristic. This change most often occurs to sebaceous-rich fingermarks on plastic substrates after 3 months post-deposition. At least one of these four possible outcomes may be observed throughout the aging process and is highly dependent on conditions and the enhancing method; however, the final common pattern is named "obliteration" as a result of either the natural course of degradation or external mechanical forces. As the name suggests, in this case the fingermark appears completely destroyed, leaving very little signs of its former presence, with a smudged area and a very low-quality or faint ridge flow. This pattern is most often observed in eccrine-rich secretions deposited on plastic beyond 3 months post-deposition.

5.6 Morphometric Parameters in 2D for Age Estimation of Latent Fingermarks

The following 2D age indicators, or parameters, have been investigated as possible time-sensitive morphometrical features to estimate the age of latent fingermarks: changes in ridge width, color contrast between ridges and furrows, presence of ridge discontinuities, decrease of minutiae count, and presence of pores (Fig. 5.6). Each of these aging parameters has its own advantages and limitations. It is important to note that no single parameter by itself may provide sufficient information for an

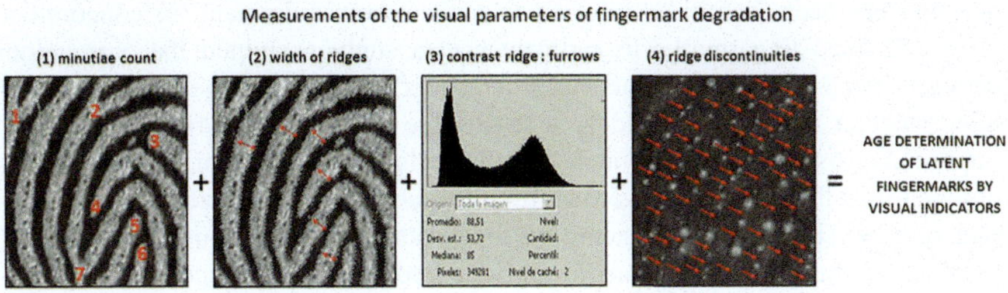

Fig. 5.6 Depiction of some physical parameters that can be used to measure fingermark degradation [3]

accurate estimate; rather, a combination of parameters would be the preferred approach as each one is affected differently by the aging process with different levels of sensitivity to degradation.

5.6.1 Ridge Width

Fluctuations of the width of ridges as the fingermark ages have been reported where a widening or narrowing of furrows occur depending on the combination of factors at play [1, 4, 9, 16, 35, 38, 39]. This is true for both fingermarks and palm marks [9, 16]. In one study, ridge and furrow widths were examined over a period of 180 days, concluding that the ridges decreased in size (measuring 30–34 mm fresh to 22–28 mm aged) while the furrows increased (measuring 24–28 mm fresh to 26–30 mm aged) [39]. These observations were supported by another study that reported a reduction in the width of ridges by 7.3%, 6.7%, and 0.14% within 7, 30, and 60 days, respectively [35].

In the experiment by Czech et al. (2018) [35], a 3% increase in the width of ridges was observed, which was also described by another study [69]. It was hypothesized that this increase may be the result of aging in higher room temperatures, stressing the importance of environmental factors [35]. Further research on nonporous surfaces determined that the most impactful factors on this aging parameter were substrate type, secretion type, and the presence of natural light, respectively [9]. In this occasion, when on plastic, the loss of ridge width was observed across all conditions examined. However, ridges on glass showed little to no degradation, with the exception of eccrine-rich depositions exposed to direct light [9]. Sebaceous-rich fingermarks were more resilient to changes regardless of natural light conditions; in addition, the presence of light appeared to "fix" the ridges when on glass. The study concluded that the width of ridges was most affected by the type of substrate and the presence of natural light had an effect only for eccrine-rich secretions [9].

Skin ridge widths are extremely variable in size and highly dependent on the donor; therefore, this technique could only be used once the fingermark in question has been identified. In order to reduce the requirement for concomitant fresh and aged fingermarks to estimate age, a study examined the correlations between inked prints (exemplars from databases) and latent fingermarks to determine whether the former could be a reasonable substitute to a freshly deposited impression [69]. The study concluded that ridge widths of flat fingerprints, as opposed to rolled ink prints, were highly comparable to ridge sizes of fresh latent fingermarks. This means that exemplars from databases could be used as a baseline to calculate the reduction/expansion in ridge width of an aged specimen, circumventing the need for fresh depositions from a suspect. Nonetheless, in regard to ridge widths, some claim it is not a good indicator for age determinations [4]. A large part of this is due to inter-donor differences and an overlap of ridge width sizes between aged and fresh fingermarks [4, 15]. Deposition pressure and sequence of deposition could also impact the width of the ridges. When examining a fingermark at a crime scene, these variables are typically unknown, introducing limitations to the aging models.

5.6.2 *Color Contrast*

Another parameter of interest is the gradual change in color contrast between the ridges and furrows of an enhanced latent fingermark. Research has demonstrated that as a fingermark ages, there is a loss in contrast with the background [7, 37, 40]. The tool of preference to measure these variations is the color histogram feature from imaging software and the respective metrics provided. The histogram displays a numerical value to the range of colors present in an image (pixels) and can be used to track minute color changes over time. In 2015, one study demonstrated the sensitivity of this method by showing no initial alteration in contrast followed by a clear, sharp decrease after approximately 10 days [40]. The histogram tool has proven sensitive enough to perceive color differences between fresh and older depositions even in cases where the fingermarks did not appear distinct to the naked eye. This is especially evident for sebaceous-rich fingermarks aged on glass in darkness [7] which are the most resilient to degradation as detected by optical means.

Lipid content may influence this age indicator. Low and moderate content produces a change in contrast with time, whereas high levels translate to no variations for an extended period of time after deposition [40]. Still, this could be a very useful parameter with a lot of potential, although lipid content should be evaluated to detect donor differences in sebum/lipid production [40].

5.6.3 Ridge Discontinuities

Ridge discontinuities are defined as the "*naturally occurring interruptions of ridges that 'appear' during the degradation process of the mark* (…)" [8]. This refers to the total or partial disappearance of a ridge within the ridge flow (Fig. 5.7). The technique is considered "semi-quantitative" because quantitative aspects of fingermark quality need to be assessed prior to determining the number of discontinuities. This is because ridge discontinuities may be observed in both aged and fresh fingermarks [8]. In this case, to circumvent this problem, a semi-quantitative value called the Discontinuity Index (DI) was created [8]. This calculation does not provide the actual number of discontinuities; rather, it combines the number of discontinuities and a quality value of the fingermark itself [8]. The level of ridge quality is important to evaluate, especially when using ridge discontinuity as an age parameter. Earlier in the chapter, the CAST grading scale was introduced where a fingermark is rated on a scale of 1 to 4 depending on its visual quality [25]. In order to use discontinuity as a parameter, the fingermark should be a medium to high level of quality. Fingermarks rated level 4, which is the highest value, would show little to no discontinuities and therefore would not be good candidates for this analysis. Those rated as a level 2 or lower would show discontinuities so large that the ridge flow would be lost and therefore would not be useful. At quality level 3, the fingermark is viable for identification and age determination purposes; discontinuities, when present, are visible, and therefore, a more accurate DI value can be calculated. DI values have shown to be higher, and more useful, for eccrine-rich fingermarks deposited on plastic than for their counterparts on glass [8]. It has been observed

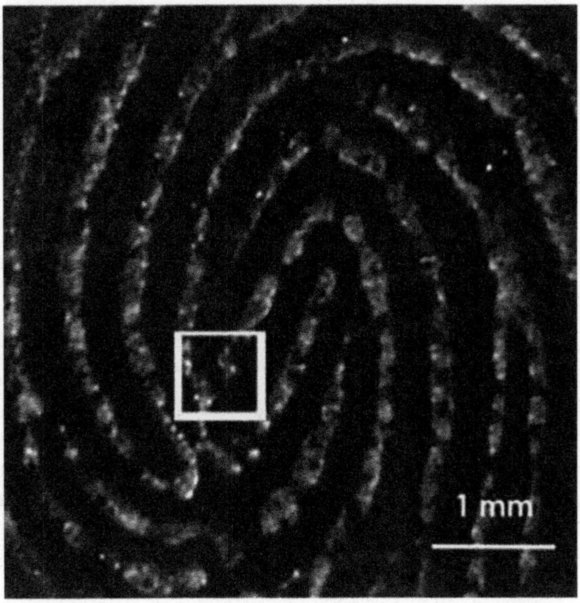

Fig. 5.7 Image depicting an enhanced latent fingermark with a ridge discontinuity (interruption) that occurs during the aging process. Ridges are visualized with TiO_2 powder (white) and photographed. (Figure reprinted with permission from Ref. [8]. Copyright 2017, John Wiley & Sons)

that secretion type and substrate have much more influence on the occurrence of discontinuities than environmental conditions; in other words, it occurs independent of light condition [8].

5.6.4 Minutiae Count

The gradual loss in the number of minutiae is another parameter explored for age estimation. This is simply the count (or ratio) of minutiae that can be unequivocally visualized over time in a particular fingermark. The number of these dactyloscopic elements has been shown to significantly decrease as the fingermark degrades [10, 39]; however, it is of little use for depositions on glass, as these have shown no substantial visual degradation, unlike on plastic [10]. Direct exposure to natural light through a window glass has shown to have a "preserving" effect on the ridges, particularly for depositions on glass, with no noticeable loss of minutiae. In dark conditions, a significant reduction of viable minutiae has been reported furthering this hypothesis, being more perceptible in eccrine-rich fingermarks than their sebaceous-rich counterparts [10].

5.6.5 Presence of Pores

The presence and shape of pores has been examined as an aging parameter. Studies have shown that after as little as 5 days, pore shapes can change [39]. It was also noted that they can disappear, appear larger, and fuse with each other displaying what is described as a "chain" [39]. The use of this feature is very limited as it requires very good visual clarity of the ridges which is not common in questioned fingermarks.

5.7 Techniques for the Analysis of Latent Fingermark Degradation in 2D

The scientific approaches to investigate fingermark degradation patterns can be broadly classified as either chemical or physical/optical, and the latter sub-classified as 2D (two-dimensional) morphometric analyses and 3D (three-dimensional) analyses. Chemical approaches specifically examine the chemical compounds within a fingermark and may use the relative composition and concentration of specific components to estimate time since deposition. This approach is covered in more detail in Chaps. 3 and 7. Chemical imaging is a "hybrid" approach that examines the chemical constituents of the fingermarks to render a 2D image. Some of these

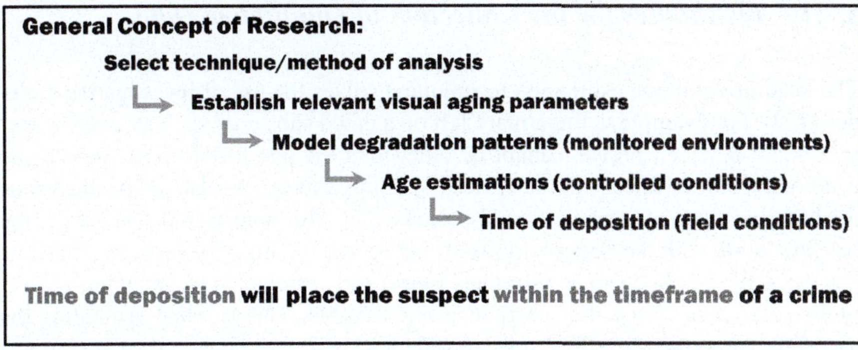

Fig. 5.8 Graphical representation of the suggested workflow to follow for research on age estimation of latent fingermarks

techniques exploit the fluorescent properties of specific chemical compounds and how these degrade overtime and include MALDI, TOF-SIMS, and fluorescence spectroscopy [1, 4, 12, 15, 27, 47, 50–55]. These are explained in depth in Chaps. 4 and 8.

On physical/optical analyses, 2D morphometric techniques refer to quantitative and qualitative methods to measure the topographical degradation of ridges. Other techniques that will be covered in this section include the use of chromatic white light and measurements of the electrostatic charge of the fingermark [7, 8, 10, 16, 37–40]. The last group of techniques uses 3D analysis to visualize and measure degradation patterns in the third dimension and may include optical profilometers to measure the topography and variations in height, area, and volume of ridges within a fingermark over time [49]. This is further discussed in Chap. 6.

Every analytical approach has its own advantages and limitations. Where one technique may fail or have stringent limitations, another technique may succeed. For example, one of the main limitations of chemical approaches is that the chemical profile of fingermarks is extremely donor specific and could be difficult to develop and "universalize" a model for age estimation. While the nature of the donor has some impact on the aging process, 2D techniques can overcome this issue because they are less sensitive at detecting differences at the molecular level. For example, it has been shown that latent fingermarks from males and females do not age significantly different on non-porous surfaces when examined by visual means (color contrast between ridges and furrows) using a powdering method [56].

Regardless of the method/technique to be used, every approach must pre-establish its own aging parameters for data measurements (Figs. 5.6 and 5.8) which will become the basis for mathematical modeling of patterns and predictions.

5.7.1 Techniques for the Collection of Qualitative Data

The most obvious age estimation technique involves the use of investigative methods [1, 2]. For example, a fingermark left on a dated substrate, such as a newspaper or a brand-new cell phone, cannot be older than the age listed on the newspaper itself or the date of release of the device [2]. A fingermark can also not be older than the last time a surface was thoroughly cleaned [1]. This approach to age estimation has little to do with the fingermark itself, rather investigating the substrate. In some instances, this method can be extremely useful and accurate; however, it has its own limitations when dating the substrate is not possible. This is when estimating the age of the fingermark itself must be considered.

Analytical techniques for the examination of fingermarks can also be categorized by the nature of the data they collect, broadly as qualitative, quantitative, and semi-quantitative. The first method provides information strictly on the visuals of a specimen (clarity); the second requires detailed measurements of the specimen in the form of physical (in 2D or 3D) and chemical changes over time. The third approach can provide both types of data. This section outlines several qualitative methods that can be used for age estimation on non-porous surfaces and, additionally, discusses their advantages and their limitations.

One of the first attempts at age estimation was by Menzel in 1992 [46]. When studying the photoluminescence using an argon laser, different color changes were observed based on the length of time since deposition [1, 15, 57]. Fresh fingermarks luminesced a greenish/yellow color, and as they aged, a red shift in color occurred [46, 57]. This property was further explored to develop a method for age estimation of fingermarks. It was later discovered that there was too much donor variation to draw reliable conclusions based on their color shift alone [4, 15, 46]. This approach has been acknowledged as the start to develop a reliable technique for dating fingermarks. This research has further inspired studies that particularly explore the fluorescent components and properties of fingermarks, such as the use of fluorescent spectroscopy.

5.7.1.1 Visual Examination of Ridge Quality: 2D Imaging

The first step to measuring quality involves an overall physical/visual evaluation to determine the clarity of the ridges. The qualitative parameters include, for example, exploring color contrast between ridges and furrows of enhanced fingermarks. The advantages of this technique are that it is simple, inexpensive, and easy to interpret the results. In this case, the histogram feature of the image's grayscale scheme is used and represents all the pixels that comprise the fingermark area of interest (Fig. 5.9). The histogram is a graph of pixel intensity (x-axis) versus number of pixels (y-axis), from the whitest pixels in the image (a numerical value of 255) to the darkest black (a numerical value of 0). As the fingermark topography degrades over time, the profile (shape) of the histogram also changes [7]. If a white powder is

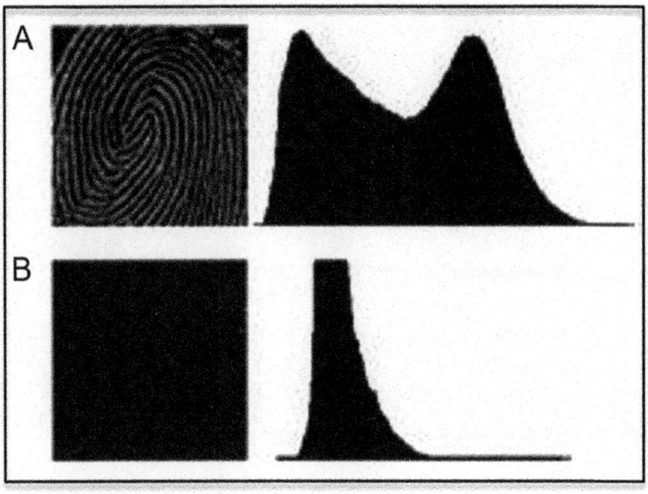

Fig. 5.9 Example of enhanced sebaceous-rich latent fingermarks with titanium dioxide powder and the respective color histogram profile. These were deposited on a plastic substrate and aged in direct natural light indoors. Over time, the histogram *A* shifts from two peaks (representing two colors, darkest and whitest of a fresh deposition) to one single peak skewed to the left of the profile of the aged fingermark image *B* (background black color)

used for enhancement (i.e., ridges in white with dark background), there will be a leftward shift in the histogram as the fingermark degrades, indicating a loss of contrast between the ridges and furrows. This analysis can be coupled with another imaging software, such as the Universal Latent Workstation (ULW) to highlight areas of the fingermark with the highest and lowest qualities by a coded map of colors. This tool is discussed in Chap. 6.

This technique is sensitive enough to distinguish patterns of degradation for different types of secretions and substrates, even with impressions appearing visually similar to the naked eye [7]. Beyond any subjective evaluations, Matuszewski (2015) and De Alcaraz-Fossoul et al. (2016) showed that the fingermarks of the same secretion type on the same substrate generally display a similar histogram profile, proving that there was some interaction and effect of the factors examined. In sebaceous-rich impressions, a much higher initial contrast value than in eccrine-rich counterparts was detected [7, 40]. Also, it took a much longer period of time for the former to show visual signs of degradation, i.e., a smaller decrease in color contrast over time [7]. This indicated that sebaceous-rich impressions had a much higher resilience to degradation indoors [40]. The effect of the type of non-porous substrate was also noted. Glass surfaces showed less reduction in contrast with time, whereas plastic showed a much more significant loss with varied patterns of aging [7]. This technique should be further explored since, at a crime scene, it is unknown to the examiner whether a residue is from a first, second, or third successive touch of eccrine or sebaceous nature. The fingermarks in these studies were all representative of the "first touch" in the sequence.

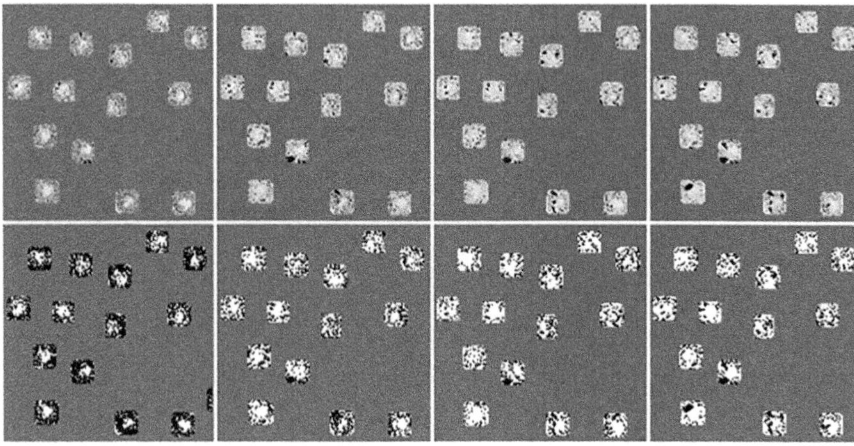

Fig. 5.10 Example of pore changes obtained using CWL over a long-term period of time (from left to right: 1 day, 35 weeks, 79 weeks, 120 weeks). Top row of images displays the pores unprocessed. The second row depicts the binarized image for better visualization (explained more in depth in Sect. 5.7.1.6). (Figure reprinted with permission from Ref. [90]. Copyright 2014 Logos)

5.7.1.2 Visual Examination of Pores: 2D Imaging

The second type of qualitative methods is porescopy (Fig. 5.10). This is the examination of pores and how they are altered over time. In fact, pores in a fingermark can "change" location and shape and may disappear altogether along the degradation of ridges [4, 39]. In one instance, microscopic observations at 1-day post-deposition have shown that some pores are closed and located in the center of the ridge while others are located on the edge of the ridge, opened or closed. Within 5 days, some of the pores on the edge of the ridges may "disappear," and the pores in the center may appear to open [4, 39]. The remaining pores fuse and form an elongated "chain" [39]. This study only examined the pores for a period of 5 days; therefore, this research alone cannot provide evidence for a successful technique [4]. Future studies would need to examine pores over longer periods of time. One of the main concerns for using pores for age estimation would be those (common) situations where they are not visible due to poorer quality of the impression, making this parameter mostly impractical [4].

5.7.1.3 Chemical Imaging: Time-of-Flight Secondary Ion Mass Spectrometry (TOF-SIMS)

2D approaches can easily be combined with other techniques. Many of these techniques use the chemical properties of a fingermark to translate this information into an image using fluorescent or non-fluorescent properties. In turn, data can be further interpreted as means for age estimations. These techniques may also use

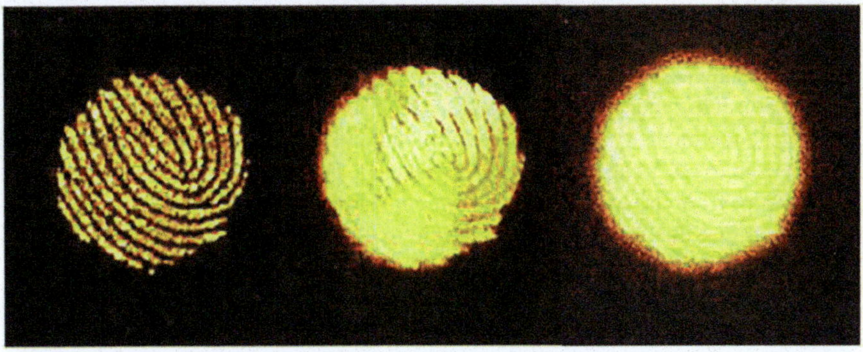

Fig. 5.11 TOF-SIMS images, examples of diffusion of fingermark residue after 1, 24, and 48 h (left to right). (Open source from Ref. [58])

quantitative values making the analyses more objective and are usually non-destructive and do not require prior enhancement processes. A further discussion on this topic is in Chaps. 4 and 8.

For example, TOF-SIMS has been used to visualize the diffusion of ridge residues over time. In one study, the diffusion of biomolecules over 1, 24, 48, 72, and 96 h was examined [58]. After the first hour, ridges were very clear and easily observed. After 24 h, a clear diffusion was detected, and by 48 h, little ridge detail could be distinguished (Fig. 5.11). However, this study only examined one substrate, which was a bare silicon wafer. This limited the examination of possible substrate effects on degradation. Additionally, this technique would only be applicable for short-term age estimation since after 48 h, all impressions became indistinguishable.

5.7.1.4 Chemical Imaging: Matrix-Assisted Laser Desorption/ Ionization (MALDI)

This technique is covered in detail in Chap. 8 and overcomes some limitations of the TOF-SIMS. For example, in [58] only small molecules could be analyzed [50, 58] which resulted in age evaluations within 48 h post-deposition [58]. By using MALDI, larger molecules such as triacylglycerol, in addition to fatty acids [50], could be examined, allowing for age estimates to be achieved for longer time periods. In addition, multiple substrates have been examined, such as plain glass microscope slides, microscope slides treated with Rain-X water repellent, stainless steel slides, and the non-adhesive side of scotch tape. It was revealed that the original problem with TOF-SIMS was not the molecular weight of the particles analyzed; instead, it was hypothesized that it was due to hydrophobic/hydrophilic compounds and surface interactions [50]. The study also noted that triacylglycerol diffuses slower than fatty acids on a hydrophobic surface and vice versa on a hydrophilic surface. Further research comparing TOF-SIMS and MALDI would have to explore the interaction between the compounds and the substrates [50].

MALDI has also analyzed lipids as a degradation marker [12]. A method named "the relative ratio approach" was developed in an attempt to circumvent the highly variable chemical signatures due to donor variation. This approach assumed lipid degradation into other molecular forms which with time could be used to create a ratio of the ion intensity of the original lipid and its degraded form. This ratio could be then used to assess age of the impression [12]. This, in theory, would make lipids independent from the original deposition concentration. However, statistical analysis showed that this hypothesis was unable to classify age groups. The data generated was then reprocessed using the "entirety of the fingermarks' lipid dataset" which then showed more promising results, indicating that this was an approach that should be further investigated. In this particular case, one of the main limitations was that real fingermarks were not used. Instead, a silicon fingertip and a sebaceous reference lipid pad were employed, reducing natural variation in the fingermark composition (i.e., donor) [12]. The "relative ratio approach" is a robust method that should be considered and applied across all age estimation techniques, especially for chemical analyses. The major limitations of this technique are its cost, the level of expertise needed for its operation, and its currently prohibitive portability to the crime scene.

5.7.1.5 UV Laser Spectroscopy

This technique uses the chemical spectra of the fingermark compounds to estimate age. In one study, fingermarks deposited on glass microscope slides and white paper were examined over 1 day, 1 week, 2 weeks, 4 months, 7 months, and 1 year under direct sunlight, lamp light, and darkness [53]. The samples were visualized with and without a UV laser, using the 340 nm peak (Fig. 5.12). At 2 months, samples stored in the dark and under a fluorescent lamp in low humidity as well as the dark in moderate humidity were visible; however, the quality of specimens in moderate humidity under lamplight and sunlight conditions decreased. At 4 months, the quality across all conditions diminished, being classified as "fair" or "poor." After 1 year, the fingermarks were considered in poor condition or not detected at all. It was observed that samples stored in the dark degraded less than those stored in sunlight and under a lamp [53]. This last observation is the opposite from other studies where more degradation is detected in the dark and better ridge detail is detected in sunlight conditions [6, 9]. Fingermarks stored in low humidity degraded less than those stored in the moderate humidity conditions, indicating that levels of moisture affect degradation [53]. The study further concluded that fingermarks at different times can be visualized using light sources at 440 nm. As the fingermark ages, the intensity of fluorescence increases, and this peak could be used to estimate the age of the deposition [53].

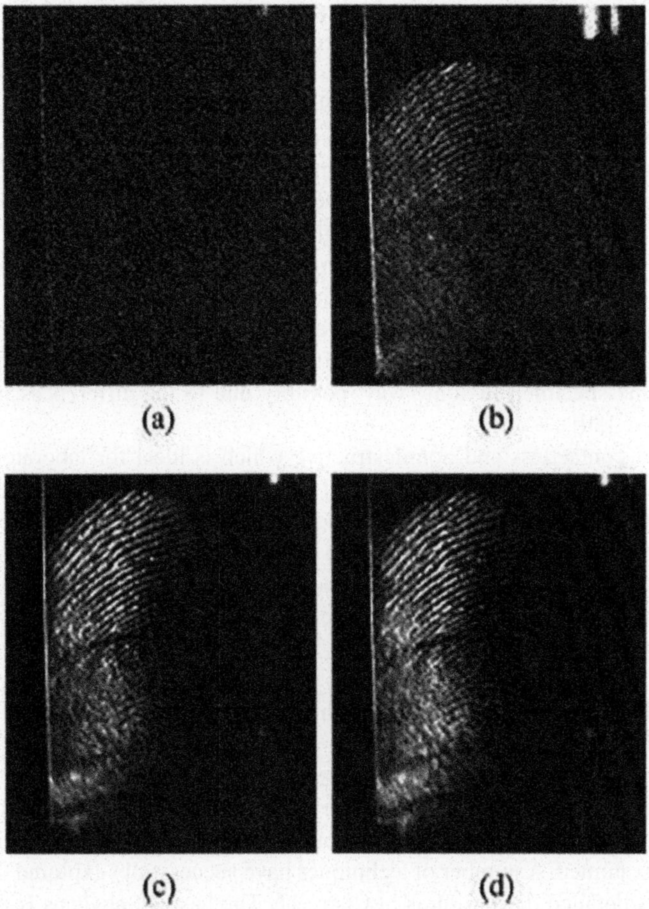

Fig. 5.12 An example of the fluorescence of fingermarks on glass using the peak at 440 nm from 0 days to 2 months. (**a**) After deposition; (**b**) 1 day; (**c**) 1 week; and (**d**) 2 months. (Figure reprinted with permission from Ref. [53]. Copyright 2018, Jon Wiley & Sons)

5.7.1.6 Chromatic White Light (CWL)

This technique exploits the visible changes in ridge/furrow contrast over time to assess the age of fingermarks. A detailed discussion is in Chap. 4. In summary, CWL provides an intensity and a topographic image and converts it into binary pixel values. The binary pixel values are used to measure the amount of contrast of a fingermark [15]. One study reported an error rate for age estimation between 13 and 40% [15] which is not an ideal aging model. Researchers have been able to successfully estimate the short-term age of specimens to discern the sequential overlapping of fingermarks by observing their degradation behavior [19]. The reported success rate for this technique was a promising 80%; however, it is still not ideal [19].

Degradation behaviors over a longer time frame, 1.5 years, have also been explored with CWL [59]. In addition to contrast, ridge width and porescopy parameters were analyzed; however, the reduction of ridge width claimed by others could not be confirmed [9, 16, 39]. It is hypothesized that a reduction in contrast in the edge of the ridges could not be detected or was overlooked when using the binary pixel feature. This could perhaps be a limitation of the technique itself [59]. A significant reduction in contrast around the pores was observed which made them appear as if they became larger, although this was possibly caused by the loss of contrast around the pore rather than changes in size [59]. The study concluded that the best age estimation feature to use with CWL is contrast which decreased over time. However, graphs of the data in this experiment demonstrated that the degradation speed of each sample was very different, indicating that some variable was affecting the specimen in some way, possibly due to the differences in chemical composition [59].

CWL is contactless and non-destructive which is ideal for laboratory applications, although field applications are still unproven. The technique appears to have an issue with the presence of dust, particularly larger particles that may influence the binary pixel feature. Further research needs to be conducted on the influence of dust particles, temperature, humidity, sweat composition, and other aging parameters that could be robust indicators of age.

5.7.1.7 Fingermark Age Estimation by Known Contaminants: Miscellaneous Approaches

A very different approach to estimate age could involve the known degradation pattern of a specific contaminant, such as blood, where the decay of the matrix itself could be examined. A number of techniques have successfully explored this method [60–62]. A detailed discussion is in Chap. 12. The basis of analysis is the color of blood changing (oxidizing) over time [60–63]. One of the most successful methods is hyperspectral image analysis [60–63]. In Li et al. (2013), a linear discriminant analysis was conducted based on the change of the absorption spectra over time. In this case, age was determined with a minimal error rate [60, 62, 64]. Other techniques have targeted the ratios of mRNA and rRNA [60, 65] or blood-related components by reflectance spectroscopy [60, 61].

5.7.2 Techniques/Methodologies for the Collection of Quantitative Data in 2D

These approaches comprise methods for the quantification of physical (topographical) features or chemical markers as means to collect measurable data to build mathematical aging models. These analyses may be more precise than qualitative counterparts as numerical data can be directly obtained with little interpretation from the observer.

5.7.2.1 Morphometrics

The quantitative measurement of the form in size and shape is called *morphometrics*. In regard to fingermarks, this technique can be used to collect information on color contrast, ridge widths, minutiae count, and number of ridge discontinuities, among other possible time-sensitive indicators of age. Scientists can use a wide array of computer imaging tools and software to measure and record this type of data.

Color contrast between ridges and furrows may be measured using the histogram tool present in many image editing software [7, 37]. One recommended method is to use the color intensity value provided by the histograms, for both the ridges and furrows [7, 37]. One particular experiment used the following formula to calculate the difference (D) in intensity between valleys (Iv) and ridges (Ir). This formula is as follows: $D = Iv - Ir$ [37]. The intensity values are calculated using the generated color intensity value from the histogram with 0 being pure black and 255 being pure white [7, 37]. Other studies have developed different formulae and procedures to successfully calculate this parameter [7, 37]. Usually, the maximum contrast is achieved with fresh fingermarks, displaying two clear color peaks in the histogram, as seen in Fig. 5.9.

Ridge discontinuities are more challenging to calculate. This aging parameter was exlored by measuring what was named the *Discontinuity Index* (DI). The DI is calculated by using both quantitative and qualitative data [8]. First, the fingermark is evaluated for its level of clarity and quality. This is due to the fact that ridge patterns with low quality (level 1 or 2) would not provide sufficient data for this parameter to work accurately since it would introcude too much instrumental error. Quality is recommended to be assessed using the CAST grading scale mentioned earlier in the chapter (Table 5.1). If suitable for this analysis, then a value for the actual number of ridge discontinuities is obtained and a DI value calculated [8].

Ridge width measurements can also be assessed with image editing software. For higher levels of accuracy, multiple random ridges distributed around the fingermark should be measured perpendicular to the flow of the ridge. These measruments would account for any differences in ridge widths throughout the fingermark, as well as pressure and distortion effects during deposition. De Alcaraz-Fossoul et al. (2019) suggests choosing at least 30 ridge points of measurement, randomly distributed throughout the impression [8]. In this study, areas of measurement are marked by drawing lines from one ridge to another, with at least four ridges in the area of interest and avoiding minutiae [8]. The ridge widths are then individually measured at each point where the line and a ridge intersect. The relative locations of these lines are used as templates and make it easier to measure the same areas on different impressions of the same fingermark. This ensured that the same points are measured across impressions of the same donor to measure intravariability (either latent to latent or latent to known) [8]. These measurements can then be compared to each other over time and/or across substrates to assess changes in ridge widths in the form of ratios. A ratio <1 implies a loss in size, whereas a ratio >1 suggests the opposite.

Calculation of minutiae count is a relatively simple concept. The recommended procedure for this analysis is to use an area with high density of minutiae, usually the geographical center of the fingermark, and count all minutiae within a predetermined section, for example, $1cm^2$ [10]. Next, calculate a ratio of the number of minutiae counted in the questioned fingermark over the number of minutiae counted in the inked (exemplar) impression from the same $1cm^2$. Once again, this ratio can be compared to ratios of the same fingermark calculated at a later time. A ratio <1 implies a detectable loss in viable minutiae.

Morphometrics can be very easily implemented as a methodology in any practical setting as it is non-destructive, easy and fast to apply, and inexpensive. However, like all methods, some limitations are noted. When a fingermark is enhanced at a crime scene, a number of additional variables go into the process of development (mode of application, type and affinity of the developer to residue, etc.) and visualization (lighting conditions, capturing device, etc.). At the crime scene, it is impossible to know what the fingermark would have looked like prior to its discovery (i.e., fresh) [4]. Due to this common eventuality and in order to provide meaningful age estimations, the environmental, storage, and deposition conditions should be known or inferred. If these conditions are met, it is possible to obtain a fingermark sample from the suspect to develop a "case-specific aging model" [4] for comparison. Alternatively, in some instances, exemplar fingermarks from previous arrests could also be used in lieu of a fresh fingermark (in the form of inked or scanned impressions) for comparison purposes. Many studies have noted that morphological parameters are sufficiently sensitive to environmental factors. These may become reliable age indicators that reflect the effect of different combinations of conditions over time [4, 9, 10, 16].

5.7.2.2 Chemical Analysis and Chemical Imaging

In addition to the described above, the chemical makeup of an aging fingermark has been examined by other quantitative and semi-quantitative techniques [4, 15, 44, 48]. Much like two-dimensional techniques, age estimation using chemical analysis explores certain chemical constituents as age indicators and describes the changes to these constituents over time [4, 15, 48]. Fourier transform infrared spectroscopy (FTIR) allows the visualization of the spectra from the chemical constituents of fingermarks [51]. As with most approaches, FTIR comes with limitations. It can only be used on non-porous surfaces and for short-term age determinations [4, 12, 51]. In this case, accurate results in age estimates have only been obtained for spectra less than 20 days old. This technique is further discussed in Chap. 7. Raman spectroscopy has also been explored for this same purpose; it is non-destructive and has identified components of the fingermark that can be the foundation for further research on age estimation. Andersson et al. (2017) used Raman spectra to identify specific compounds and how they degraded over time [4, 54]. Fluorescence spectroscopy explores different levels of fluorescence of sweat components. Under laboratory conditions, it has been possible to correctly estimate the age of 55% of fingermarks from male donors with a 1.9 day error rate for up to 3 weeks [1, 27, 47, 55]. Although this technique was promising and showed a low error rate, it was not useful on most of the fingermarks analyzed, particularly those deposited by females due to inadequate fluorescence [15, 55]. It was further noted that 77% of females and 27% of male fingermarks had insufficient fluorescence for the technique to be utilized [15, 55]. Gas chromatography/mass spectrometry (GC/MS) and high-performance liquid chromatography (HPLC), among other techniques, are commonly employed in research to quantify and identify changes in residues over time [66–68]. These techniques fall out of the two-dimensional spectrum and are only briefly mentioned here. More detailed explanations of these techniques are in Chaps. 3 and 7.

5.7.2.3 Electric Charge Approach

Another quantitative technique explores the electric charge of the fingermark and how it varies over time [2, 15, 39]. This method is donor independent and is only affected by the physical properties of the fingermark and the environment [2, 15]. For example, electrochemical impedance spectroscopy (EIS) measures the impedance charge of the fingermark [39]. This parameter varies and decays as the fingermark ages and provides estimates for recently deposited impressions [2, 15, 39]. This technique is able to overcome some limitations of other approaches; for instance, the surface charge can be quantified, and the fingermark charge would not be affected by the surface [1]. However, measurement of electric charge is limited to metallic (conductive) substrates and is not useful on older depositions [15, 39]. This method is explained in more detail in Chap. 9.

5.8 Criticisms to and from the Community of Experts

Fingerprinting (or more widely, friction ridge skin pattern analysis) in the USA has been thoroughly scrutinized for the entirety of processes involved in identifications as well as for all the scientific principles which it is based on, especially since the 2009 National Academy of Sciences (NAS) report and even prior to that in several court hearings [29]. The main concern surrounding this discipline still is the relative degree of subjectivity of the examination process, potentially influenced by bias and context effects [70–73]. It has been extensively reported that experts can make errors induced by contextual information which may become an influencing factor in decision-making [71–76].

Until very recently, rough estimates of the time of deposition (or age) of a latent fingermark could be provided by experts, if inquired in court. This information was based on the (educated) subjective opinion of an examiner with little to no scientific basis on factual aging indicators that could “support” such opinion. In the near future, adding a routine “age estimation” step to the fingerprint identification workflow may further increase the objectivity of the overall analysis and results. Then, it is important that age determination as a process is also scrutinized just as the rest of the examination processes are. To this date, the decisions of the examiners are subject to (expert) opinion, expertise, experience, and even bias. However, it is important to ensure that these decisions are founded on reliable scientific research and the most updated information available. All the research that has been conducted on age estimations, especially since the 2000s, provide a solid basis to divert from subjective interpretations and (educated) opinions, more so with novel technological adaptations that can help automize this process and minimize the "human bias effect." And despite all efforts, as long as we have humans performing or involved in friction ridge examinations, this forensic discipline will be partially subjective in nature, as with many others.

In principle, age estimation can assist with minimizing potentially erroneous identifications. For example, if the time of the crime does not fall within the predicted time of deposition of an identified fingermark, then it could be excluded as "primary evidence." This can be useful in some cases, but not for those circumstances in which the age of the evidence falls within the estimated time of the crime. For this reason, other avenues for reducing erroneous identifications should be explored [62, 72], whether that is addressing contextual effects on examiners [72, 73] or re-defining how fingermark identification results are interpreted, such as viewing results as a continuum with likelihood probabilities [62]. Furthermore, scientific research has emphasized the large variability of factors involved in natural aging processes, making it necessary to standardize mathematical models of aging (if possible) with a level of certainty and error. Variables such as inter- and intra-donor variability, deposition conditions, substrate types, environmental conditions, and visualization techniques may affect the interpretation of age estimates [1, 14, 62].

The numerous scientific studies addressed throughout this chapter have involved the analysis of latent fingermarks in situations where time of deposition and environmental factors are known. In many of these experimental settings, degradation patterns can be discerned and the relative influence of variables inferred. However, it has been demonstrated that under certain combination of conditions, aged fingermarks can visually appear relatively "fresh" [7–10, 62]. In addition, it has been noted that fingermarks can persist almost intact for long periods of time when using traditional optical and enhancement methods [1, 77]. It is known that the clarity of a latent fingermark over time is highly dependent on its initial quality [77]. This is a relevant fact from a crime scene analysis perspective because the visual quality of a fingermark is based on how fresh (or degraded) it "looks", which is information that is the basis for age determinations supported by time-sensitive parameters. It is recommended the forensic community to not simply correlate good visual quality to indicate freshness [62]. Studies do, however, show promise in co-examining the chemical composition and three-dimensional features for age estimations [62, 78]. A two-step process, primary and secondary screenings, is suggested (Fig. 5.13) in which optical and chemical techniques could provide a comprehensive estimate of the time since deposition.

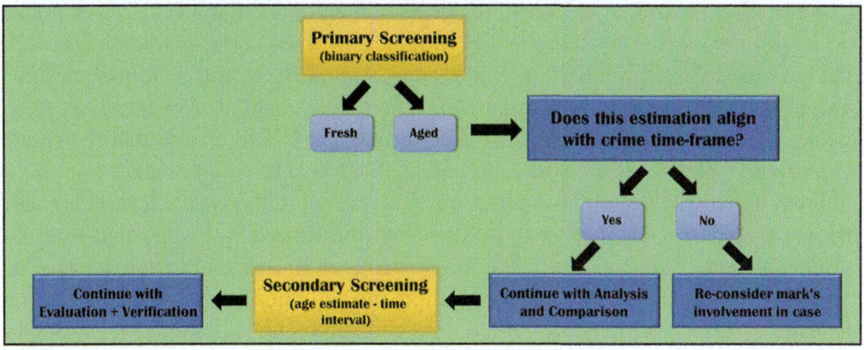

Fig. 5.13 Decision-making representation of the recommended screening stages and steps within ACE-V when estimating the age of a latent fingermark

5.9 Future Challenges and Limitations of 2D Analyses

In practice, prior to attempting to estimate the age of a fingermark by optical means, a series of basic aspects need to be considered to succeed in such an endeavor. First, the fingermark in question must contain sufficient ridge detail (visual quality) for a potential identification, usually level 2 ridge detail. Next, a number of variables from the crime scene may be considered which include, among others, the type of substrate (always known), the fingermark enhancing technique (if any), the type of secretion (rarely known), the surrounding environmental conditions (usually known), and the unique interactions and outcomes of these taken together (the main goal of research). Each of these variables affects one another and in turn affects the fingermark itself. This can be simplified by the triangle of interaction: substrate <–> donor <–> environment, with other sub-variables influencing each one of these categories (e.g., biological sex and age of donor) as shown in Fig. 5.3. Both in experimental and operational (field application) settings, two types of data can be obtained from every fingermark: qualitative and quantitative in nature. These can be collected by "measuring" or "categorizing" parameters such as changes in ridge width, color contrast between ridges and furrows, presence of pores, level of fluorescence, clarity scores, and electrostatic charge, among others, that can be used to build mathematical models of fingermark degradation.

5.9.1 Crime Scene Applications and Limitations

One of the core limitations to age estimations is extrapolating the laboratory-based models to crime scene applications. Indeed, models may be difficult to apply as every crime scene is influenced by a myriad of variables, and it is unlikely that the exact combination and interaction of all variables will be ever twice the same.

However, there is still hope. One of the goals in fingermark aging research is, in fact, to identify the most influencing or key variables involved in the degradation process and their mode of action; first, in indoor conditions with the anticipation to design a simplified mathematical model that could include as many as reasonably possible factors. Later, the most difficult obstacle to overcome is to ensure that the results from these experiments are still valid and applicable to the crime scene.

From a detective/criminalist perspective at a scene, it may be unclear which variables a fingermark has been exposed to from the time it was deposited until the moment it was discovered. In fact, it may be difficult or impossible to determine the type of secretion, the sequence of deposition, the amount of pressure applied during deposition, and the presence of certain contaminants, among other variables. Research has shown that degradation models are affected by the type of secretion [1, 2, 6–8, 10, 14, 37, 40], a relevant "undeterminable" variable to consider in real scenarios; yet, separate degradation models could be developed to include the respective type of secretion (i.e., eccrine-rich vs sebaceous-rich vs natural). Other variables could be easily recorded in situ or inferred in indoor situations, such as the degree of exposure to natural light, levels of RH%, temperature, type of substrate, and type of fingermark enhancer used (if any). These "determinable" variables are very informative and easy to collect unless the fingermark was aging for a very extended period of time (months to years). Once the questioned fingermark is identified, further information can be obtained about the donor, such as biological sex, ethnicity, type of diet, use of cosmetics, etc., providing more key data that could help select a more specific mathematical model that could be applicable to these specific circumstances.

Using 2D topographical features alone for age estimations could lead to potentially undesirable high rates of error and/or a large window of statistical uncertainty (standard deviation). It has already been noted that "[…] *a method for age determination using physical characteristics is less viable, as ridge detail is highly dependent on the original fingerprint*" [15, 87]. Chemical composition is also highly variable, and it could be argued that the initial state of the fingermark could have more influence on degradation and quality than time [1]. Perhaps, a model that encompasses a combination of the two approaches would be ideal as not all chemical changes over the life span of a fingermark necessarily translate to an immediate topographical change; thus, time-sensitive morphometric parameters could be considered less sensitive than chemical approaches, or, in other words, more robust to inter- and intra-donor variability.

5.9.2 Limitations of the Aging Parameters in 2D

Minutiae count, pore size, and ridge width are prime examples of time-sensitive parameters that may be impacted by the angle, pressure, and duration of transfer. The number of minutiae is dependent on the quality of the deposition itself and at the same time may also be affected by the effectiveness of the enhancement method

[42]. Differences in the width of ridges based on biological sex and/or ethnicity have also been documented [30, 79–81]. This feature is also highly impacted by deposition pressure [13, 20–22] where the more pressure exerted the wider the ridge becomes and more distortion is introduced [13, 20]. At the crime scene, for age estimation purposes, it could be assumed (for modeling simplicity) that the pressure applied is medium [13]. However, a fresh fingermark that is deposited with low pressure would initially be smaller in size and could be easily mistaken for an aged fingermark when the analysis relies on topographical (visual) parameters alone. The levels of pressure could mislead the examiner and have an effect on the age estimate, so additional parameters should be co-analyzed to minimize this effect. Therefore, these aging parameters may only be used in the secondary screening step after the identification of its source (Fig. 5.1).

Color contrast between ridges and furrows may be a good indicator to estimate fingermark age because it is sensitive to elapsing time, to environmental changes and the type of secretion but robust to biological sex. However, contrast may be affected by the sequence in deposition as well. It is not known if color contrast can provide meaningful information on whether the deposition is truly "fresh"; in other words, a first deposition in an iteration of depositions. For example, the donor could touch a number of items prior to depositing the fingermark of interest without giving sufficient time for the fingers to naturally recharge with sweat secretions. Research is needed to explore how much of the secretion deposited can affect the level of color contrast in the fingermark. Research is currently being conducted by De Alcaraz-Fossoul and his team to answer this question. Most studies to date have assumed that the fingermark analyzed is the first in a sequence of depositions, meaning that the finger is fully charged with a mix of both types of secretions, eccrine and sebaceous (Fig. 5.14).

Another example of an aging parameter with limitations would be Ridge Discontinuities, as it is dependent on the initial quality of the fingermark itself. The lower the initial quality, the more of the ridge flow may be lost or undetectable which makes evaluating ridge discontinuities of little use. This parameter alone would be clearly insufficient for age estimations and should be combined with others.

5.9.3 Variables and Interactions of Variables

Fingermark degradation studies have investigated plausible variables that may influence the natural aging process. The list of variables examined, and the interactions among them, may not be able to explain all possible cases of degradation but could be considered as a starting point for future research and still prove very useful in certain real scenarios. Realistically, it will never be possible to include all influencing factors in an age estimation model, but it may be feasible to select those that play a major role in degradation and that are present in many practical situations.

Several studies have examined and described donor inter- and intra-variability, confirming that there are fingermark differences between donors and within the

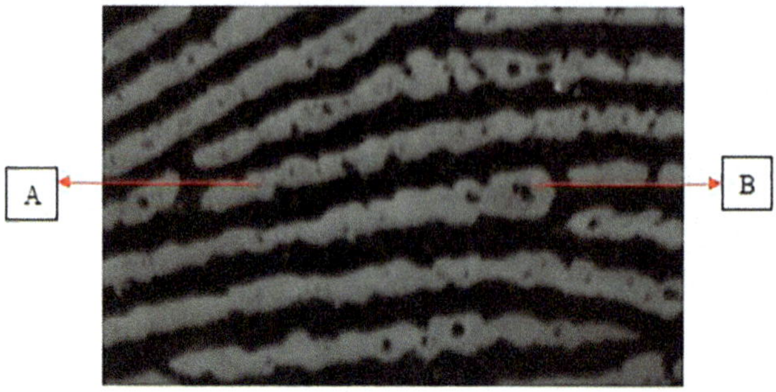

Minimum pressure (75 grams). A convex feature (A) and a large group of pores (B) are labeled.

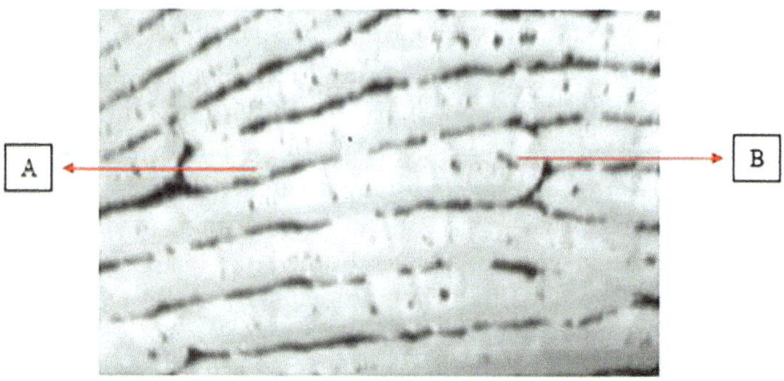

Maximum pressure 5214 grams. Note that (A) is still slightly rounded. The pores (B) appear smaller, but are still obvious.

Fig. 5.14 Example of visual differences of the width of ridges of latent fingermarks as a result of varying pressures [13]. (Open source from Ref. [13])

same donor [1, 4, 11, 12, 14, 15, 27, 28, 30, 33, 34, 36, 39, 79–81]. This alone already implies that physical, chemical, and biological parameters such as ridge density, sweat composition, or levels of detectable DNA could be used as possible fingermark age indicators. Others have demonstrated differences in the quality and quantity of the deposition depending on secretion types, pressure, and angle, among other factors [13, 14, 20–22, 26–28, 88]. Thus, it is of paramount importance to discern which variables are relevant from those that have a smaller impact on aging. To date, research has shed some light towards identifying pertinent factors involved in degradation. For instance, the interaction of substrate and environment has particularly strong effects on the patterns and outcomes of degradation. An example is seen with latent fingermarks on glass that tend to degrade to a lesser extent, and much more slowly or not at all, compared to their counterparts on plastic or tile [6, 8, 10]. Glass substrates, in many cases, appear to have a "preservation" effect, especially for fingermarks exposed to direct natural light. It has also been shown that on plastic, ridges tend to diffuse to the furrows, particularly those stored in the dark

Table 5.4 Table summarizing the sensitivity and effect of time on four 2D aging parameters used as parameters of degradation

Environmental variable			Fingermark aging parameter			
			Loss of minutiae	Loss of color contrast	Increase in discontinuity	Variation in ridge width[a]
Glass	Eccrine-rich	Light	*No effect*	*Strong effect*	*No effect*	*Strong effect*
		Shade	*Weak effect*	*Strong effect*	*Strong effect*	*Weak effect*
		Dark	*Strong effect*	*Weak effect*	*Strong effect*	*Strong effect*
	Sebaceous-rich	Light	*No effect*	*No effect*	*No effect*	*No effect*
		Shade	*Weak effect*	*Strong effect*	*Weak effect*	*No effect*
		Dark	*Weak effect*	*No effect*	*Weak effect*	*No effect*
Plastic (Polystyrene)	Eccrine-rich	Light	*Strong effect*	*Weak effect*	*Strong effect*	*Strong effect*
		Shade	*Strong effect*	*Weak effect*	*Strong effect*	*Strong effect*
		Dark	*Strong effect*	*Strong effect*	*Strong effect*	*Strong effect*
	Sebaceous-rich	Light	*Weak effect*	*Weak effect*	*Weak effect*	*Weak effect*
		Shade	*Strong effect*	*Strong effect*	*Strong effect*	*Weak effect*
		Dark	*Weak effect*	*Weak effect*	*Weak effect*	*Weak effect*

Combinations of experimental variables (substrate + secretion + light exposure) are included with only three possible results: no effect, weak effect, and strong effect
[a]At 170 days post-deposition or before obliteration

[6, 8, 10]. High relative humidity and temperature accelerate the decaying process, both chemically and physically, possibly as a result of the activity of microorganisms and the longer deleterious presence of water affecting other organic/inorganic components [2, 14, 15]. Above all, the unique interaction among all the variables (i.e., donor, substrate, and environment) affects and determines the initial quality and composition of fingermarks, how degradation progresses, and how these dictate their life expectancy. It is a challenging endeavor for forensic researchers to comprehend and mathematically model these interactions and their impact on aging (Table 5.4).

5.10 Where Do We Go from Here?

Research in latent fingermark age estimation has come a long way since it was first explored in the 1950s; however, a tumultuous road is ahead, especially at developing robust mathematical models for age predictions. The first hurdle to overcome relates to the fact that most of the techniques discussed in this chapter have only been experimented under laboratory conditions and therefore are not yet fully applicable to crime scenes [4]. It would be imperative to conduct wider-scale studies with larger pools of donors, as most experiments have included a very limited number of participants and experimental conditions [62]. In addition, black-box studies would have to be conducted to test the accuracy of the methods in uncontrolled environments. Further consideration into operational deployment and use is certainly needed which may require substantial financial support and/or strong multidisciplinary collaborations in the years to come.

Research efforts should focus on identifying and testing time-sensitive parameters of age and providing reliable age estimates by as many efficient and effective methods as possible. Evaluating other common enhancement methods such as cyanoacrylate fuming (which is very sensitive and non-destructive) would open the door to yet unexplored aging parameters that could be assessed. Most studies to date have examined non-porous substrates, but the degradation patterns of fingermarks on porous items, although much more complex, should also be investigated. Chemical developers on porous substrates (such as ninhydrin and DFO) would hopefully provide more information on fingermark aging in these cases.

From the array of available choices, some researchers favor the analysis of chemical constituents over morphometrics to model aging patterns for allegedly being more accurate and objective [1, 78]. In the chemical arena, forensic scientists should consider exploring the relative ratios of chemical residues to develop "chemical profiles" for fingermark aging patterns [12, 62]. Many have examined chemical composition using only a small sample size, and, therefore, a larger population of donors is still needed for these results to become relevant [62]. For example, promising experimental results have demonstrated the degradation of squalene and cholesterol [4]. For this type of examination, the following could be further considered: the reference and age-dependent components need to be present in all donor fingermarks, be time-sensitive, and be differentiated from contaminants. Ideally, these chemicals should be detectable and measurable using common techniques available to most laboratories and potentially portable to the scene. Lastly, the target components would have the lowest possible variability and show reproducible degradation patterns over time [1]. Lipids meet almost all of these criteria; however, it is a known exogenous substance that can be found in many cosmetics [1]. Lipids are stable overtime and would be a potential candidate to further investigate in a larger-scale study [4].

Regardless of the methodological approach, the last step in research would be to develop a mathematical model and software application that could include a combination of aging parameters and environmental conditions to make predictions of age [70–76, 82–86]. For example, one particular case that could be applied for this purpose is the use of a software application called "FRStat" which allows fingerprint examiners to demonstrate the validity and reliability of their results [89]. This software calculates a "similarity characteristic" which is further evaluated against a database that compares impressions made by the same and different sources, creating a similarity statistic value [89]. A similar software could be developed using the similarity of fingermarks based on the visual parameters they exhibit as they age. Additionally, the software should be able to account for the various environmental, substrate, and donor variables and potentially even provide image examples of typical fingermarks of the same age that could be visually correlated to the questioned fingermark being examined [62].

References

1. Girod A, Ramotowski R, Lambrechts S, Misrielal P, Aalders M, Weyermann C (2016) Fingermark age determinations: legal considerations, review of the literature and practical propositions. Forensic Sci Int 262:212–226. https://doi.org/10.1016/j.forsciint.2016.03.021
2. Bleay SM, Croxton RS, Puit MD (2018) Fingerprint development techniques: theory and application. Wiley, Hoboken. https://doi.org/10.1002/9781119187400
3. De Alcaraz-Fossoul J, Barrot Feixat C, Mestres Patris C, McGarr L, Stow K, Tasker J, … Gene Badia M (2015) Determining the age of fingermarks: no longer myth. In European Academy of Forensic Science conference. Prague, Czech Republic, p 371
4. Hagan AO, Green S (2018) Crime scene to court: a study on finger-mark aging Forensic Res Criminol Intl J 6(6):491–503. https://doi.org/10.15406/frcij.2018.06.00249
5. De Alcaraz-Fossoul J, Barrot Feixat C, Roberts KA (2015) The paradigm of fingerprint age determination. Jacobs J Forensic Sci 1(1), 006th series, 1–5
6. De Alcaraz-Fossoul JD, Patris CM, Muntaner AB, Barrot Feixat C, Gené Badia M (2012) Determination of latent fingerprint degradation patterns – a real fieldwork study. Int J Legal Med 127(4):857–870. https://doi.org/10.1007/s00414-012-0797-1
7. De Alcaraz-Fossoul J, Barrot Feixat C, Tasker J, Mcgarr L, Stow K, Carreras-Marin C et al (2016) Latent fingermark aging patterns (Part II): color contrast between ridges and furrows as one Indicator of degradation. J Forensic Sci 61(4):947–958. https://doi.org/10.1111/1556-4029.13099
8. De Alcaraz-Fossoul J, Barrot Feixat C, Carreras-Marin C, Tasker J, Zapico S, Gené Badia M (2017) Latent fingermark aging patterns (Part III): discontinuity index as one indicator of degradation. J Forensic Sci 62(5):1180–1187. https://doi.org/10.1111/1556-4029.13438
9. De Alcaraz-Fossoul J, Barrot Feixat C, Zapico SC, Mcgarr L, Carreras-Marin C, Tasker J, Gené Badia M (2019) Latent fingermark aging patterns (part IV): ridge width as one indicator of degradation. J Forensic Sci 64(4):1057–1066. https://doi.org/10.1111/1556-4029.14018
10. De Alcaraz-Fossoul J, Mestres Patris C, Barrot Feixat C, Mcgarr L, Brandelli D, Stow K, Gené Badia M (2016) Latent fingermark aging patterns (part I): minutiae count as one indicator of degradation. J Forensic Sci 61(2):322–333. https://doi.org/10.1111/1556-4029.13007
11. Menzel ER (2001) Recent advances in photoluminescence detection of fingerprints. Sci World J 1:498–509. https://doi.org/10.1100/tsw.2001.76
12. Duff JM, Menzel ER (1978) Laser-assisted thin-layer chromatography and luminescence of fingerprints: an approach to fingerprint age determination. J Forensic Sci 23(1). https://doi.org/10.1520/jfs10661j
13. Richmond S (2004) Do fingerprint ridges and characteristics within ridges change with pressure? Australian Federal Police, Forensic Services
14. Girod A, Ramotowski R, Weyermann C (2012) Composition of fingermark residue: a qualitative and quantitative review. Forensic Sci Int 223:10–24. https://doi.org/10.1016/j.forsciint.2012.05.018
15. Cadd S, Islam M, Manson P, Bleay S (2015) Fingerprint composition and aging: a literature review. Sci Justice 55(4):219–238. https://doi.org/10.1016/j.scijus.2015.02.004
16. Barros RM, Faria BE, Kuckelhaus SA (2013) Morphometry of latent palmprints as a function of time. Sci Justice 53(4):402–408. https://doi.org/10.1016/j.scijus.2013.08.002
17. Sodhi G, Kaur J (2001) Powder method for detecting latent fingerprints: a review. Forensic Sci Int 120(3):172–176. https://doi.org/10.1016/s0379-0738(00)00465-5
18. Merkel R, Gruhn S, Dittmann J, Vielhauer C, Bräutigam A (2012) On non-invasive 2D and 3D chromatic white light image sensors for age determination of latent fingerprints. Forensic Sci Int 222:52–70. https://doi.org/10.1016/j.forsciint.2012.05.001
19. Schott M, Merkel R, Dittmann J (2012) Sequence detection of overlapping latent fingerprints using a short-term aging feature. 2012 IEEE international workshop on information forensics and security (WIFS). https://doi.org/10.1109/wifs.2012.6412630
20. Maceo AV (2009) Qualitative assessment of skin deformation: a pilot study. JFI 59(4):390–440

21. Fieldhouse SJ (2015) An investigation into the effects of force applied during deposition on latent fingermarks and inked fingerprints using a variable force fingerprint sampler. J Forensic Sci 60(2):422–427. https://doi.org/10.1111/1556-4029.12661
22. Fieldhouse S (2011) Consistency and reproducibility in fingermark deposition. Forensic Sci Int 207:96–100. https://doi.org/10.1016/j.forsciint.2010.09.005
23. Pulsifer DP, Muhlberger SA, Williams SF, Shaler RC, Lakhtakia A (2013) An objective fingerprint quality-grading system. Forensic Sci Int 231(1–3):204–207. https://doi.org/10.1016/j.forsciint.2013.05.003
24. Hicklin RA, Buscaglia JA, Roberts MA, Meagher SB, Fellner W, Burge MJ et al (2011) Latent fingerprint quality: a survey of examiners. JFI 61(4):385–417
25. Sears VG, Bleay SM, Bandey HI, Bowman VJ (2012) A methodology for finger mark research. Sci Justice 52(3):145–160. https://doi.org/10.1016/j.scijus.2011.10.006
26. Chadwick S, Moret S, Jayashanka N, Lennard C, Spindler X, Roux C (2018) Investigation of some of the factors influencing fingermark detection. Forensic Sci Int 289:381–389. https://doi.org/10.1016/j.forsciint.2018.06.014
27. Lambrechts SAG, van Dam A, de Vos J, van Weert A, Sijen T, Aalders M (2012) On the autofluorescence of fingermarks. Forensic Sci Int 222:89–93. https://doi.org/10.1016/j.forsciint.2012.05.004
28. Williams DK, Brown CJ, Bruker J (2011) Characterization of children's latent fingerprint residues by infrared microspectroscopy: forensic implications. Forensic Sci Int 206:161–165. https://doi.org/10.1016/j.forsciint.2010.07.033
29. Yamashita B, French M (2011) Latent print development. In: Holder EH, Robinson LO, Laub JH (eds) The fingerprint sourcebook. National of Justice, US Department of Justice, Washington, DC, pp 7.1–7.67
30. Sánchez-Andrés A, Barea JA, Rivaldería N, Alonso-Rodríguez C, Gutiérrez-Redomero E (2018) Impact of aging on fingerprint ridge density: anthropometry and forensic implications in sex inference. Sci Justice 58(5):323–334. https://doi.org/10.1016/j.scijus.2018.05.001
31. Ashbaugh DR (1999) Quantitative-qualitative friction ridge analysis: an introduction to basic and advanced ridgeology. CRC Press, Boca Raton. https://doi.org/10.1201/9781420048810
32. Thomas GL, Reynoldson TE (1975) Some observations on fingerprint deposits. J Phys D Appl Phys 8(6):724–729. https://doi.org/10.1088/0022-3727/8/6/017
33. Nayak VC, Rastogi P, Kanchan T, Yoganarasimha K, Pradeep Kumar G, Menezes RG (2010) Sex differences from fingerprint ridge density in Chinese and Malaysian population. Forensic Sci Int 197:67–69. https://doi.org/10.1016/j.forsciint.2009.12.055
34. Dhall JK, Kapoor AK (2016) Fingerprint ridge density as a potential forensic anthropological tool for sex identification. J Forensic Sci 61(2):424–429. https://doi.org/10.1111/1556-4029.12959
35. Czech A, Szabelak A, Sowiński A (2018) Changes in fingerprints depending on physiological factors. J Forensic Sci 64(3):711–716. https://doi.org/10.1111/1556-4029.13937
36. Acree MA (1999) Is there a gender difference in fingerprint ridge density? Forensic Sci Int 102(1):35–44. https://doi.org/10.1016/s0379-0738(99)00037-7
37. Matuszewski S, Szafałowicz M (2013) A simple computer-assisted quantification of contrast in a fingerprint. J Forensic Sci 58(5):1310–1313. https://doi.org/10.1111/1556-4029.12224
38. Baniuk K (1990) Determination of age of fingerprints. Forensic Sci Int 46(1–2):133–137. https://doi.org/10.1016/0379-0738(90)90151-n
39. Rosa R, Giovanardi R, Bozza A, Veronesi P, Leonelli C (2017) Electrochemical impedance spectroscopy: a deeper and quantitative insight into the fingermarks physical modifications over time. Forensic Sci Int 273:144–152. https://doi.org/10.1016/j.forsciint.2017.02.016
40. Matuszewski S (2015) Age-dependent changes of contrast in fingermarks of various lipid content. Problems Forensic Sci 101:5–13
41. Popov K, Sears V, Jones B (2017) Migration of latent fingermarks on non-porous surfaces: observation technique and nanoscale variations. Forensic Sci Int 275:44–56. https://doi.org/10.1016/j.forsciint.2017.02.015

42. De Alcaraz-Fossoul J, Roberts KA, Barrot Feixat C, Hogrebe GG, Gené Badia M (2016) Fingermark ridge drift. Forensic Sci Int 258:26–31. https://doi.org/10.1016/j.forsciint.2015.11.003
43. Dorakumbura BN, Becker T, Lewis SW (2016) Nanomechanical mapping of latent fingermarks: a preliminary investigation into the changes in surface interactions and topography over time. Forensic Sci Int 267:16–24. https://doi.org/10.1016/j.forsciint.2016.07.024
44. Olsen RD Sr (1987) Chemical dating techniques for latent fingerprints: a preliminary report. The Identification News:10–12
45. Kent T (2016) Water content of latent fingerprints – dispelling the myth. Forensic Sci Int 266:134–138. https://doi.org/10.1016/j.forsciint.2016.05.016
46. Menzel, ER (1992) Fingerprint age determination by fluoresence. J Forensic Sci. Letters to the Editor: 1212–1213
47. Van Dam A, Aalders MCG, Todorovski T, Van Leeuwen TG, Lambrechts SAG (2016) On the autofluorescence of aged fingermarks. Forensic Sci Int 258:19–25. https://doi.org/10.1016/j.forsciint.2015.11.002
48. Girod A, Spyratou A, Holmes D, Weyermann C (2016) Aging of target lipid parameters in fingermark residue using GC/MS: effects of influence factors and perspectives for dating purposes. Sci Justice 56(3):165–180. https://doi.org/10.1016/j.scijus.2015.12.004
49. De Alcaraz-Fossoul J, Mancenido M, Soignard E, Silverman N (2018) Application of 3D imaging technology to latent fingermark aging studies. J Forensic Sci 64(2):570–576. https://doi.org/10.1111/1556-4029.13891
50. O'Neill KC, Lee YJ (2017) Effect of aging and surface interactions on the diffusion of endogenous compounds in latent fingerprints studied by mass spectrometry imaging. J Forensic Sci 63(3):708–713. https://doi.org/10.1111/1556-4029.13591
51. Girod A, Xiao L, Reedy B, Roux C, Weyermann C (2015) Fingermark initial composition and aging using Fourier transform infrared microscopy (μ-FTIR). Forensic Sci Int 254:185–196. https://doi.org/10.1016/j.forsciint.2015.07.022
52. Francese S, Bradshaw R, Denison N (2017) An update on MALDI mass spectrometry based technology for the analysis of fingermarks – stepping into operational deployment. Analyst 142:2518–2546. https://doi.org/10.1039/c7an00569e
53. Akiba N, Kuroki K, Kurosawa K, Tsuchiya K (2017) Visualization of aged fingerprints with an ultraviolet laser. J Forensic Sci 63(2):556–562. https://doi.org/10.1111/1556-4029.13588
54. Andersson PO, Lejon C, Mikaelsson T, Landström L (2017) Towards fingermark dating: a Raman spectroscopy proof-of-concept study. Chemistry Open 6:706–709. https://doi.org/10.1002/open.201700129
55. Van Dam A, Schwarz JCV, De Vos J, Siebes M, Sijen T, Van Leeuwen TG et al (2014) Oxidation monitoring by fluorescence spectroscopy reveals the age of fingermarks. Angew Chem Int Ed 53(24):6272–6275. https://doi.org/10.1002/anie.201402740
56. Einfalt M (2020) The influence of biological sex on latent fingermark aging. Unpublished Master's thesis. University of New Haven
57. Dalrymple BE, Duff JM, Menzel ER (1977) Inherent fingerprint luminescence—detection by laser. J Forensic Sci 22(1):106–115. https://doi.org/10.1520/jfs10373j
58. Muramoto S, Sisco E (2015) Strategies for potential age dating of fingerprints through the diffusion of sebum molecules on a nonporous surface analyzed using time-of-flight secondary ion mass spectrometry. Anal Chem 87(16):8035–8038. https://doi.org/10.1021/acs.analchem.5b02018
59. Merkel R, Otte K, Clausing R, Dittmann J, Vielhauer C, Bräutigam A (2013) First investigation of latent fingerprints long-term aging using chromatic white light sensors. In: Proceedings of the First ACM workshop on Information Hiding And Multimedia Security – IH&MMSec, p 13. https://doi.org/10.1145/2482513.2482519
60. Singh P (2018) Age estimation of a dried bloodstain using different techniques – a review article. Acad J Forensic Sci 1(1):23–27

61. Li B, Beveridge P, O'Hare WT, Islam M (2011) The estimation of the age of a blood stain using reflectance spectroscopy with a microspectrophotometer, spectral pre-processing and linear discriminant analysis. Forensic Sci Int 212(1–3):198–204. https://doi.org/10.1016/j.forsciint.2011.05.031
62. Francese S (2019) Emerging Technologies for the Analysis of forensic traces. Springer, Cham. https://doi.org/10.1007/978-3-030-20542-3
63. Edelman G, van Leeuwen TG, Aalders MCG (2012) Hyperspectral imaging for the age estimation of blood stains at the crime scene. Forensic Sci Int 223:72–77. https://doi.org/10.1016/j.forsciint.2012.08.003
64. Li B, Beveridge P, O'Hare WT, Islam M (2013) The age estimation of blood stains up to 30days old using visible wavelength hyperspectral image analysis and linear discriminant analysis. Sci Justice 53(3):270–277. https://doi.org/10.1016/j.scijus.2013.04.004
65. Anderson S, Howard B, Hobbs GR, Bishop CP (2005) A method for determining the age of a bloodstain. Forensic Sci Int 148(1):37–45. https://doi.org/10.1016/j.forsciint.2004.04.071
66. Weyermann C, Roux C, Champod C (2010) Initial results on the composition of fingerprints and its evolution as a function of time by GC/MS analysis. J Forensic Sci 56(1):102–108. https://doi.org/10.1111/j.1556-4029.2010.01523.x
67. Archer NE, Charles Y, Elliott JA, Jickells S (2005) Changes in the lipid composition of latent fingerprint residue with time after deposition on a surface. Forensic Sci Int 154(2–3):224–239. https://doi.org/10.1016/j.forsciint.2004.09.120
68. Koenig A, Girod A, Weyermann C (2011) Identification of wax esters in fingermark residues by GC/MS and their potential use as aging parameters. JFI 61(6):652–676
69. De Alcaraz-Fossoul JD, Barrot Feixat C, Zapico SC, Mancenido M, Broatch J, Roberts KA et al (2017) Ridge width correlations between inked prints and powdered latent fingerprints. J Forensic Sci 63(4):1085–1091. https://doi.org/10.1111/1556-4029.13656
70. National Research Council (2009) Strengthening forensic science in the United States: a path forward. The National Academies Press, Washington, DC
71. Cole SA (2005) More than zero: accounting for error in latent fingerprint identification. J Crim Law Criminol 95(3):985–1075
72. Dror IE (2017) Human expert performance in forensic decision making: seven different sources of bias. Aust J Forensic Sci 49(5):541–547. https://doi.org/10.1080/00450618.2017.1281348
73. Dror IE, Charlton D (2006) Why experts make errors. JFI 56(4):600–616
74. Dror IE (2013) Practical solutions to cognitive and human factor challenges in forensic science. Forensic Science Policy Manage Intl J 4(3–4):105–113. https://doi.org/10.1080/19409044.2014.901437
75. Dror IE, Thompson WC, Meissner CA, Kornfield I, Krane D, Saks M, Risinger M (2015) Letter to the editor- context management toolbox: a linear sequential unmasking (LSU) approach for minimizing cognitive bias in forensic decision making. J Forensic Sci 60(4):1111–1112. https://doi.org/10.1111/1556-4029.12805
76. Dror I, Rosenthal R (2008) Meta-analytically quantifying the reliability and biasability of forensic experts. J Forensic Sci 53(4):900–903. https://doi.org/10.1111/j.1556-4029.2008.00762.x
77. Midkiff CR (1993) Lifetime of a latent print how long? Can you tell? JFI 43(4):386–392
78. Wertheim K (2003) Fingerprint age determination: is there any hope? JFI 42(53):42–49
79. van Dam A, van Beek FT, Aalders MCG, van Leeuwen TG, Lambrechts SAG (2016) Techniques that acquire donor profiling information from fingermarks – a review. Sci Justice 56(2):143–154. https://doi.org/10.1016/j.scijus.2015.12.002
80. Kapoor N, Badiye A (2015) Sex differences in the thumbprint ridge density in a central Indian population. Egypt J Forensic Sci 5(1):23–29. https://doi.org/10.1016/j.ejfs.2014.05.001
81. Gutiérrez-Redomero E, Alonso C, Romero E, Galera V (2008) Variability of fingerprint ridge density in a sample of Spanish Caucasians and its application to sex determination. Forensic Sci Int 180(1):17–22. https://doi.org/10.1016/j.forsciint.2008.06.014

82. Mnookin J, Kellman PJ, Dror I, Erlikhman G, Garrigan P, Ghose T, Metler E, Charlton D(2016) Error rates for latent fingerprinting as a function of visual complexity and cognitive difficulty. Report. NIJ Award 2009-DN-BX-K225
83. Ulery BT, Hicklin RA, Roberts MA, Buscaglia J (2016) Interexaminer variation of minutia markup on latent fingerprints. Forensic Sci Int 264:89–99. https://doi.org/10.1016/j.forsciint.2016.03.014
84. Ulery BT, Hicklin RA, Buscaglia J, Roberts MA (2012) Repeatability and reproducibility of decisions by latent fingerprint examiners. PLoS One 7(3):1–12. https://doi.org/10.1371/journal.pone.0032800
85. Ulery BT, Hicklin RA, Roberts MA, Buscaglia J (2016) Data on the interexaminer variation of minutia markup on latent fingerprints. Data Brief 8:158–190. https://doi.org/10.1016/j.dib.2016.04.068
86. Ulery BT, Hicklin RA, Buscaglia J, Roberts MA (2011) Accuracy and reliability of forensic latent fingerprint decisions. Proc Natl Acad Sci 108(19):7733–7738. https://doi.org/10.1073/pnas.1018707108
87. Hildebrandt M, Dittmann J, Pocs M, Ulrich M, Merkel R, Fries T (2011) Privacy preserving challenges: new design aspects for latent fingerprint detection systems with contact-less sensors for future preventive applications in airport luggage handling. Lecture Notes Comput Sci Biomet ID Manage:286–298. https://doi.org/10.1007/978-3-642-19530-3_26
88. Merkel R, Dittmann J, Vielhauer C (2011) How contact pressure, contact time, smearing and oil/skin lotion influence the aging of latent fingerprint traces: first results for the binary pixel feature using a CWL sensor. In: 2011 IEEE international workshop on information forensics and security. https://doi.org/10.1109/wifs.2011.6123153
89. Swofford H, Koertner A, Zemp F, Ausdemore M, Liu A, Salyards M (2018) A method for the statistical interpretation of friction ridge skin impression evidence: method development and validation. Forensic Sci Int 287:113–126. https://doi.org/10.1016/j.forsciint.2018.03.043
90. Merkel R (2014) New solutions for an old challenge: chances and limitations of optical, non-invasive acquisition and digital processing techniques for the age estimation of latent fingerprints. Doctoral dissertation, Otto-von-Guericke-Universität Magdeburg. Logos, Berlin. ISBN 978-3-8325-3823-3

Chapter 6
Latent Fingermark Aging in 3D: Uncovering Hidden Degradation Patterns

Josep De Alcaraz-Fossoul and Meredith A. Narowski

Abstract Usually, in fingermark aging studies, two-dimensional (2D) morphometric features are examined in order to obtain information on how, and by how much, fingermarks visually degrade over time. These include parameters such as the distance between ridges, color contrast between ridges and furrows, and the width of ridges. However, in this process, there are other aging features that are being overlooked because fingermarks are rather three-dimensional (3D) objects. These features are the height and area occupied by the ridges, as well as the volume of the fingermark secretion. This chapter discusses novel 3D features with the objective to aid in establishing a multi-parameter methodology for determining time since deposition by visual means. Fingermark age estimation is an expanding subfield of forensic science research that will undoubtedly provide meaningful insights on the aging process of this common physical evidence. In addition, optical profilometry, a nondestructive 3D imaging technology, is discussed in detail to explain how it can be employed to reveal 3D features not explored before in the field of friction ridge pattern analysis.

Keywords Three dimensional · Volume · Area · Height · Fingermark · Fingerprint · Aging · Degradation · Optical profilometer · Morphometrics · Topography · 2D · 3D

J. De Alcaraz-Fossoul (✉)
Henry C. Lee College of Criminal Justice and Forensic Science,
Forensic Science Department, University of New Haven,
West Haven, CT, USA
e-mail: jfossoul@newhaven.edu

M. A. Narowski
Raleigh/Wake City-County Bureau of Identification,
Raleigh, NC, USA

J. De Alcaraz-Fossoul (ed.), *Technologies for Fingermark Age Estimations: A Step Forward*, https://doi.org/10.1007/978-3-030-69337-4_6

Acronyms

2D	Two dimensional
3D	Three dimensional
ALS	Alternate light source
AFIS	Automated fingerprint identification system
CWL	Chromatic white light
MALDI	Matrix-assisted laser desorption/ionization
ToF-SIMS	Time-of-flight secondary ion mass spectrometry
OP	Optical profilometer
ULW	Universal Latent Workstation
SED	Silver electroless deposition
TiO_2	Titanium dioxide
FTIR	Fourier transform infrared spectroscopy
GC-MS	Gas chromatography/mass spectrometry
CM	Confocal microscopy

6.1 Introduction

Despite the advent of DNA identification as an aiding tool for solving crimes, (latent) fingermarks are still one of the most common forms of physical evidence worldwide [1]. Since its inception, fingerprint examinations have involved the comparison and identification of general patterns and minutiae[1] [2] from a manual mode of operation to semi-automatic systems known as Automated Fingerprint Identification System (AFIS) [3]. AFIS allow for the search/comparison of millions of stored fingerprints from persons of interest and provides a list of top candidates for every query. This comparison involves the (still) manual location of the minutiae within the fingerprint and the identification of the type of minutiae (bifurcation, ridge ending, dot), as well as determining their frequency and orientation [3]. These combined characteristics of minutiae vary among donors and fingers, and, to this day, no two fingers have been shown to share the exact same combination of microscopic features [4]. Ultimately, it becomes the responsibility of the AFIS operator to confirm/reject the identity of a "hit" provided by the computer software. The development of AFIS in the 1970s illustrates how technology can have a significant impact on the way fingerprint evidence can be analyzed. Aside from the distribution of minutiae, fingermarks will have an inherent different chemical composition based on the exogenous and endogenous substances on the skin, such as sweat, moisturizers, drugs, or cosmetics. The presence of these substances also differs

[1] Microscopic distinguishable and discriminating features located along the ridges of friction ridge skin, from human fingers, toes, palm of the hands, and sole of the feet.

from person to person depending on their unique lifestyle [5], as discussed in Chap. 7.

In addition to the identification power of friction ridge skin, primarily from fingers and palms, many studies are exploring the potential of fingermarks in relation to their "aging" on a variety of substrates, such as glass, metals, and an array of plastics, exposed to different environmental conditions, thus widening the scope of friction ridge skin research from the original goal of discovering and improving methods of enhancement. This is a paradigm shift from attempting to answer the "who" of evidence to the more challenging "when." Scientists are intrigued by how factors such as the donors themselves, the type of substrates, and other environmental conditions play a role in the visualization (i.e., ridge quality) and proper recovery of latent fingermarks at a crime scene [6–8]. A better understanding of the influencing factors pre-, peri-, and post-deposition and the natural degradation processes involved will help to develop fingermark aging models that can, subsequently, aid in estimating the time since deposition. Fingermark age estimation is a subdiscipline of research that can greatly aid in the criminal justice system, especially when the identification of suspects by conventional minutiae comparisons may fail, for example, in cases of misattributions [9]. By estimating the age of a (latent) fingermark, a person can be included or excluded from the time frame of a crime simply based on the moment they touched any given object.

Some of the most novel microscopic, analytical, and imaging technologies are being adapted for fingermark visualization, composition analysis, and aging studies (as discussed in Chap. 5). These include technologies such as Chromatic White Light (CWL), Matrix-Assisted Laser Desorption/Ionization (MALDI), and Time-of-Flight Secondary Ion Mass Spectrometry (ToF-SIMS) [10–13], just to mention a few. These have been adopted by forensic researchers to assist in expanding the probative value of evidence by determining time since deposition. For example, morphological changes in ridge detail could be examined, such as micro-variations in ridge width and detection of certain chemical compounds (as discussed in Chap. 7).

Although most technologies can provide information on topographical or chemical alterations, they are unable to measure variations in the height and volume of ridges over time. Because of this, research on latent fingermark topography has recently encompassed 3D imaging technologies such as the Optical Profilometer (OP) that can measure with great accuracy any modifications in height and volume in addition to obtaining 3D renderings of the ridges [8, 14]. An OP microscope is a nondestructive technology that collects data on the x, y, and z axis to form a 3D topographical image of an object. In the realm of fingermark age research, the OP is used to measure the height, area, and volume of fingermark ridges. Most relevant, it can monitor the loss of ridge material of the same exact specimen over time, overcoming a common limitation from other methods. Furthermore, researchers have used the OP to investigate whether fingermark constituents (i.e., water, proteins, fatty acids, etc.) either evaporate during the natural aging process or diffuse onto the substrate, or a combination of both processes [8]. It has been hypothesized that if fingermark components were to evaporate post-deposition, then a significant volume of the impression would also be lost. However, if most of the residues were to diffuse over the substrate's surface, then the volume would remain about the same [15]. If both evaporation and diffusion were to occur simultaneously, then scientists would be able to determine their relative influence under certain environmental conditions. Changes in volume can also be assisted by measuring the weight of the fingermark over time. If the weight diminishes, then the constituents are likely evaporating [16]. In order to successfully estimate the age of a fingermark, it is essential to fully comprehend first the complex process of degradation and the influence of the environment under controlled and/or monitored conditions.

After being transferred on a surface, the unique interaction between donor characteristics (e.g., age, sex, diet, eccrine- vs sebaceous-rich secretions, etc.); environmental conditions (e.g., temperature, relative humidity, exposure to natural light, the activity of microorganisms, etc.); and the type of substrate (e.g., differences in porosity, roughness, wettability, etc.) will play a role in latent fingermark aging. All fingermarks will evolve differently based on the combination of these factors until they reach total obliteration, either by the intrinsic natural degradation process or caused by an external acting force (e.g., mechanical or chemical). Several investigations into the morphometric changes of latent fingermarks post-deposition are focusing on and have the broad goals to:

1. Understanding how environmental factors influence latent fingermark aging.
2. Identifying the best topographical features that can reliably characterize the aging process. Ideally, these must be consistent across donors and environmental conditions but sensitive enough to be affected by, and be indicators of, changes over time.
3. Developing a reliable mathematical model for age estimation with minimal error rates that could be applied in as many practical scenarios as possible.

This chapter aims to introduce and discuss novel aging parameters in 3D that could be used to model latent fingermark degradation patterns and, ultimately, to estimate time since deposition. Additionally, the potential of 3D imaging technologies will be compared with the more commonly used 2D technologies (i.e., scanners and photography); studies on the aging of fingermarks will be critically discussed and how they can be improved by utilizing certain methodologies. Throughout the sections, fingermark aging patterns will be uncovered, driving forces hypothesized, and research approaches in 3D introduced (Fig. 6.1).

6.2 Quality Assessment of Latent Fingermarks

The first step before attempting to collect measurable data for fingermark identification, as well as age estimation based on morphometric data, is to evaluate the (sufficient) quality and clarity of ridges. This is extremely important during the preliminary examination process to correctly associate (match) a questioned fingermark to its possible source (donor) [17]. Clarity is based on the overall quality of the

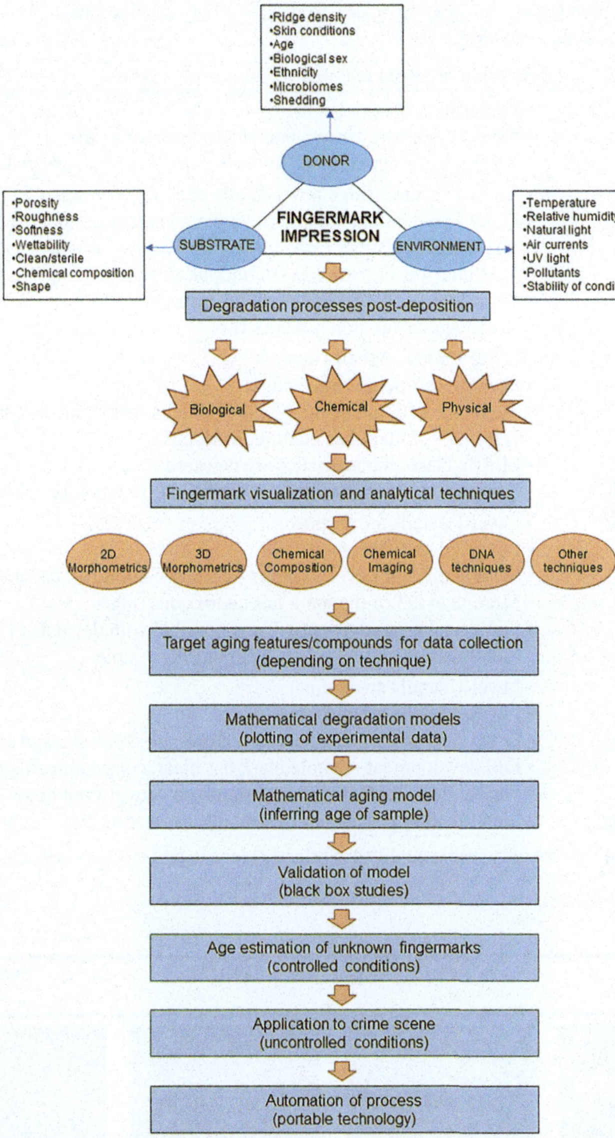

Fig. 6.1 Schematic diagram of factors influencing fingermark aging and steps to developing an age estimation technology

ridge detail that can be distinguished in the (enhanced) fingermark. There are a variety of grading scales that rank fingermark's visual quality dependent on the degree of ridge detail and the overall clarity of the impression [18–20]. These scales can be combined to form a cohesive grading scale to classify fingermarks on their visual quality (Table 6.1 and Fig. 6.2). These rankings are typically categorized into a relatively subjective scale from 0 to 4, with "0" score showing no visible ridges

Table 6.1 Combination of the grading scales from [18–20] to display one, cohesive scale for classifying levels of fingermark quality

Grading level	Description of visual features
0	• Fingermark is not visible • No development of any general structures or forms • No signs of finger contact with surface – only background
1	• Level 1 (overall ridge flow) details may not be distinct • Level 2 (individual friction ridge paths) details are not distinct • There are no distinct Level 3 (ridge structures) details • Less than 1/3 of mark has continuous ridges • Some signs of finger contact with surface • Visual age estimation not possible
2	• Level 1 details are distinct • Few Level 2 details are distinct • There are minimal distinct Level 3 details (debatable minutiae) • 1/3 to 2/3 of mark has continuous ridges • Mostly class characteristics are detected • No definite identification but visual age estimation possible
3	• Level 1 details are distinct • Level 2 details are distinct • Several distinct Level 3 details (definitive minutiae, debatable pores) • More than 2/3 of the mark has continuous ridges • Some background noise/artifacts possible with debatable ridge edges • Identification and visual age estimation possible
4	• Level 1 details are distinct • Level 2 details are distinct • Level 3 details are distinct (all features are definitive and clear) • Full development – whole mark has clear, continuous ridges • Very high contrast with background, no background noise • Identification possible but visual age estimation may not be possible

Scores "3" and "4" are the only viable qualities to reliably identify and estimate age of an impression

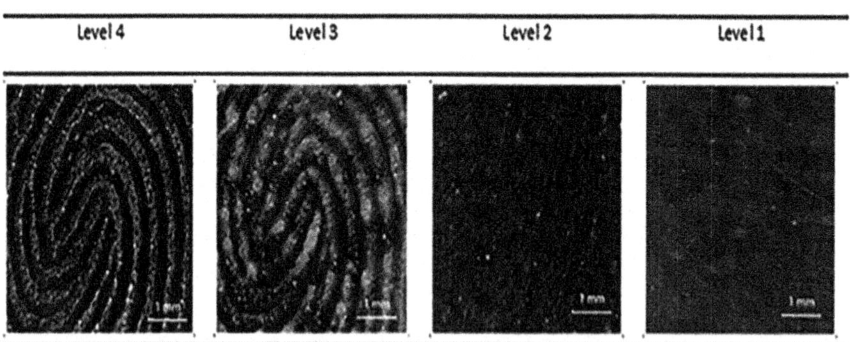

Fig. 6.2 Examples of latent fingermark images classified by the level of quality 1–4. Note the large difference in quality between Levels 2 and 3, while differences between Levels 1/2 and 3/4 are relatively minor. These correspond to titanium dioxide-enhanced impressions, showing ridges in white; adapted and reprinted with permission from John Wiley & Sons [33]

and “4” having clear, continuous ridges and unambiguous pores [18]. Impressions that are given a quality score of “3” or “4” have the potential to be matched to a known exemplar, whereas scores of “0” to “2” imply no sufficient visible ridge detail to confidently render an identification [19]. To determine the grading score, a fingermark should also be classified as having Level 1, 2, or 3 ridge details. Level 1 detail implies being able to observe the overall ridge flow of the impression, such as an arch, loop, or whorl. Level 2 detail includes the detection of microscopic friction ridge characterisitics, i.e., minutiae. Level 3 consists of other ridge structures in the fingermark, such as visible pores.

An alternative and more objective method to grade the quality and clarity of a fingermark is by using computer-based software, such as the Universal Latent Workstation (ULW) developed by the US Federal Bureau of Identification (FBI). The ULW is able to color-code and map the clarity of a fingermark and display the selected areas in four colors: red, yellow, green, or blue [18]. These colors correspond to the background, debatable ridge flow, debatable minutiae, and definitive minutiae, respectively. ULW also provides a series of metrics relevant to the degree of quality. Briefly, after an image is imported, the software will detect ridge characteristics and display the original image overlapped with the corresponding color-coded regions (Fig. 6.3). The clarity of the image is then automatically calculated [21].

The level of quality becomes especially pertinent when attempting to estimate fingermark age. That is, if the fingermark is not suitable for identification, there is no necessity to proceed with dating the evidence. In addition, a high degree of visual

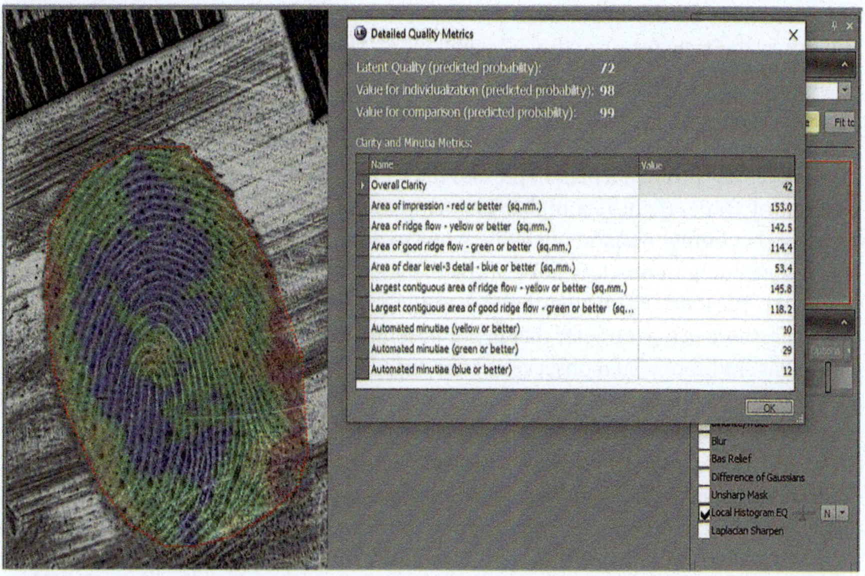

Fig. 6.3 Example of a latent fingermark examined with ULW. This specimen was powdered with titanium dioxide (color shown is inverted from original) and aged in the dark on a plastic substrate. Note the color-coded scheme on the fingermark and the quality metrics on the right-hand side

degradation, whether the "true" age is just a few days or a few years, can introduce too much error in the calculated age estimate. In all cases, examiners should evaluate the number of viable minutiae and other ridge characteristics to determine whether an identification could even be possible before attempting to date the fingermark. There is no established minimum number of minutiae that need to be present for age estimates or identifications. However, a survey conducted by INTERPOL in 2011 showed that out of 73 countries, 44 used a point standard, and 24 of the 44 required a minimum of 12 minutiae for a reliable identification [22]. A similar practice could be applicable to age estimations, but in this case, the dimensions of the area with viable ridges also becomes relevant for dating fingermarks. It is common to observe different levels of quality simultaneously in the same mark as a result of an uneven transfer or degradation across the surface. In this case, a conservative approach applies where the best "preserved" areas would be used for identification, as well as for age estimation.

6.3 Factors affecting the quality of Latent Fingermarks Peri-deposition

There are many factors that affect the visual quality of a fingermark at its deposition and, therefore, the fingermark's overall quality as it ages in response to environmental conditions. As discussed in previous chapters, pre-deposition factors, mostly donor conditions and the presence of exogenous substances on the skin, will also have an effect on the fingermark life expectancy, but not necessarily on the initial quality at the transfer of a "fresh" deposition. These factors are discussed in more detail in Chap. 5.

One of the peri-deposition (transfer) factors affecting the initial fingermark quality, and thus influencing its life expectancy, is the pressure exerted and duration of contact [23, 24]. When a person presses their finger onto a substrate with a very high amount of force, there is a higher chance that the imprinting ridges will merge or smudge, becoming indistinguishable when examined microscopically. On the other hand, if a person lightly touches a surface, there might be little useful area for analysis or the presence of discontinuities in the ridges unrelated to degradation [25]. Since pressure is a parameter that is not possible to be measured in real crime scene situations, indirect visual observations are used to help with its class categorization. For simplicity, pressure can be broken down into low, medium, and high amounts of force (Fig. 6.4). Low amounts of force are considered to be less than 2 kilograms (kg), or approximately 19 Newtons (N), while medium and high amounts are, respectively, between 2 and 7 kg (19–68 N) and greater than 7 kg (68 N) [26]. It has been demonstrated that the maximum contact surface area between a finger and a substrate occurs at 5.26 N of force [27] and that 69% of the fingertip's area will be visible on a substrate with just under 1 N of force, as the fingertip is quite flexible and flattens onto the surface easily [28]. In fact, the visual quality of a fingermark

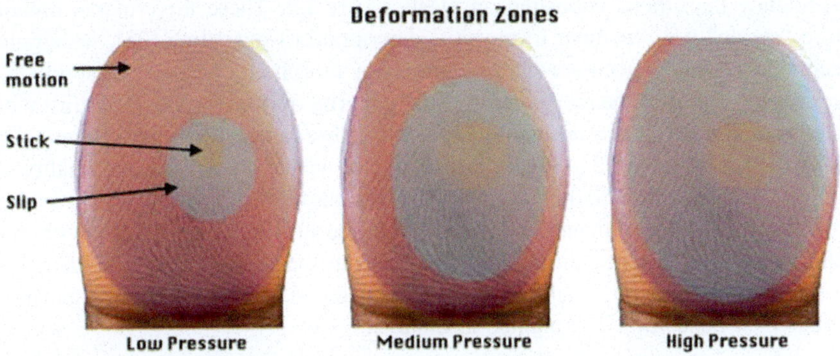

Fig. 6.4 Images depicting the deformation areas of a fingertip that will come into contact with a surface at low, medium, and high pressure; reprinted with permission from the International Association for Identification [28]

from the same finger can appear slightly different when examined microscopically based on the pressure exerted during deposition as well as the substrate the fingermark is deposited on [29]. Impressions best suitable for age estimation will fall mostly within the category of medium pressure, usually showing little distortion and smudging. This range of medium pressure (between 2 and 7 kg) is also used for most research with fingermarks, significantly reducing any potential variations in ridge quality and morphometric measurements.

The type of substrate also has a significant impact on the initial quality of the impression. Different substrates possess different characteristics that influence the deposition process and subsequent degradation, such as the texture, shape, and porosity. Other factors related to, but not part of, the substrate may include the presence of contaminants and temperature. In latent fingermark research, experimentation is typically performed on porous and nonporous surfaces.[2] The nonporous substrates have been primarily glass and different types of plastics and metals, since these are surfaces on which latent fingermarks are often discovered on at crime scenes. Chadwick et al. [30] examined the effect that the substrate had on the aging of latent fingermarks on two porous (Reflex 100% recycled A4 white paper and Reflex Ultra white A4 premium paper) and two nonporous (Marbig A4 polypropylene sleeves and Hercules large polyethylene Ziploc bags) substrates. They discovered a small difference in the degree of visual degradation between the two types of papers; however, there was a clear difference between the quality of the marks on the Ziploc bags and the plastic folder sleeves. Fingermarks on the Ziploc bags received higher quality scores. Research on nonporous substrates has also shown that fingermarks on glass appear to have a higher overall quality and longer life

[2] Semiporous surfaces such as finished wood and glossy cardboard must be treated with a combination of techniques for fingermark development which adds a layer of complexity to aging studies. Little is yet known on aging patterns on this type of surfaces.

expectancy than those deposited on plastics [31, 32]. These fingermarks remain more "stable" (i.e., resilient to physical deterioration) over time [33], which can become a caveat for accurate age estimations by morphometrics. The "stability" of fingermarks refers to the preservation of the integrity of ridges, (color) contrast with background, and the presence/absence of ridge discontinuities (interruptions). De Alcaraz-Fossoul et al. [31–34] have investigated visual changes in the quality of latent fingermarks in indoor conditions. These studies have shown that ridge quality was affected most significantly by the nature of the substrate itself (plastic vs glass) rather than the different natural lighting conditions they were exposed to (darkness vs shade vs direct natural light). Further discussion on the effects of environmental factors post-deposition is explained in Chap. 5.

6.4 Visual Parameters of Degradation: Meaningful Quality Metrics for Estimating Time Since Deposition

Little is known on the total number and degree of impact of (un)known donor, substrate, and environmental factors involved in latent fingermark degradation. Despite this limitation, there are numerous visual parameters that can be used to determine the level of visual quality of a fingermark image, both in 2D and 3D. As mentioned earlier, quality scores are of a subjective nature (except when using ULW) which implies a certain degree of bias and error [35, 36]. The visual parameters of reference may include the number of viable minutiae, (color) contrast between ridges and furrows, clarity of ridge edges, the general flow of the ridges, visible pores, presence of ridge discontinuities, and ridge inversions. Examples of different 3D qualities and parameters to be evaluated are shown in Fig. 6.5 and Table 6.2, respectively. Some of these parameters can later be used as metrics to examine the aging process of the fingermark, such as the color contrast between ridges and furrows.

(a) *Viable Minutiae*

One of the primary indicators of visual quality is the presence of unambiguously identifiable minutiae. Every individual, except for rare cases of adermatoglyphia [37], has distinct fingermarks as result of the unique distribution of minutiae based on the number, location, orientation, and type of these microscopic features which are determined by genetic and environmental factors during fetus gestation [38]. If minutiae are not recognizable, the fingermark is not suitable for identification purposes beyond a possible exclusion based on class characteristics. This fingermark will receive a lower score on the quality grading scale. In cases with no viable minutiae, the age of the impression will not be accurately estimated by morphometrics, but other alternative methods could still be used, such as the chemical analysis of residues as described in Chap. 7.

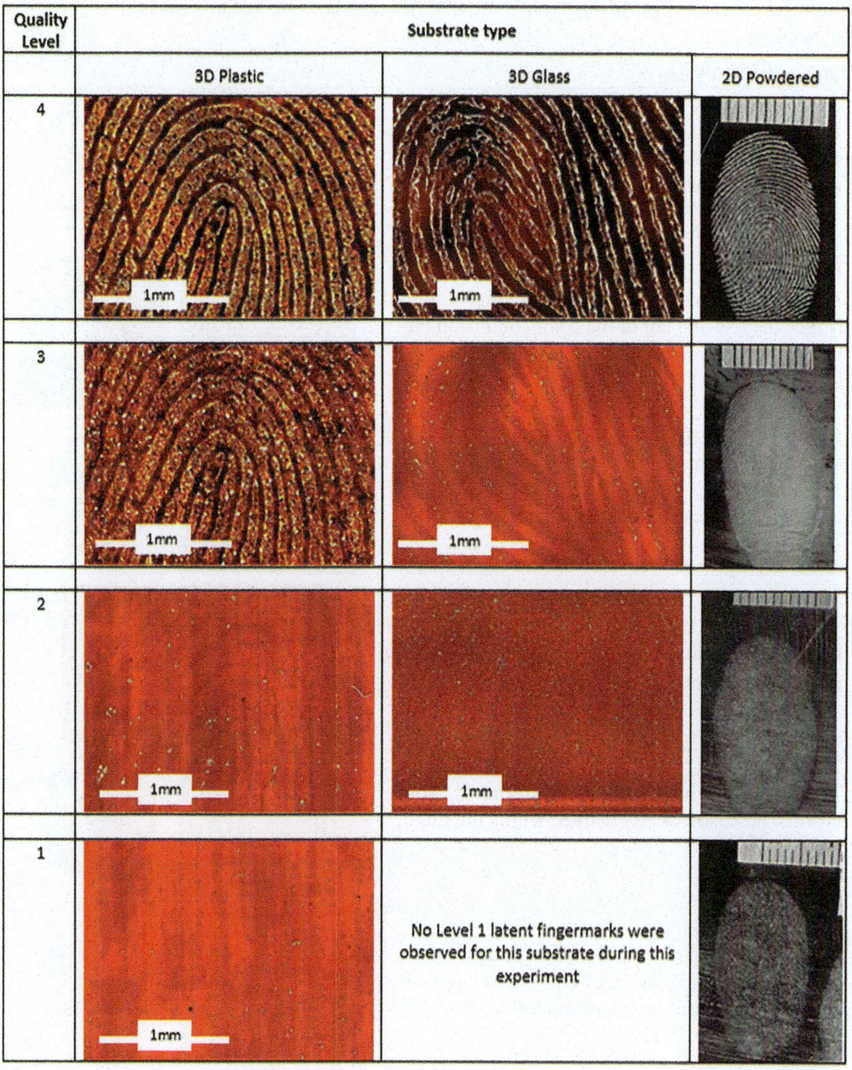

Fig. 6.5 Examples of visual quality scores of latent fingermarks as flattened 3D images on glass and plastic substrates. 3D images are renderings that depict the heights of ridges of non-enhanced (non-powdered) specimens. For comparison purposes, a 2D fingermark image enhanced with titanium dioxide (ridges in white) is shown at the same level of quality

(b) *(Color) Contrast Between Ridges and Furrows*

This parameter is another important indicator of quality. It is primarily used in conventional 2D examinations and also serves as a means of estimating age. Over time, as the quality of a latent fingermark decreases, the color contrast of the enhanced (e.g., powdered) specimens will also change [6]. This is the result of the

Table 6.2 Summary of qualitative parameters that may be collected to assess latent fingermark degradation

Qualitative analysis	
Viable minutiae This fingermark is one of highest quality with many viable minutiae. Notice the ridge edges are very well-defined, including the core and deltas	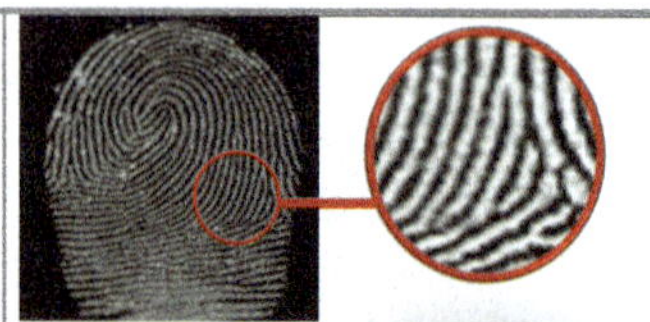
Color contrast between ridges and furrows This is a fingermark next to its histogram color profile. Notice the rise and fall of the color contrast in the histogram, representing all pixels from darkest furrows (left) to brightest ridges (right)	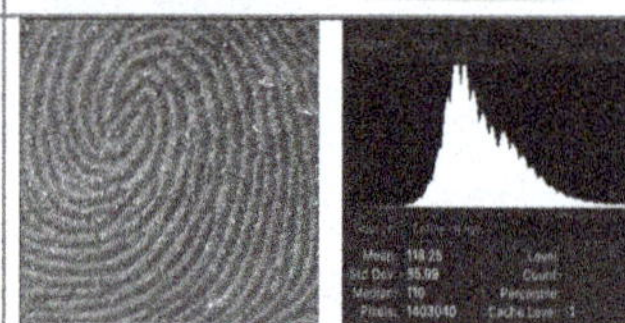
Clarity of ridge edges This is a fingermark with clearly defined ridge edges. Notice a clear distinction between the ridges (white) and the furrows (black)	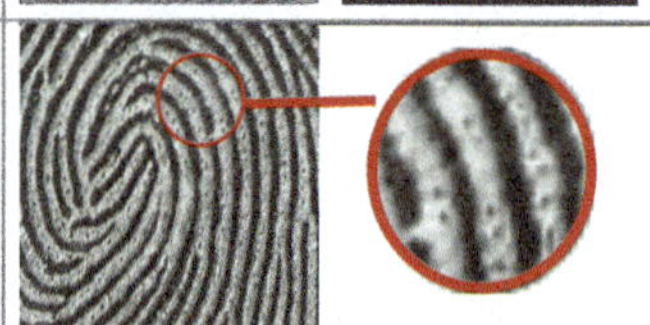
Ridge flow This fingermark shows an excellent ridge flow. Notice how the ridges do not have any interruptions and are smooth and easy to follow	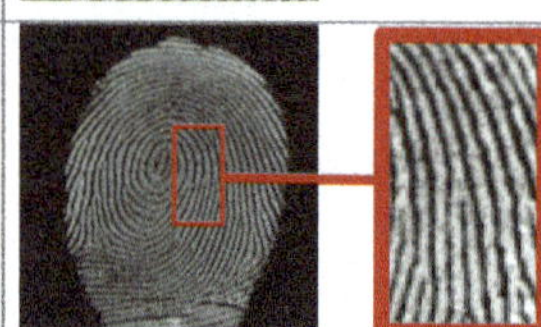
Visible pores This fingermark has visible pores magnified. Notice how the pores are clearly defined within the ridges	
Ridge discontinuities This is a fingermark with visible ridge discontinuities. Notice how some of the ridges that would normally be smooth and defined have interruptions in them	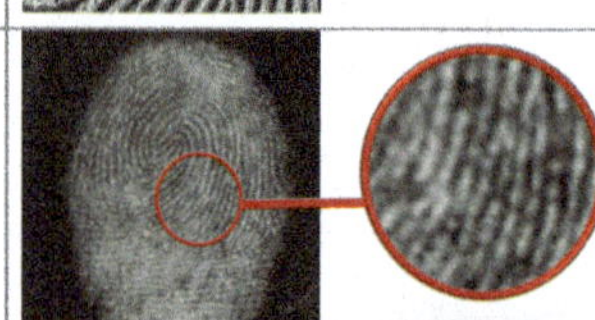
Ridge inversions These are two fingermarks showing "before and after" a ridge inversion, a sign of aging. Notice how in the left image (fresh) the ridges are white and the furrows are black. However, in the image on the right (72 days old), only the edges of the ridges are white. These were both enhanced with the same TiO_2 powder	
Relative surface areas This is a fingermark visualized with ULW showing areas with viable minutiae (highlighted in blue and green) and others not as clear (yellow, and red). This is an example of how a mark may contain different levels of quality	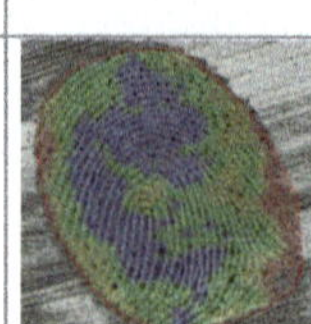

All the fingermarks shown were enhanced with TiO_2 powder (ridges are in white) and photographed gradual loss of adhesive properties of ridges to the developer (and vice versa) and/ or the lower quantity of residue present as the fingermark ages [11]. If the developer is not adhering in a sufficient amount, then the contrast with the background will decrease and the impression will receive a lower quality score. Moreover, the residue that forms the ridges can also diffuse into the furrows causing the developer to adhere to the furrows as well, resulting in contrast being less noticeable. In 3D, the (color) contrast is not dependent on the enhancement capabilities of the developer as none is required; however, contrast is interpreted as the heights of the ridges with respect to the substrate. The diffusion of the residue as well as the height of the ridges is discussed further in Sects. 6.5 and 6.9. Especially in 3D examinations, the contrast of the ridges is extremely important. If the ridges are not of sufficient visual quality, their heights, volume, and widths will not be measured accurately as the values may fall beyond the detection limit (sensitivity) of the microscope.

(c) *Clarity of Ridge Edges*

The clarity of ridge edges loosely relates to the concept of (color) contrast between ridges and furrows. As the fingermark ages, and the ridge residue diffuses and/or evaporates, the edges of the ridges will begin to appear "fuzzy," with a softer transition to the furrows (Fig. 6.5 and Table 6.2). This is likely due to the ridges losing their physical integrity and its components degrading and/or diffusing onto the furrows. The onset of this phenomenon can be observed as early as 2 - 24 h after deposition, but it is highly dependent on the technique used for detection, as explained in Sect. 6.9.1. As the edges of the ridges lose visual definition, the ridges and furrows will begin to blend together, making it more difficult for examiners to locate viable minutiae. A fingermark that does not have clear ridge edges will receive a lower score on the quality grading scale. In 3D examinations, unclear ridge edges can become a challenge, especially when measuring small ridge heights beyond the z resolution limit of the microscope, which falls below 2–20 nm in height depending on instrument capabilities. In this circumstance, accurate measurements of heights, of both ridges and diffused residues, may not be possible at all.

(d) *Ridge Flow*

The flow of ridges is one of the first aspects that examiners evaluate when grading the quality of a fingermark. If ridges do not have a continuous flow, it will be more difficult to distinguish minutiae from ridge breakages caused by degradation (Table 6.2). Ridge breakages (or discontinuities) usually present a jagged and irregular contour at their tip compared to the rounder ends of the minutia type known as ridge endings [33]. The detection of breakages is dependent on the level of sensitivity of the developer. A fingermark without a continuous ridge flow will receive a lower score on the quality grading scale. However, it is possible, and common, to have areas with good ridge flow while others are poorer in the same fingermark. If there are sufficient viable minutiae in the one "good" area, an identification and age estimation can still be possible. In extreme cases, if no flow is visible, the fingermark becomes unidentifiable and impossible to date with morphometrics alone. In

3D examinations, the quality of ridge flow is not as important as other metrics could be used. As stated previously, poor quality in ridge detail may result in a less reliable measurement of the height of ridges; however, other tools are available for comparison, such as the total ridge area and volume.

(e) *Visible Pores on Ridges*

Fingermarks that have visible and well-defined pores are often scored the highest quality on the grading scale. The presence of clearly visible pores usually implies a very high ridge quality and, therefore, potentially viable minutiae. As with ridge flow, there can be certain areas that have visible pores, whereas other parts might not. In this case, the fingermark may receive a lower score of "3" since most of it is not of the highest quality. In 3D examinations, the visibility of pores is not pertinent, since the pores are contained within the ridges that are ultimately being measured. If there are visible pores, the 3D examination would most likely be successful.

(f) *Ridge Discontinuities*

Ridge discontinuities, or interruptions, are naturally occurring breakages in the ridge flow that appear as either a result of the natural aging process [39] or because of the initial low quality of the fingermark due to, for example, low pressure applied during transfer. There are many factors that can affect ridge discontinuity, such as the type of substrate, the pressure applied, the presence of contaminants, the environment, and the type of secretion (eccrine-rich vs sebaceous-rich). As an aging feature, it may occur primarily during the degradation process and is dependent on the interaction of certain conditions. In general, a higher number of discontinuities will imply lesser fingermark quality. At a certain point, entire fingermark areas will disappear as a result of merging ridge interruptions; thus, the counted number of discontinuities will decrease in stages of advanced decay. In 3D examinations, ridge discontinuities, like with ridge flow, have a relative impact on the measurement of average height as other parameters could be measured (area and volume) until the resolution limit of the instrument is reached.

(g) *Ridge Inversions of Powdered Fingermarks: A Case of Interest*

Ridge inversion is a phenomenon that occurs when the developer (typically a powder) attaches to the edges of the ridges and, to some degree, to the furrows, leaving the "center" of the ridge mostly unpowdered. This could be the result of the loss of affinity of the powder to the ridge residue, or the residue diffusing from the ridge onto the furrow as it ages. The powder may adhere to the residue that has diffused rather than the ridge itself. The fingermark may be seen as a "negative" of a freshly powdered one (as seen in Table 6.2); however, this phenomenon is not applicable to 3D imaging as it uses no developer. This occurrence is a clear (indirect) sign of the loss of ridge integrity at the "height" level which may be overlooked in 2D. Due to the z resolution limitation of the 3D technology, the very thin layer of material diffused will be also difficult to detect by this method.

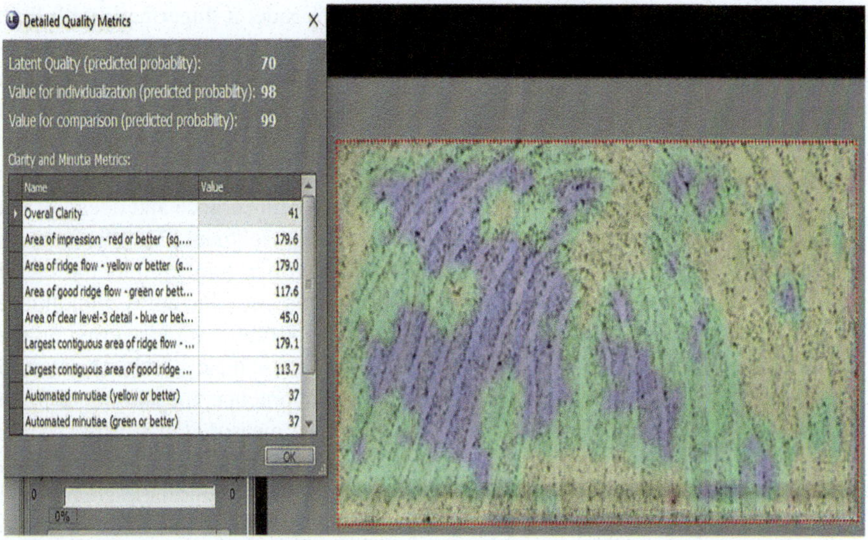

Fig. 6.6 Example of a latent fingermark examined with ULW. This image corresponds to a flattened 3D fingermark that depicts the heights of ridges of a non-enhanced (non-powdered) specimen aged in the dark on glass. Note the color-coded scheme and the quality metrics on the left-hand side

(h) *Relative Surface Areas*

In many instances, fingermarks will not degrade homogenously across the entire surface; therefore, certain areas will show different levels of ridge degradation. The *relative surface area* could be defined as the size of a fingermark's region that is of a specific ridge quality with respect to the total surface of the impression (mm^2). Depending on how much detail is visible and with discernable minutiae, a positive identification may be possible even with certain areas of poor quality as long as they are not overwhelmingly extensive. The area of the fingermark that is clear will affect the overall quality score given by the examiner. Alternatively, the ULW provides estimates of topographical ridge qualities using different metrics and a color-coded scheme that can also be used for images captured in 3D (Fig. 6.6).

6.5 Quantitative Analysis of Latent Fingermarks: Meaningful Metrics for Estimating Time Since Deposition

The quantification and objectivation of ridge detail modifications measured by metrics (i.e., morphometrics) are more pertinent, and necessary, for aging analyses in both research and field applications. A fingermark cannot exist without an intrinsic quality and quantity of features; each trait is dependent on the other [17]. Therefore, qualitative and quantitative parameters work side by side; a fingermark with high clarity is expected to have a high quantity of details and vice versa. The quantification refers to the measurable morphometric information contained within a friction ridge impression [17]. Some of the quantitative measurements that can be obtained include the width, area, volume, and height of ridges.

As previously mentioned in terms of quality, donor, substrate and transfer conditions are factors that will have an impact on the initial appearance and decay of the fingermark post-deposition in combination with the environment [23, 40]. In terms of quantity, the amount of force (pressure) used can affect, for example, the initial measurable width and height of ridges [8, 24]. A study conducted by Fieldhouse et al. supported the idea that ridge widths increase with higher force exerted on the substrate [25] and that the use of lighter pressure will cause residue to be transferred intermittently instead of leaving a continuous ridge. In practice, the amount of force has a relatively low relevance because it can never be measured at a crime scene; thus, it does not need to be precisely known to estimate the time since deposition. Pressure can be inferred based on the visual quality of the fingermark, such as the presence of smudges, discontinuities, interpapillae ridges, etc. Time of contact also becomes relevant [23] because the longer a finger is in contact with a surface, the higher the probability that the resulting impression will have continuous ridges. However, these two factors are dependent on the type and quantity of the secretion that is on the fingertips pre-deposition. If there is a high volume of secretion or contaminants, the ridges might merge together during transfer. In controlled research, mathematical aging models should include the ratio between a fresh and the respective aged fingermark to discover the significance, and minimize the effect, of initial conditions of transfer in volume and height metrics. This way, degradation patterns can be revealed and an estimate of age calculated. In practical applications in the field, this would require the identified suspect to "donate" a series of fresh fingermarks at different pressures for comparison purposes. If this were not an option, fingerprint exemplars from databases could be used to collect alternative reference data on ridge width and number of minutiae [41].

The type of substrate that a fingermark is deposited on will also affect the quantitative analysis [17]. Indeed, the type of substrate may affect the viable number, orientation, and location of the minutiae detected on the surface as the impression decays [6, 42]. Meyers et al. discovered that between a plastic bag, a compact disc (CD), a ceramic plate, glass, and paper, the ceramic tile displayed the greatest number of minutiae [42]. The interaction with donor characteristics, including type of secretion, is important for the quantitative analysis. In relation to secretion type, De Alcaraz-Fossoul et al. concluded that eccrine-rich fingermarks aged on plastic over six months will show fewer ridge discontinuities than their counterparts on glass [32] after visualization with titanium dioxide (TiO_2) powder.

There are other factors that can have an impact on the metrics (i.e., data from aging parameters) collected during a quantitative analysis [17]. These metrics can be used for identification purposes but also assist in determining the time since deposition. In the latter case, common parameters that are examined for fingermark aging purposes are the widths of ridges, distances between ridges, heights of ridges, and the area and volume of the transferred secretion, among others [8, 43]. These aging parameters are shown in Table 6.3 and are further discussed in Subsections A to E.

(a) *Ridge Width Measurements*

The width of a ridge is the distance from side to side of this friction skin structure and perpendicular to the ridge flow. This distance is variable person to person, finger to finger, ridge to ridge, and within a ridge depending on a number of biological factors [44–46]. Once deposited, and as the fingermark ages, the width of ridges can expand or reduce in size depending on factors such as exposure to natural light, temperature, relative humidity, substrate, and secretion type [7, 32]. Ridge width is a morphometrical feature that can be easily measured when conducting aging studies. The sizes of the imprinted ridges may differ from the skin pattern depending on the conditions of transfer. There is also an intra-person ridge width variation depending on the receiving surface. For example, if a person transfers a fingermark on a glass window indoors and then on a plastic bin indoors, the width of the fresh ridges

Table 6.3 Summary of quantitative parameters that can be collected to assess latent fingermark degradation

Quantitative analysis	
Ridge width This 2D fingermark contains red markings in circles depicting areas where ridges would be measured. Changes in the width of ridges over time can indicate a sign of degradation. This parameter can be automatically measured by an OP instrument	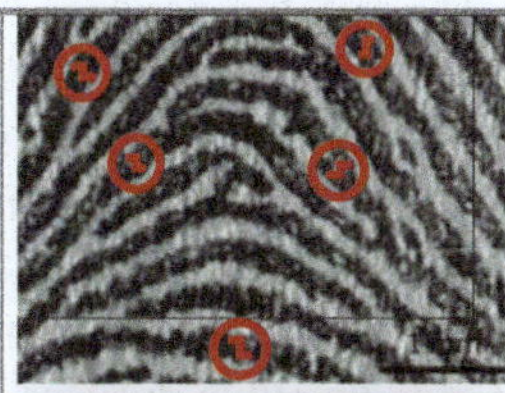
Distance between ridges This is a detailed 2D image of a powdered fingermark (white ridges). The arrows show where to measure the distance between two ridges. This parameter can be automatically obtained by an OP instrument	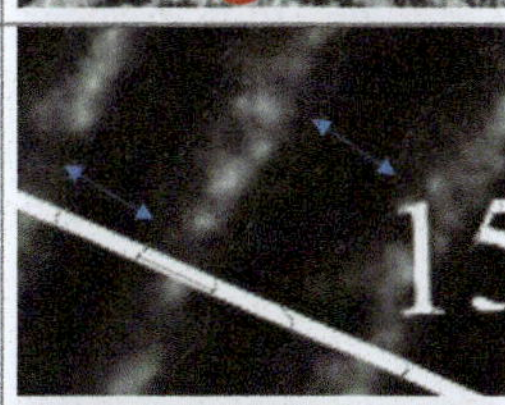
Height of fingermarks and ridges A 3D image of a raw latent fingermark captured with an OP, where a cross section of the mark is selected for height analysis. A height profile of this selected section is provided by the instrument	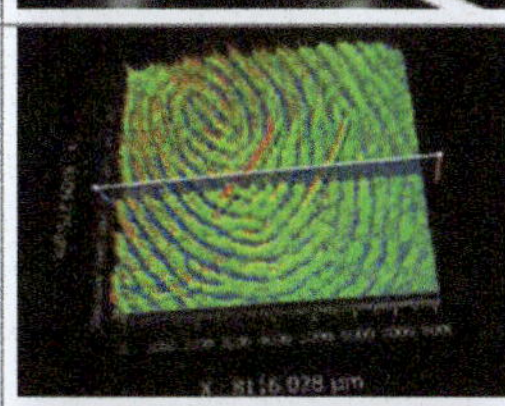
Area/volume of fingermarks and ridges A 3D image (flattened) of a raw latent fingermark obtained by an OP. Several metrics on volume and ridge area on any given surface are provided by the instrument	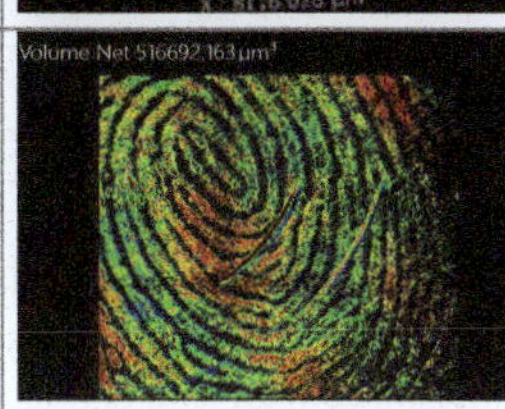

of these two impressions may differ and, more so, as they age because of their substrate and the degree of exposure to natural light which also influences temperature and relative humidity. The fingermark on the window may have more exposure to direct sunlight, whereas the impression on the plastic bin might be in complete darkness. Then, it becomes relevant to collect sufficient ridge measurements from every fingermark and calculate ratios over time rather than comparing measurements ridge to ridge. The disparity of ridge width measurements within a fingermark can be overcome, to some degree, by averaging 30 ridge width points at exactly the same locations across the fingermarks analyzed [32]. In 2D morphometric studies, the degradation process can be measured by changes in the ridge width after enhancement at predetermined discrete times, e.g., 1 day, 3 days, 1 week, etc. This can provide pertinent information on the evolution of the width of ridges and how distinctively environmental conditions impact this aging parameter [32]. Due to the variety of influencing aging factor combinations (donor, substrate, environment), different mathematical models for age estimation may then be applied to practical scenarios. For 3D studies, the same principles apply, with the exception that no enhancement methods are needed, thus cancelling the effect of the developer on data interpretation (i.e., level of sensitivity of enhancer, proper application, etc.).

(b) *Distance Between Ridges*

The distance between two adjacent ridges is another parameter that can be quantitatively examined in 2D/3D analyses [29] and can become larger or smaller as the constituents in the ridges evaporate and/or diffuse onto the substrate [45, 47] (Fig. 6.7). For example, in certain conditions, the ridges may dry out and decrease viscosity, almost to a point of "mummification" [48], sligthly enlarging or maintaining the distance between ridges. The distance between ridges can also lessen when the ridge residue diffuses onto the furrows. This phenomenon can be observed, for example, after a latent fingermark is powdered and the developer clings to the ridges as well as to any secretion in the furrows. In these cases, the diffused residue could be mistaken for being a part of the ridges [6].

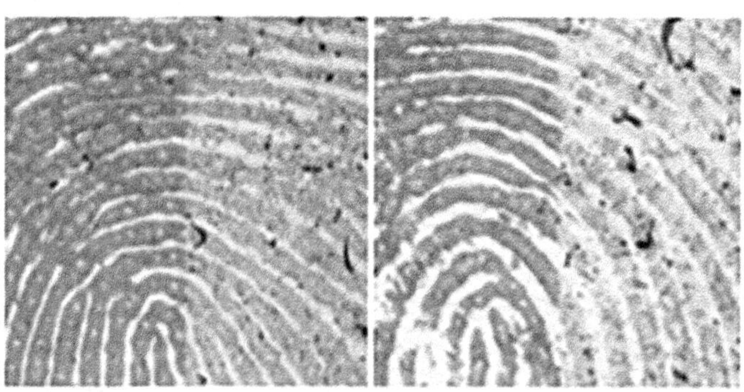

Fig. 6.7 Examples of grayscale images demonstrating the difference in the distance between ridges after 1 day (left) and 1.5 years (right) of untreated latent fingermarks deposited on hard disk platters; reprinted with permission from author [11]

(c) *Heights of Ridges*

The height of fingermark ridges is a parameter that can be measured only when performing a 3D analysis on nonporous surfaces using, for example, an optical profilometer (OP) [49]. This instrument can measure the ridges on the x, y, and z axis and provide topographical information with high degree of precision [8]; its capabilities are described in Sect. 6.9. In fact, the loss of ridge height is currently being evaluated as an aging parameter to date fingermarks and might be a reliable indicator to determine whether ridge constituents are evaporating or, alternatively, diffusing onto the substrate [50]. Further research needs to be conducted to confirm either of these hypotheses by this technology. In order to decipher whether evaporation or diffusion is the leading force in aging processes, a quantitative analysis combining the volume, height, and area of the secretion would be ideal (see Section "d"").

The loss of ridge height over time could be used in conjunction with other aging parameters, such as the aforementioned ridge width. Depending on the type of substrate, the environmental factors, and the type of secretion, the height of the ridges will diminish accordingly and can provide valuable additional information that 2D analyses alone cannot offer. Ideally, a simple ratio between a fresh and aged deposition at predetermined times and known temperatures can be used to develop mathematical models of degradation that can eventually be adapted for age estimation at the scene.

On the height of ridges, cross sections can also be examined in 3D (Table 6.3). These can be created when a handful of ridges are selected and their profiles compared over time with themselves. These measurements can be used to obtain an average ridge height, the distance between the ridges, or the distance between the peaks of the ridges. Cross sections allow for the visualization of how sweat secretions "travel" over time between adjacent ridges on a much smaller area, rather than evaluating the entire fingermark.

(d) *Area and Volume of Fingermarks and Ridges*

The volume of the fingermark, and its changes post-deposition, is a metric that has not been widely researched as a possible aging parameter and can only be measured in 3D. This metric can provide critical information on what occurs to a latent fingermark over time. Fingermarks are composed partially of water [1, 50], and it has been suggested that most water will evaporate shortly after deposition [24, 51]. Little is known on the impact of the other chemical constituents on the topography of the fingermark. Some argue that residues will also evaporate post-deposition [51] or will diffuse onto the substrate [15]. These hypotheses could be assessed by measuring changes in the surface area and volume of the transferred material on nonporous materials. If the initial measurement of volume is drastically higher than subsequent measurements, it could indicate that evaporation is the acting force. However, if the subsequent volume measurements are approximately the same as the initial measurement but the area increases, then diffusion could be the main acting force, as long as measurements fall within the limits of detection of the measuring instrument. The area and volume of the secretion, as well as the height of ridges, will be discussed in detail in the following sections. All these aging parameters can be easily measured [8, 14] by the OP and are obtained automatically by metrics this instrument provides, such as the average roughness of the surface (Sa), kurtosis of the areal surface (Sku), maximum peak height (Sp), mean roughness of the surface (Sq), and the skewness of the areal surface (Ssk) [52].

6.6 Understanding the (Physical) Aging Process of Latent Fingermarks

The ability to accurately estimate the age of a latent fingermark could be of great service to criminal investigations and in court. Being able to scientifically set a time frame onto a fingermark could aid in objectively including, excluding, or prioritizing individuals from a pool of suspects beyond (potentially biased) reports from witnesses. For example, a homicide is committed in the living room of a home. The scene is processed 24 h later, and a fingermark is identified and estimated to be between 0 and 5 days old by the time the analysis is conducted, falling within the possible time frame of the event. However, the prime suspect claims the fingermark was left when they were visiting the victim 1 month ago. In such scenario, there are grounds for debate during an investigation, and in court, since the predicted age of the fingermark contradicts the suspect's statement.

The physical (visual) aging parameters being assessed to model degradation patterns, and eventually to estimate the age of a latent fingermark, are the (color) contrast between ridges and furrows, the loss of ridge detail (minutiae), the broadening or reduction of ridge width, the continuity of the ridges, the overall clarity of the fingermark, and more recently the height, area, and volume of the fingermark, just to name a few. These parameters mirror the qualitative and quantitative features that can be extracted from an image and were discussed previously. In simpler terms, age estimation begins by evaluating the general quality of the mark and continues with the data collection and combination of the aging parameters that are most pertinent and least redundant to develop a mathematical model.

In the methodology proposed, once a sufficient quality fingermark is located at a scene or on an object (e.g., by oblique lighting or ALS) on a nonporous object, it should first be photographed and scanned in 3D, so that its appearance at the time of discovery is thoroughly documented. Only after this step, traditional enhancing methods should be applied. Any qualitative and quantitative analyses would be automatically obtained on a handheld device or performed in a laboratory setting, so that the data is recorded and could be used for comparison. Once an alleged suspect is identified, reference samples of their fingermarks should be taken at different pressures, preferably on a substrate similar to the one recovered from the crime scene, although there are likely to be significant legal challenges to performing such tasks. A known fingermark provides critical qualitative and quantitative information that can be compared with the questioned evidence to determine whether it is fresh, aged, or falls within the time frame of the crime. Most importantly in age estimations, an examiner should not mistake an aged fingermark of high visual quality with a fresh one. For example, a fingermark that is 6 months old might visually appear similar to another one that was deposited just 6 h prior to its discovery. This circumstance has already been described in a laboratory setting using conventional 2D imaging tools [43]. In scenarios like this, the height and volume of the ridges might be crucial additional information in discerning whether the evidence is indeed "fresh." It must be noted that a 3D examination must be conducted prior to applying any enhancing techniques (e.g., by powders) on nonporous surfaces (plastic, glass, metal, tile, etc.). Based on experimental data [8], an older fingermark will have lower ridge heights and less volume than a freshly deposited one of equivalent 2D visual quality. The reference average heights of a donor's fingermarks can be obtained from a selection of known fresh samples from the same individual who has been already identified using common identification methods.

For a reliable age estimation model to work, time-sensitive morphometrical parameters (i.e., morphological features that change with time) should be combined and supported by science. These include monitored and/or controlled observations of natural degradation patterns of enhanced and untreated fingermarks [31–34, 43] which will be influenced by the type of secretion and its composition (sebaceous- vs eccrine-rich), the environmental conditions (exposure to natural light, temperature, humidity), and the substrate (nonporous) indoors (Fig. 6.8), as described in more detail in Chap. 5.

Indeed, the triad of interactions between donor, environment, and substrate is crucial in elucidating patterns of degradation and, subsequently, estimating the age of a latent fingermark [5, 45]. Donor characteristics may include sex, age, ethnicity, and diet, among others. Numerous studies have investigated differences in fingermarks between males and females [2, 4, 48] and have shown that females typically have finer ridges than males and male's ridges tend to be coarser. Females also typically have a greater ridge density than males [45, 46]. It should be noted that the measurement of uncertainty (when accurately presented) tends to show that average ridge widths tend to overlap. In regard to secretion, the fingermarks of males tend to be "greasier" [44]. All these characteristics might be relevant to fingermark age estimation for numerous reasons. Since females typically have finer ridges, the

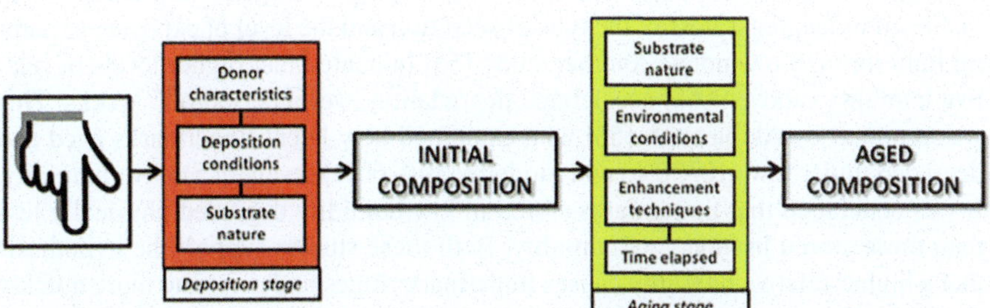

Fig. 6.8 The factors that affect the initial and aged composition of a latent fingermark; reprinted with permission from Elsevier Science & Technology Journals [66]

ridge width measurement will likely be smaller. This should be taken into consideration when determining whether the ridge is naturally thinner or the result of aging. Although the size and density of ridges appear to be different, studies available today have not proven any significant difference between sexes to aging patterns using physical parameters of degradation alone. This could be an indication of the relatively lower sensitivity of optical approaches (e.g., similar sensitivity of the powder to male/female secretions) compared with chemical examinations. As a donor ages, their ridge groove pattern becomes coarser and puberty affects the overall chemical profile of the impression [5]. It has been shown that children's fingermarks are difficult to locate and enhance if they had been deposited longer than 24 h prior on nonporous surfaces. Children's ridges are more prone to degradation because they contain a higher level of volatile constituents, such as unesterified free fatty acids, unlike adults that contain more stable constituents such as long-chain fatty acid esters [5]. Although there are differences between the composition of a child and adult's fingermark because of puberty, there are no current methods that can definitively distinguish between a child and adult fingermark. However, if the composition of the secretion is tested and contains a high concentration of long-chain fatty acid esters, there is a higher likelihood that the donor is an adult rather than a child [5]. Studies on an individual's diet have shown that certain compounds found in fingermarks can be traced back to dietary habits [5]. For example, it is been reported that caffeine and other substances can be detected in fingermark secretions [53, 54]. In regard to ethnicity, a study examined one Spanish and two Argentinian populations [2, 4]. The results showed that South American Indians (one of the Argentinian populations studied) have greater variation in their minutiae frequency than any other American population. The Spanish population presented a significantly higher number of ridge endings and a lower number of bifurcations and convergences than either of the Argentinian populations studied. However, at this time, the effects of both diet and ethnicity on the aging of fingermarks are unknown.

Experiments have consistently shown a significant impact of the type of substrate over the type of secretion and exposure to natural light. For instance, fingermarks on glass will typically show fewer physical signs of aging than those on plastic, regardless of other factors [32]. In research conducted by De Alcaraz-Fossoul et al. [32, 34], it was concluded that fingermarks on glass will be less prone to physical degradation, such as the loss of integrity of ridges and the number of viable minutiae, regardless of the type of secretion and the level of exposure to natural light through a window. Another study [55] indicated that higher levels of relative humidity (above 90%) accelerated degradation over a period of 7 weeks. This was reported by Akiba et al. [56] who examined how latent fingermarks aged and degraded under low (10–30%) and moderate (40–60%) levels of relative humidity. It was concluded that fingermarks stored in low humidity degraded physically less than those stored in moderate humidity. Both these studies support the hypothesis that a higher relative humidity causes fingermark ridges to deteriorate more quickly over time. Despite these findings, no substantial influence of the presence/absence of natural light has been detected to influence physical degradation.

Aging studies of latent fingermarks have primarily focused on nonporous surfaces, such as glass and plastics, since these are two common nonporous surfaces on which fingermarks are recovered at crime scenes (see Chap. 5 for detailed discussions). Although fingermarks are also deposited and developed on porous surfaces, these are not typically used because they absorb the residue at an undetermined rate, making some ridge features, especially the heights, difficult to analyze in 3D.

6.6.1 Experiments on Other Nonporous Surfaces

Besides glass and plastics, it is important to address other nonporous surfaces that fingermarks can be deposited on to understand how other surfaces might impact the aging process.

(a) *Steel*

Steel as receiving surface has been briefly researched in terms of fingermark aging. This type of substrate has shown to significantly impact the deterioration of fingermarks in as little as 24 h [1]. A study on a hand grenade handle using ungroomed fingermarks from 50 male donors showed that only after 24 h no ridge detail was detected using conventional methods. However, after 48 h, some ridge detail (such as ridge endings, bifurcations, etc.) was seen using ToF-SIMS. Another study demonstrated that the degradation on steel was much faster than on Teflon [12] and supported the theory that the iron in steel has an influence on the lactic acid degradation process that occurs over time. A third study explored whether ridge detail was discernable on stainless steel before enhancement [14], and it was discovered that ridges could be detected before enhancement, though the quality was not sufficient for analysis. Several studies have examined the effect of fingermark residue on metallic surfaces, but due to the wide array of alloy compositions, more research is needed since many objects such as weapons, ammunition, tools, and explosives can be made out of metals and alloys that can be key pieces of evidence at crime scenes.

(b) *Aluminum Foil*

Aluminum foil is another nonporous substrate that has minimally been researched in terms of fingermark aging. One study aged a fingermark on aluminum foil for 2 weeks and was partially developed using cyanoacrylate fuming [13]. The impression was then analyzed using ToF-SIMS, and ridge detail was clear and continuous in the negative ion images created. However, the ridges were showing as a series of dots, rather than continuous lines, when examined solely after cyanoacrylate fuming (Fig. 6.9). More studies should be conducted to document how a fingermark would age on aluminum foil (and other aluminum-based objects) without the cyanoacrylate fuming and staining procedures.

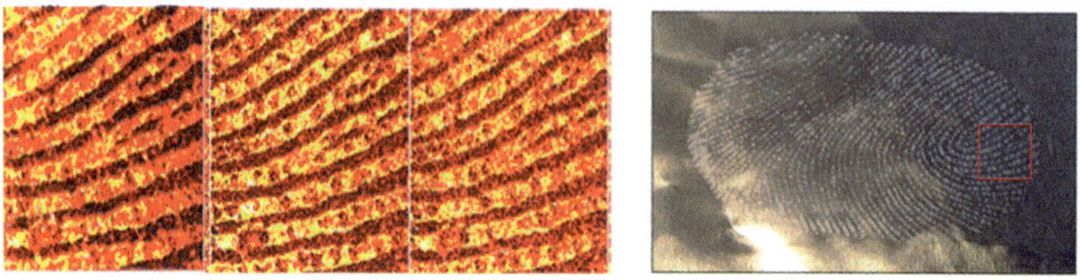

Fig. 6.9 Images depicting the difference between a fingermark deposited on aluminum foil enhanced with cyanoacrylate fuming and ToF-SIMS analysis (**a**) and the same fingermark only enhanced by cyanoacrylate fuming (**b**). The red box in image B represents the area of the fingermark that was further enhanced with ToF-SIMS; reprinted with permission from the Royal Society of Chemistry [13]

(c) *Ammunition*

Methods for the detection and enhancement of fingermarks on ammunition have been extensively researched in forensic science; however, aging patterns have not been widely explored. In one study, fingermarks were aged on brass squares in a locked cupboard and in glass-filtered natural light for up to 6 weeks [57]. These were then analyzed using silver electroless deposition (SED) which develops the fingermark by releasing silver ions from an aqueous ionic liquid such as ethaline or choline (which is composed of a 1:2 ratio of ethylene glycol to choline chloride) onto the brass surface as a silver metal. The silver deposits onto the part of the metal surface where there is no residue, therefore creating a "negative" of the fingermark. The results revealed that on brass surfaces, fingermarks will degrade much more rapidly when stored in the dark. Although this study did not examine ammunition directly, many types of cartridges and cartridge cases are manufactured from brass. Therefore, the information gained from this study could be useful in the analysis of fingermark evidence on ammunition as well.

6.7 The Basics of 2D Analysis for Latent Fingermark Aging Studies

There are several approaches to study latent fingermarks post-deposition, for identification and/or aging purposes. These include chemical imaging, chemical analysis, physical analysis by 2D and 3D visualization methods, and the analysis of touch DNA and microbial composition. The potential of these approaches is discussed in detail in other chapters of this book. Age determination of fingermarks should begin with the use of nondestructive 2D/3D methodologies and instruments that aid in visualization, allowing for the examination and recording of morphometrical aspects such as ridge width, the distances between ridges, and ridge discontinuities [29]. The most basic form of 2D analysis uses a powder developer, such as carbon black or titanium dioxide (TiO_2) that enhances color contrast with the background [6]. These can be used on their own or after chemical imaging on the same fingermark as long as the chemical imaging method is nondestructive. Aside from the common use of photography and image editing software, some of the instruments for 2D analysis include ToF-SIMS, MALDI, CWL, and FTIR. All these technologies have provided valuable information on ridge detail alterations post-deposition and new data on the impact of the environment over time. These are largely covered in Chaps. 4, 5, and 8. Some of the technologies, like MALDI and CWL, have characteristics that can complement optical 3D analysis. For example, MALDI can be used to explore how compounds diffuse on a substrate over time. 2D and 3D techniques provide pieces of information that can easily complement each other, especially when attempting to develop a comprehensive mathematical model of fingermark degradation.

6.8 The Use of 3D Technologies for the Analysis of Forensic Evidence

Although 2D methodologies can provide critical information about fingermark topography and describe degradation patterns post-deposition, there are certain intrinsic limitations to these approaches. For example, these cannot give information about the height of the ridges or the volume of the secretion on the substrate. However, these data can be collected by 3D analysis, providing a novel insight into the changes that occur to latent fingermarks over time. The use of 3D imaging in the field of fingerprinting is fairly recent, with very few studies published, but has been successfully applied in other areas of forensic science such as firearms, ammunition, and toolmark evidence [58, 59].

(a) *Firearms and Ammunition*

Traditional 2D analysis utilizes technologies such as comparison microscopes to associate an unknown sample to a known specimen. In 3D, a more in-depth examination of these items can be conducted. Research has explored the depth of indentations on fired bullets, toolmarks, and cartridge cases [59]. 3D technologies such as the stylus profilometer [60] and the 3D virtual comparison microscope [61] have been used for this purpose. The stylus profilometer provides traditional 2D roughness surface characterization and step height as well as 3D mapping and film stress analyses [62]. The 3D virtual comparison microscope compares virtual images of two specimens side by side and replicates the mode of operation of a traditional comparison microscope. However, it has the additional advantage of a wide range of 3D enhancement tools such as rotating and translating scanned images and annotating topographical surface data that extends the comparison further than a traditional comparison microscope could perform [61]. If the depth of the striations post-firing is of the same or similar depth as the lands and grooves in the rifling of the bore, then there is a higher probability that a pattern match could be made based on factual measurements rather than simple visualization [59]. For example, Fig. 6.10 depicts the height measurements of a portion of land impressions taken with white light interferometry that could be compared to striations post-firing.

(b) *Toolmarks*

Similar to ammunition, toolmark evidence is examined in 2D with the goal of finding pattern correlations based on striations left by a tool. A 3D analysis can examine toolmark evidence quantitatively and compare the depths of the toolmark impression and its striations to a known specimen (Fig. 6.11). Toolmark evidence has successfully been quantified using technologies such as confocal microscopy (CM) [63]. CM is an imaging technique for the observation of surface topographies in 3D. The microscope produces two different images, one is an in-focus image of

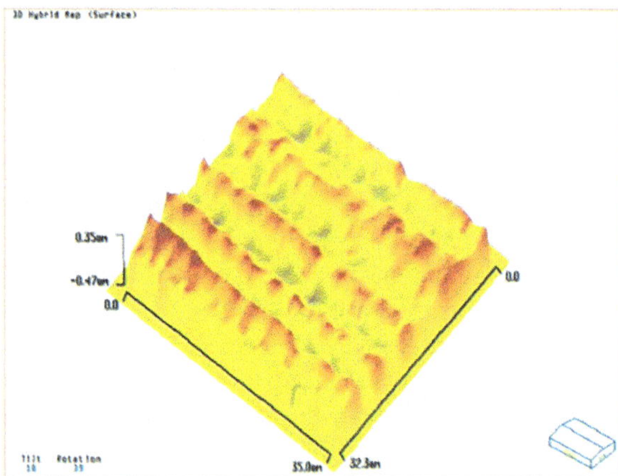

Fig. 6.10 Image showing the measurements of a portion of a land impression captured with white light interferometry; reprinted with permission from John Wiley & Sons [60]

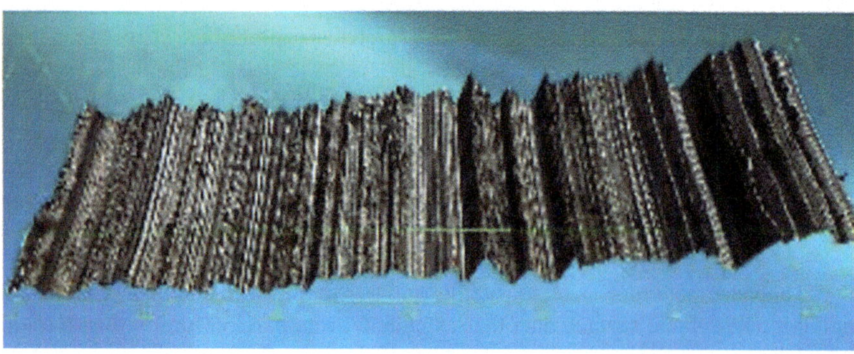

Fig. 6.11 A three-dimensional image of a striated toolmark; reprinted with permission from John Wiley & Sons [63]

the sample, and another shows the height information in varying gray scales [63]. Accurate images could be obtained and inputted into the database to later be used for comparison purposes. In a study on screwdriver and chisel impression patterns, a match was made based on height information that was previously inputted into a toolmark software. The algorithmic error rate when using the 3D analysis from CM was only about 1% [63].

6.9 The Use of 3D Technologies for Latent Fingermark Aging Studies

3D examination of physical evidence has been demonstrated to be extremely useful in revealing novel information regarding topographical details. One technology that has been tested in fingermark aging studies is the OP. This is a nondestructive technology that allows for the visualization and collection of data on the x, y, and z axis. The OP has proven to be successful in detecting, among other data, the loss of fingermark ridge heights over time [8].

The OP works by using light to create a 3D image of the specimen that is placed on the instrument's stage. It uses the various wavelengths of light to compare the difference between the test specimen and the reference mirror that is built into the OP itself. Since the reference mirror in the OP is as close to being perfectly flat as possible, the topographical data that is displayed in the resulting image are the differences in heights between the reference mirror and the test sample. The OP measures the specimen vertically. As the material on the stage passes through the focal plane of the OP, it creates interference, i.e., the difference between the flatness of the reference mirror and the sample itself. The resulting image will include light and dark bands; the light bands are considered constructive interference and the dark bands as destructive interference. From this, a 3D surface map of the sample is produced [64] (Fig. 6.12).

The OP has been adopted to examine (latent) fingermarks because of its ability to reveal key information that cannot be otherwise obtained by 2D analysis. Standard metrics on height, area, and volume of ridges can be collected as well as for the entirety of the ridge pattern transferred to the surface. In fact, the OP can measure a wide variety of physical features of a fingermark which include, but are not limited to, average roughness, kurtosis of the areal surface, maximum peak height, root mean square roughness, skewness of the areal surface, mean peak profile height, Swedish height, volume up, and volume net [52]. These suggested parameters are explained below and summarized in Table 6.4.

(a) *Average Roughness (Sa)*

This value refers to the average area roughness that is evaluated over the entire 3D surface. On a fingermark, this would correspond to the average height of all of the ridges examined compared to the arithmetical mean of the surface.

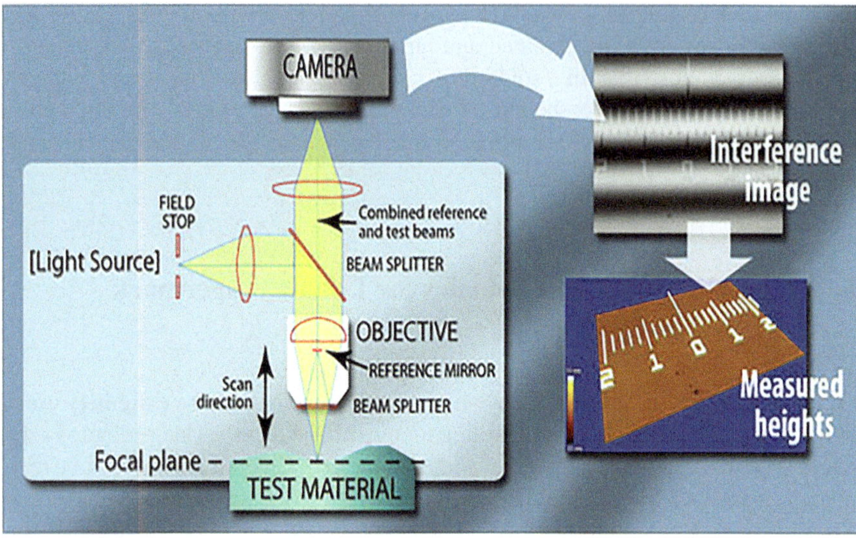

Fig. 6.12 Diagram depicting the working mechanisms of an optical profilometer to provide a 3D image of a specimen; image courtesy of Zygo Corporation [64]

Table 6.4 Summary of the suggested optical profilometer parameters that may be used to characterize a latent fingermark as it ages; images courtesy of Zygo Corporation [51]

SUGGESTED OP PARAMETERS FOR FINGERMARK AGE ESTIMATIONS*	
Sa 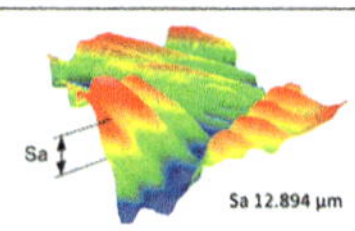 Average height of the surface	***Ssk*** 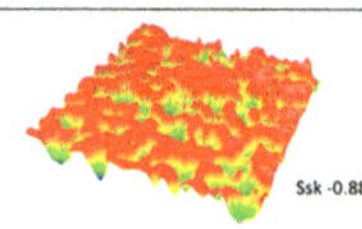 Level of symmetry of the surface heights
Sku 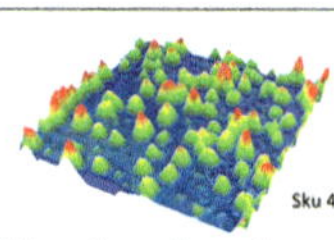 Inordinately high peaks on the surface	***Sdq*** 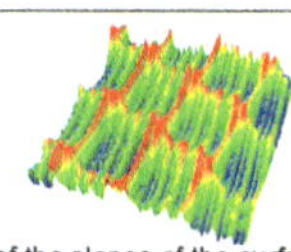 Measurement of the slopes of the surface
Sp 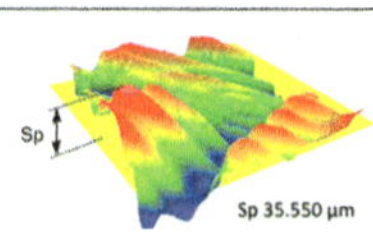 Maximum peak height of areal surface	***Rpm*** 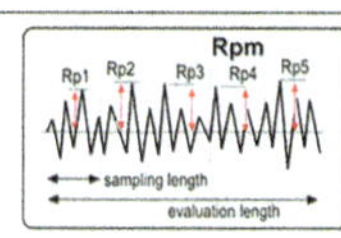 Average heights of the highest peaks
Sq 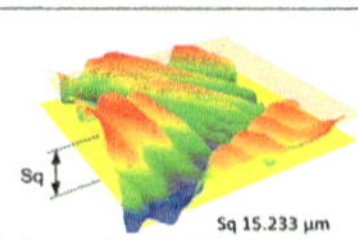 Standard deviations of the heights	***H*** 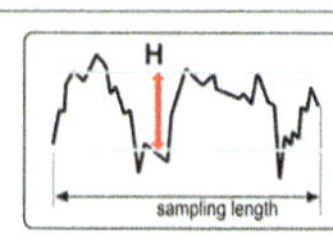 Height between two reference lines
Volume Up 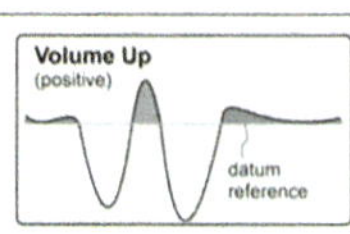 Volume higher than the reference area	***Volume Net*** 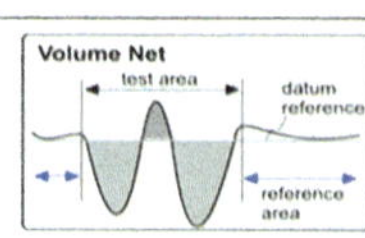 Overall volume of the sample being tested

(b) *Kurtosis of the Areal Surface (Sku)*

This value indicates the presence of any inordinately high peaks (sharpness) or valleys making up the surface. On a fingermark, this would indicate the unexpected predominance of high ridges or deep scratches compared to other ridges in relation to the plane of the surface. This value would be especially useful when examining fingermarks for signs of degradation.

(c) *Maximum Peak Height (Sp)*

This value is the maximum peak height of the areal surface examined. On a fingermark, this would be the highest ridge or the highest point in a specific ridge, in other words, the height of the highest peak within the defined area.

(d) *Root Mean Square Roughness (Sq)*

This value is the square root of the mean height of the surface. Essentially, it is equivalent to the standard deviation of heights, similar to Sa values but more robust to the effect of unusual heights. On a fingermark, this value would provide an average of the heights of the ridges that are being measured.

(e) *Skewness of the Areal Surface (Ssk)*

This value represents the level of symmetry of the surface heights in relation to the plane they are on. It indicates the predominance of the peaks or valleys comprising the surface. On a fingermark, this value would provide insight into the ratio of ridges and any surface scratches in relation to the plane of the substrate.

(f) *Root Mean Square Gradient of the Surface (Sdq)*

This value is a general measurement of the slopes that comprise a surface. It can be used to differentiate surfaces with a similar average of roughness (the Sa value). On a fingermark, this value would indicate how spaced the ridges are on the surface.

(g) *Mean Peak Profile Height (Rpm)*

This value represents the mean height of the peaks of a selected sampling length. It is determined by averaging the five highest peaks of a selected region. On a fingermark, this would be calculated by a cross section of five ridges and then averaging them together.

(h) *Swedish Height (H)*

This value is the roughness between two predetermined reference plane lines; the upper line exposes 5% of the data and the lower line 90%. It is less sensitive to spikes in the height than other values. On a fingermark, this value would be the height of the ridges in relation to the reference lines.

(i) *Volume Up*

This value is the volume of the sample that is higher than the reference area. A positive volume up value is the space that is occupied by bumps on the test area. On a fingermark, this would correspond to the volume of the mark's secretion that is present on the substrate regions examined.

(j) *Volume Net*

This value is the overall volume of the sample being tested. It is equal to the volume up minus the volume down (below the reference area). Essentially, it is the average volume of the sample being measured. On a fingermark, this value represents the overall volume of the test area.

The OP has been successful in revealing latent fingermark aging-related processes, providing meaningful, and also novel, data on time-sensitive morphometrics. A few of these aging parameters can also be measured using 2D methods, such as ridge width and ridge discontinuity with high precision and little error. However, the OP can simultaneously measure ridge width and heights by taking cross sections of ridges at once over the exact same fingermark (Fig. 6.13). This opens the possibility of recording videos by elapsing images of the same fingermark as it ages,

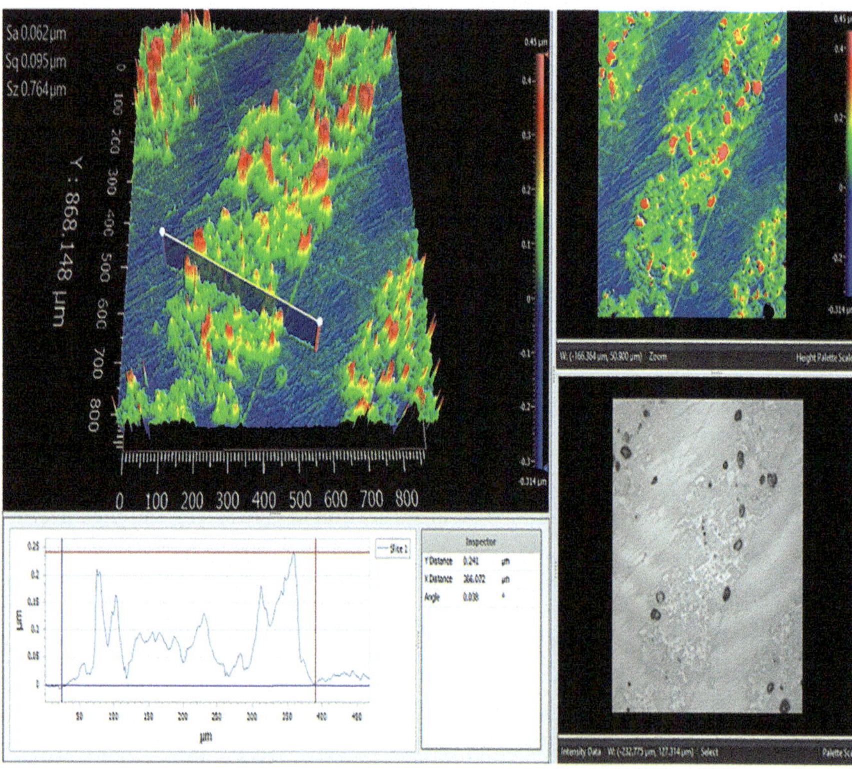

Fig. 6.13 Captured composite figure of a raw sebaceous-rich latent fingermark by OP, showing a cross section of a single ridge (top left) and the respective data profile of heights over length (bottom left). A flattened 2D image (top right) of the same fingermark area and the unprocessed image (bottom right) are also shown

providing a new insight to the degradation process that may reveal information never observed before. Changes in ridge width may be a relevant indicator of fingermark aging as the width can increase or decrease depending on the combination of environmental factors, substrate, and donor types. In 2017, an OP was utilized by Alexandra Košir to verify droplet volume of nucleic acids previously measured by digital polymerase chain reaction (dPCR) [65]. The OP was employed as a reference method to verify the heights of the droplets that were predetermined. In this study, the level of accuracy of the OP method was higher than 99.2% and was comparable to the other methods employed.

Ridge discontinuities can also be studied with an OP by measuring the heights of the ridges; if there is an area where the height drastically decreases and then increases again, it might be an indication of a discontinuity. Since the fingermark does not need to be enhanced before examination, the data are a true and accurate representation of the heights of the ridges without being skewed by the method of application and sensitivity of a developer. The OP reduces the potential detection of false discontinuities on nonporous materials which might be created by powdering.

6.9.1 Applications of OP Technology

There are numerous hypotheses, and a few clues, as to what occurs to a latent fingermark as it naturally ages, chemically and physically. One of these, named evaporation-driven degradation, supports the idea that many of its components will gradually evaporate [51] and that this will translate into a physical change in the ridges. As stated previously, it is argued that fingermarks are estimated to be about 20% water after deposition [16, 50]. This percentage of water will evaporate as the fingermark ages, but it is unclear to what may occur to the remaining 80%. Some researchers argue that these other components will eventually also evaporate [24]. As a fingermark loses its moisture, the ridges will become increasingly brittle and may result in a lower affinity to chemical or physical enhancement methods [66]. In cases of loss of adherence or efficiency, the OP clearly becomes a beneficial alternative for visualization since no pre-treatment is necessary. However, it is yet to be confirmed whether the rest of the volatile constituents evaporate as the fingermark ages. The evaporation of volatile constituents can be examined by chemical analyses to estimate the time since deposition. It has been shown that the levels of organic and inorganic compounds, such as squalene and lipids, decrease over time [66]. However, this approach is difficult to standardize because the quantities of compounds in fingermark secretions are largely dependent on donor-related factors such as diet and genetics [16]. Despite donor differences, there are some compounds that decrease in concentration almost entirely over a certain period of time. For example, some types of fatty acids, as well as wax esters, triglycerides, cholesterol, and squalene, decrease significantly over 30 days [66]. If a fingermark were to be tested for these substances and not detected in significant quantities, it would imply that the fingermark is likely more than 30 days old. Although there is some research that

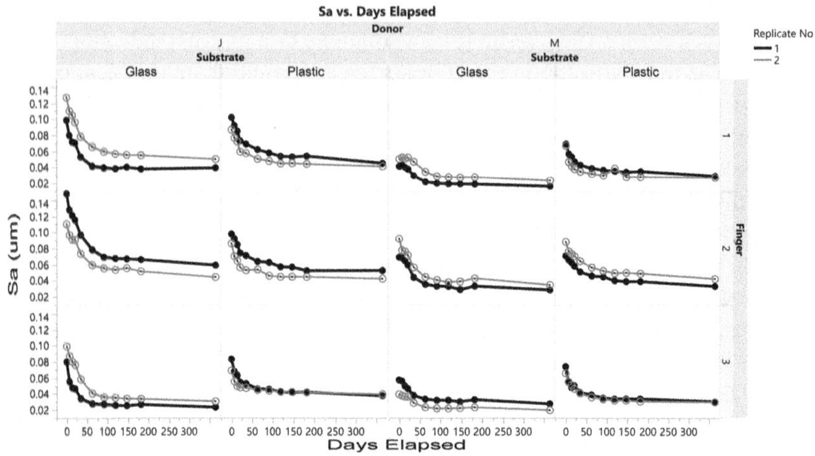

Fig. 6.14 *Sa* values over time of two replicate latent fingermarks (heights in μm) from a male (J) and female (M) deposited on glass and plastic. The values represent total average area height analyzed; reprinted with permission from John Wiley & Sons [8]

supports the hypothesis that fingermark residue evaporates with time, there is also the possibility that the ridge residues diffuse onto the furrows, or a combination of the two may as well occur (Fig. 6.14).

The hypothesis of diffusion-driven degradation of ridges is one that has not been heavily researched. Some reports have shown that the residue contained within a ridge will diffuse onto the furrows and substrate over time, as demonstrated in Fig. 6.15 [15, 16, 67, 68]. This process would result in the overall volume of the fingermark to remain very similar, or close to its original deposition. Research with ToF-SIMS showed that biomolecules could travel from the ridges into the furrows in as little as 24 h [16]; after 48 h, the ridges were completely indistinguishable once the compounds diffused into the furrows and onto the substrate [67]. Although this technology is 2D, these findings could be further supported by a 3D technology like the OP. In fact, the OP has the capability of measuring the volume of a sample; however, if the layer of diffused material is smaller than 0.02 μm in height, it could not be measured. De Alcaraz-Fossoul et al. observed a plateau in the average height of ridges after 60 days of degradation once the OP reached its 0.02 μm height detection limit [8] (Fig. 6.14).

The interaction between the fingermark and the surface has been investigated by optical and chemical approches, for example, on how fatty acids and triglycerides behave during the aging process [43, 66]. It was discovered that both compounds diffuse with no significant difference between their diffusion rates and that the substrate had the largest impact. Some researchers hypothesize that the molecular weight plays a role in the diffusion of residue and the speed of the constituents' spread [15, 16]. However, others disagree and support the idea that the interaction between the endogenous compounds of an individual's secretion and the substrate

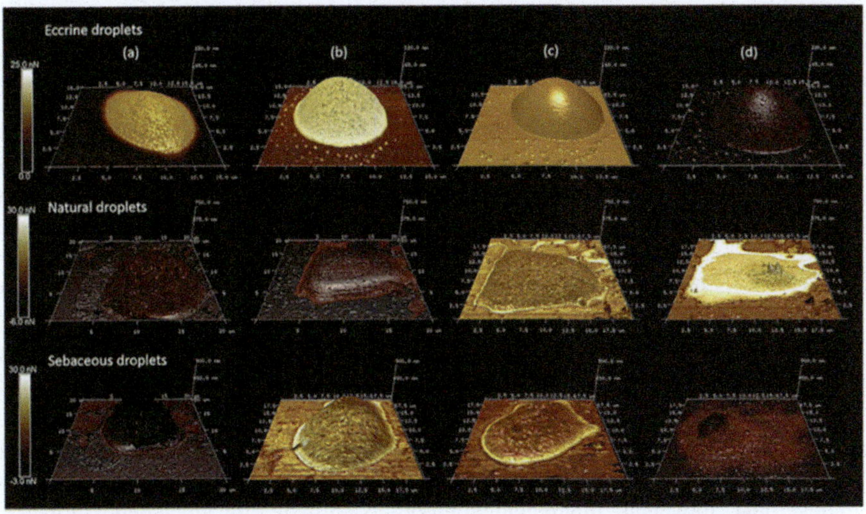

Fig. 6.15 Images of the change in adhesion and surface topography of fingermarks over time: (**a**) fresh, (**b**) 1 week old, (**c**) 2 weeks old, and (**d**) 4 weeks old. The "z" height axis has been set to be the same in each image so the change in height over time can be demonstrated; reproduced with permission from Elsevier Science & Technology Journals [15]

has the most impact [68]. Further research should be conducted to test whether the molecular weight of residue constituents play a role on the diffusion of ridge compounds.

The hypotheses of evaporation- vs diffusion-driven degradation can be further researched by exploring all the capabilities of the OP. It has already been proven that the height of ridges diminish over time [8]; thus, the next step would be to investigate whether the fingermark experiences evaporation, diffusion, or a combination of both processes. If both phenomena occur simultaneously, it will be necessary to determine which constituents are evaporating and which are diffusing and at what rate and circumstances. This could be analyzed by coupling the OP with chemical analyses by first calculating the overall volume of the sample and then measuring how much of various compounds remain in the residue, such as fatty acids, triglycerides, and squalene. OP, and other 3D technologies, can provide novel data on fingermark aging processes that could be the foundation to develop mathematical models for age estimations.

6.10 Latent Fingermark Aging: Kinetics of the Degradation Process

Recent developments, as well as past literature, have shown that the natural aging of latent fingermarks might not be a linear process but rather can be categorized, as with the decomposition of human bodies, into three discreet and distinct aging stages: fresh, intermediate, and advanced decay. Each one of the three discrete stages may be characterized by specific chemical, biological, and physical phenomena (i.e., dehydration, oxidation, bacterial growth, diffusion, etc.) that translate to a particular observable morphological change (i.e., loss of ridge edge definition, changes in ridge widths, loss of ridge continuity, etc.) that, in turn, can be described by measuring different morphometric parameters (ridge width, ridge height, total volume, etc.). These stages are applicable to latent fingermarks on nonporous surfaces and have been identified by combining existing knowledge on chemical and topographical degradation patterns. This categorization is relevant when attempting to develop a "universal" mathematical model to estimate time since deposition as a different model might have to be applied at each stage.

Chemical imaging has revealed very valuable information on the nature of fingermarks. It is a technology that can be used to increase the contrast and visual qualities of fingermarks [69]. It is usually nondestructive and helps to remove unwanted background noise such as fluorescence and texture. It requires little to no sample preparation, but it tends to take long periods of time to capture even small areas of the mark. Chemical imaging includes visualization techniques such as alternate light sources (ultraviolet light, infrared light, etc.) and optical microscopy [70]. Conventional imaging systems can collect data from a specimen in one color using a barrier filter or up to three colors using a red, green, and blue (RGB) camera. Images are recorded as a function of wavelength, and as a result, millions of spectra are created and correspond to spatial locations. The resulting image will display the sample with various color contrast [69]. Chemical imaging can be highly useful when studying the degradation of fingermarks because it may provide data about what organic and inorganic compounds are detectable during the aging process [51].

With chemical analyses, scientists might be able to support qualitative and quantitative data by the detection of certain sweat components, such as proteins, lipids, keratins, squalene, and carotenoids which can be affected over time by environmental conditions [66, 70]. Therefore, these products can become good indicators of fingermarks aging. Common analytical techniques include, but are not limited to, gas chromatography/mass spectrometry (GC-MS), high-performance liquid chromatography (HPLC), and ToF-SIMS. Unfortunately, many of these techniques are desctructive to the specimen.

In a practical setting, it is suggested age estimation be a two-step analytical process. During the first step, similar to class characterization of evidence, a fingermark would be classified into one of the aforementioned three aging stages. This classification would be dependent on the limited known information from the scene, such as the type of substrate, environmental conditions, and the donor. During the second step, a specific age estimation model for that stage would be applied to the questioned fingermark, providing a time since deposition estimate with an associated error. The three stages are distinguishable based on the visual examination of powdered and untreated fingermarks (clarity) over time (as described in Chap. 2), although more comprehensive studies are needed. These changes are physically (on the 2D and 3D visual spectrum), chemically, and biologically concurrent and may be described as follows based on current data:

1. *Fresh stage*: a "freshly" visible fingermark, enhanced or not, could be defined as the one being just deposited and up to 3–7 days old. It may show minimal morphological changes when using traditional 2D capturing methods, i.e., by cameras or scanners. However, a few minute modifications are still detectable; for example, ridges might begin to "drift," as a sign of aging, on plastic substrates [6], as soon as 3 days after deposition. At this stage, the most robust aging parameter indicating degradation (in 2D) is the number of minutiae [34]. Transfer of the sweat secretion into the furrows has been shown to occur in as little as 2 h [15] after deposition; however, this does not necessarily translate into visible topographical changes of the ridges. On the other hand, when observed in 3D, other noticeable morphological alterations can be detected. In most cases analyzed, ridge heights may decrease significantly within the first 7 days [29] after deposition, and in some instances, the loss may represent up to 50% of the initial height. This physical alteration may be overlooked if employing 2D methods alone [8].

At a chemical level, the observed modifications in 2D and 3D may be correlated with compound alterations. The quantity of squalene detected in the fingermark will fluctuate within the first 5 days [71]. Also, the concentration of cholesterol will decrease over time, and a large number of oxygenated products will be formed including cholesterol oxides and oxysterols [66]. A study suggested that a latent fingermark may lose up to 98% of its original weight within the first 3 days; this drastic loss of weight is thought to be from the evaporation of water content [72].

At a biological level, microbes' survival and growth are expected to not last beyond this initial stage due to dehydration of the fingermark and consumption of compounds (endogenous and exogenous) [50], although this would be dependent on the type of substrate and other environmental conditions. A case of interest that could be extrapolated to fingermark aging is the study of biofilms. Few studies have shown how biofilms evolve and for how long they survive on various substrates when exposed to different incubation times (1 week, 1 month, and 2 months) and light conditions (dim and ambient) [73]. Pinto et al. demonstrated that after 1 week, a biofilm could be formed and detectable on PVC, glass, and plastic substrates in both dim and ambient lighting. This discovery may provide indirect evidence on the effect of microbes in aging patterns based on how developed a biofilm is on a fingermark. Along with the ability to detect biofilms, human DNA can also be recovered within the initial 3-day period [74]. Touch DNA has been the source of major case breakthroughs in criminal investigations since it can be isolated from many different substrates after transfers from person to person, surface to surface, person to surface, or surface to person [74, 75]. It has been reported that DNA can be even recovered from a fingermark deposited on a victim's neck up to 48 h after deposition [74]. After these initial days, DNA may become too degraded to obtain full profiles due to enzymatic or other degradation processes.

At a crime scene, for a fingermark expected to be in the fresh stage, chemical or 3D analysis are the best methods to examine it. Fingermarks will not visibly appear with major significant morphological changes on a 2D spectrum within the first week; therefore, 2D analysis alone would not be sufficient for age estimation, beyond class characterization. Chemically, major molecular changes occur as soon as the fingermark is deposited, including evaporation, oxidation, or diffusion of constituents, so the chemical analysis and 3D analysis could be the best methods of examination.

2. *Intermediate stage*: a fingermark in this case would be approximately 7–30 days old based on data gathered from 2D + 3D morphometrics. This one would be visibly and noticeably different from "fresh" impressions at the 2D and 3D topographical level. In traditional 2D examinations (i.e., powdering), ridges may experience "drift"[3] [6]; over 10% of latent fingermarks appear to experience this phenomenon. The ridges may also begin to lose their defined edges and decrease affinity to powder developers, i.e., loss of color contrast [16]. In 3D, the integrity of ridges will continue to deteriorate as the volatile constituents evaporate and/or diffuse onto the surface [5]. Especially on plastic substrates, the heights of the ridges will decrease rapidly. If examined with the OP, a plateau may be observed after 25 days due to the z resolution limits of the instrument [8]. In some instances, between 8 and 32 days after deposition, the height of ridges may fluctuate before eventually plateauing [24]. This can be explained primarily by the rate of evaporation, rehydration, and diffusion of certain areas of the fingermark as it degrades. On the other hand, the width of ridges will start to drop after 7 days, although fluctuations (increases or decreases) might also be observed under certain environmental conditions [44]. The number of viable minutiae will also begin to decrease, with a notable change at 30-day post-deposition [16].

At a chemical level, the concentration of short chain fatty acids will increase over the first 15 days as longer chain fatty acids are degrading [76]. The lipid components within fingermarks, such as fatty acids, wax esters, triglycerides, squalene, and cholesterol, will also decrease significantly within the first month [77]. In fact, recent findings have shown that squalene decreases rapidly and a very minute amount will be detectable at the 7-day mark [78]. Although chemical and physical degradation continues, much less is known about biological changes.

At a biological level, although difficult to extrapolate to fingermarks, research has shown that biofilms may still be detectable at 1 month [73]. In fact, it has been reported that the biofilms are more developed than at 1 week and certain types of

[3] Refers to the phenomenon where, under certain environmental conditions, a single ridge can randomly "move" from its original position over time while the adjacent ridges may remain unaltered. As a result, a minutia at a specific location may change in appearance.

biofilms grew more than others dependent on the substrate and lighting conditions. This demonstrates that biofilms actually might survive at least a month, presumably making them also detectable on a latent fingermark. The presence and growth of microbes is indirect evidence that degradation of fingermarks may also be influenced, in part, by biological action. During this stage, any human DNA continues to degrade, dependent on the substrate type and environmental conditions. Although DNA will degrade differently due to these factors, DNA can remain on a surface even after it has been handled by another individual. It has been observed that DNA on a wallet handled by one user for a week and then handled by another individual for one more week will result in a mixed profile; the original user's DNA will still be isolated [79]. The presence of human cells (and DNA) may also have a role in fingermark degradation because of the by-products of chemical/biological degradation. The roles of human DNA and microbes in fingermark degradation are discussed in detail, respectively, in Chaps. 10 and 11.

At a crime scene, 2D + 3D examinations are the best methods for age estimation within this time frame. In this intermediate stage, the ridges will noticeably begin to change, and degradation can become more apparent depending on the type of substrate. These changes will be detectable in 2D and even more so in 3D. Chemical changes will continue to occur; however, these will not be as drastic as in the fresh stage. Therefore, it is not recommended to perform chemical analyses as the primary form of examination in this stage for dating purposes.

3. *Advanced decay stage*: these visibly aged fingermarks would be characterized for being over 25–30 days old based on 2D and 3D data. In this stage, the analysis becomes increasingly difficult as the fingermark loses identifiable features. In traditional 2D methods (e.g., powdering), many physical changes will be easily observed, including the loss of ridge definition and integrity. Regardless of the light condition and substrate type, fingermarks of the same secretion type (eccrine-rich or sebaceous-rich) may show similar histogram profiles of color contrast [31]. This may indicate that this 2D aging parameter is not sufficiently sensitive, or the most adequate, to distinguish fingermarks degrading under different conditions but, at the same time, be a good candidate for a "universal" aging parameter which is robust against certain variables. Most changes in ridge width may occur after 3 months [6]. Over a 6-month period, sebaceous-rich fingermarks may show a higher minutia count and a slower rate of degradation than their eccrine-rich counterparts, especially when deposited on glass [34]. The number of ridge discontinuities may also increase as time increases, with more discontinuities being noticed when aged on plastic [33]. In penumbra (shade), the highest number of discontinuities and the fastest rate of physical degradation may occur on plastic [33]. However, some of these morphometrical alterations may not appear on glass up to 170 days post-deposition. In 3D, after 50 days, the loss of heights of ridges on glass could plateau [8], possibly due to the z detection limit of the instrument used. The number of viable minutiae may continue to decrease beyond the 30-day mark [16]. The visual quality (clarity) of the fingermark may naturally deteriorate until the impression is no longer useful for iden-

tification and thus not of value for aging purposes. Life expectancy of the impression will be highly dependent on environmental conditions, and keeping a good record of the scene's conditions will be necessary for any attempts at age estimations.

Chemically, fingermarks with a high content of lipids may reveal no changes in color contrast after a 60-day period [80]. This indicates that the contrast between the ridges and valleys of a fresh fingermark containing a high lipid content might be the same as the contrast of an aged fingermark at 60 days. Chloride may remain in fingermarks and still be detectable after 236 days, while the quantity of saturated fatty acids might become stable after 60 days [66].

As stated previously, living microbes are suspected not to be viable/detectable very long post-deposition because of the lack of nutrients. However, Pinto et al. discovered that after 2 months of incubation, all the substrates studied were patchily covered with a biofilm and diatoms were found in a greater abundance on plastic surfaces [73]. It appears that biofilms may continue to develop long beyond the 28-day mark and therefore could assist in the interpretation of time since deposition of latent fingermarks. According to Oorschot et al. (2019) literature review, DNA can still be recovered on certain surfaces after many years, as long as the DNA was stored in ideal conditions [74]. In one study, 12 fingermarks collected at a crime scene were re-tested after 14 years. Of these 12 fingermarks, DNA was recovered in low quantity from two [81]. This experiment may indicate that the quality and quantity of DNA available for recovery may decrease over time.

At a crime scene, for a fingermark in advanced decay, 2D and 3D analyses are the most effective methods to obtain meaningful data on the degree of degradation. The loss of heights of the ridges, the increase of ridge discontinuities, and the putative diffusion of compounds are just a few of the changes that may occur in this stage. 2D + 3D examinations may also detect morphological alterations like the width of ridges, distance between ridges, minutiae count, and ridge volume, among others. There are few detectable, reported to date, chemical changes occurring at this advanced decay stage, so this type of analysis would not be recommended.

6.11 Limitations, Examiner Error, and Examiner Bias

The adaptation of new technologies is advancing research on the age determination of latent fingermarks, although operational trials, validation studies, and collaborative work with forensic laboratories to verify the recent observations and assumptions need further attention. Much more has been discovered on the factors that affect ridge degradation and on the aging parameters that could potentially be used, in a combined fashion, to estimate time since deposition. Yet, one of the major limitations to overcome is the uncertain impact of the large number of (influencing) variables, including little control over the donor. To simplify research, it is then of paramount importance to discern the relevant factors from the redundant ones. This is in an attempt to develop a "universal" mathematical model (or models) that could be applied to most conditions and donors.

When a fingermark is located at a scene, little information about the donor is available, such as biological sex, age, diet, or even possible contaminants that he/she came into contact with. Therefore, in theory, it is difficult to evaluate the pre-deposition conditions, especially in outdoor situations [82]. Despite this caveat, once a fingermark is associated to a person, some valuable information may be revealed, including age, sex, and ethnicity. Still, data on diet or the presence of contaminants will be a challenge that must be addressed when formulating mathematical models of fingermark aging. Another apparent limitation is the little information about the environmental conditions that the fingermark was exposed to post-deposition and before its discovery. However, in indoor conditions, environmental factors are relatively stable and may be easier to presume. Fortunately, factors such as the substrate, the level of exposure to natural light, the relative humidity, and the temperature can be easily traced. Several experiments have demonstrated that these are some of the most impactful factors in aging. Based on this premise, it would be feasible to design tentative formulae of aging processes. This/These model/s would have to be later improved and refined, through experimentation and application, by incorporating or excluding other influencing factors specific to the scene and donor.

In principle, and in most circumstances, fingermark age determination does not need to be precise to the minute or hour. Providing an age estimate of the questioned deposition with the appropriate measurement of certainty could be greatly useful to investigators in solving a crime. It has been discussed that the chemical examination of fingermarks in a laboratory might provide more accurate estimates than the analyses in 2D or 3D [83]. However, donor-specific characteristics may have a significant deleterious effect on the chemical composition of a fingermark and much more prone to external influences than 2D or 3D examinations. Perhaps, a combination of chemical and 2D + 3D morphometric approaches may likely provide the most accurate data on time since deposition estimates.

It has been discussed in this chapter that 2D analysis can provide critical information about a fingermark status by examining its aging parameters. However, data about the height of ridges, the volume of the sample, or other 3D features may be overlooked. Traditional 2D methods also often require enhancement methods (e.g., by powdering) in order to obtain optimal visualization. The need for fingermark enhancing is necessary in some cases, but it is highly dependent on the expertise of the examiner. Too much or too little powder and too much or too little "brushing" may have a detrimental impact on the interpretation of the age status.

Some of the limitations in 2D are overcome by 3D analyses. However, there are also limitations to 3D. As previously described, the OP has a z detection limit of about 0.02 μm in height [8]. This restricts the ability to detect any occurring degradation below this point. Because of this, it is not known whether the thin layer of secretion (ridges) plateaus at this height or degradation progresses until complete obliteration. Interestingly, beyond this point of decay, is when 2D (powdering) becomes useful, since the powder will adhere to any minute layers of material that the OP cannot detect. An example is the so-called ridge "haloes," which are caused by thin layers of residue diffusing over the furrows that may be masking the actual measurements of height of the ridges. These "haloes" are detected by powdering but often overlooked by the OP [8].

As with any scientific developments and applications, examiner error and bias may still occur with fingermark age estimations. Error is defined as an act, assertion, or belief that unintentionally deviates from what is correct, right, or true [84]. In the area of fingerprinting, practitioner error has a high potential to affect the results of an analysis. Practitioner error is a mistake or human error that can either be random or systematic and could be related to negligence or incompetence [84]. It is important to acknowledge the limitations of the comparison process during fingerprint identifications, which relate back to the examiner [84]. Especially in cases of low-quality fingermarks, two examiners may differ in their conclusions [85]. This is why it is crucial to have more than one examiner assess the same fingermarks to ensure that there are minimal chances for error or bias. Bias has been largely discussed in the literature and can occur, for example, when case information is provided to an examiner [28]. An analyst may be unintentionally influenced by knowing specific details about a case or the sample itself. Factors that will affect an examiner's bias are the case context, their emotional state, their expectations, and their hope for the outcome[4] [35]. Not only can an examiner be biased by information that is given to them or by their mentality, but fingermark analysis is also somehow subjective since it requires the final approval of a human being [36]. Examiners may erroneously support the "freshness" of a fingermark based on inspecting its rough visual quality rather than by an in-depth examination of aging parameters such as the ones suggested for age estimation [47].

Despite the limitations, error, and bias within fingermark age determinations, this subfield of forensic science can aid in reducing misattribution errors and in the prevention of wrongfully sentenced innocent people [86]. Misattribution is referred to a fingerprint that is correctly identified to a person but is inconsistent with the time frame of the crime [86]. It is unknown exactly how many people may have been incarcerated based on fingerprint evidence alone. However, it is expected that a small percentage of these individuals might have been wrongfully imprisoned due to misattribution. Some cases have already been reported in the literature where convicted suspects, such as Stephan Cowans, have been later exonerated by DNA evidence [87]. With an established fingermark aging standard, this number could be greatly decreased because we could determine whether the fingermark was deposited before, during, or after the crime. There is no doubt that the development of a mathematical standard model/s will have both public and investigative impacts. The "aging models" will strengthen the probative value of latent fingermark as evidence; speed up the judicial process involving fingermark evidence; reduce the number of evidence needed to secure or overturn a conviction; decrease the annoyance to victims (or witnesses) by not requiring them to testify; reduce the public administration

[4]A detailed NIST report "Latent Print Examination and Human Factors: Improving the Practice through a Systems Approach" was published in 2012 https://nvlpubs.nist.gov/nistpubs/ir/2012/NIST.IR.7842.pdf

spending because fewer individuals will be wrongfully incarcerated; and ultimately reduce errors in the identification of suspects [86].

6.12 Conclusion

The age determination of fingermarks is a slowly expanding subdiscipline in forensic science. Factors such as the substrate, the environment, and donor characteristics are known to affect how and by how much latent fingermarks degrade; however, much more research and funding opportunities are needed to investigate and discover the impact, interaction, and relevance of each combination of factors. The chemical signature of latent fingermarks, more than their physical appearance, is highly dependent on the donor as well as the contaminants they come in contact with, so perhaps chemical approaches could become more challenging and difficult to standardize for field applications. If a "universal" mathematical model for age estimation is to be developed, it may have to include more than one aging parameter, and from different methods of analysis, as no method alone has proven ideal for all stages of fingermark degradation: fresh, intermediate, and advanced decay.

The visual aspects of a fingermark that are altered over time include characteristics such as ridge width, ridge discontinuity, contrast between ridges and furrows, distance between ridges, ridge height, surface area, and volume. Data from these parameters can be collected with 2D technologies including MALDI, ToF-SIMS, and CWL or by simpler and more inexpensive methods like powdering and capturing an image with a conventional camera. However, ridge height and volume are two parameters that can only be measured with 3D technologies, such as the OP. 3D analysis can provide more detailed information about how ridges evolve post-deposition based on the interaction of the aforementioned parameters.

Future experimentation should primarily focus on population studies of diverse donors, such as individuals from varying ethnicities as well as from different age groups. This can help researchers to develop fingermark aging models by gathering the much needed data from individuals of different backgrounds and lifestyles. Studies should also be directed into the degradation of fingermark ridge height lower than the 0.02 μm height detection limit of the OP. Although powdering allows for the visualization of ridge heights lower than 0.02 μm, further investigations should examine how the ridges continue to degrade below this height point. Ultimately, black-box studies should be included to test the accuracy of the models developed before deployment.

Fingermark aging studies and their findings will have a great impact on criminal investigations and society. It may help to reverse wrongful convictions of individuals by incorporating the factor "time" in physical evidence [6]. Many scientific research support the basic principle that latent fingermarks follow certain degradation patterns that can be mathematically modeled; but no standard methodology yet exists for age estimations. The (un)reliability of fingermarks in court may arise when a criminal cannot be directly linked to the exact moment a crime is committed

[86]. In cases like that, the criminal justice system would save an abundance of resources on unnecessary court cases and testimonies if time since deposition could be implemented as a screening step in the fingerprint examination workflow. With new studies and more advanced technologies constantly being developed, the age estimation of latent fingermarks will become a reality sooner than later.

References

1. De Paoli G, Lewis SA, Schuette EL, Lewis LA, Connatser RM, Farkas T (2010) Photo- and thermal-degradation studies of select eccrine fingerprint constituents. J Forensic Sci 55(4):962–969. https://doi.org/10.1111/j.1556-4029.2010.01420.x
2. Gutierrez-Redomero E, Rivalderia N, Alonso-Rodriguez C, Martin LM, Dipierri JE, Fernandez-Peire MA, Morillo R (2012) Are there population differences in minutiae frequencies? A comparative study of two Argentinian population samples and one Spanish sample. Forensic Sci Int 222(1–3):266–276. https://doi.org/10.1016/j.forsciint.2012.07.003
3. Puertas M, Ramos D, Fierrez J, Ortega-Garcia J, Exposito N (2010) Towards a better understanding of the performance of latent fingerprint recognition in realistic forensic conditions. In: Proceedings of the twentieth international conference on pattern recognition, IEEE, pp 1638–1641. https://doi.org/10.1109/ICPR.2010.405
4. Gutierrez-Redomero E, Alonso-Rodriguez C, Hernandez-Hurtado LE, Rodriguez-Villalba JL (2011) Distribution of the minutiae in the fingerprints of a sample of the Spanish population. Forensic Sci Int 208(1–3):79–90. https://doi.org/10.1016/j.forsciint.2010.11.006
5. van Dam A, van Beek FT, Aalders MCG, van Leeuwen TG, Lambrechts SAG (2016) Techniques that acquire donor profiling information from fingermarks – a review. Sci Justice 56(2):143–154. https://doi.org/10.1016/j.scijus.2015.12.002
6. De Alcaraz-Fossoul J, Roberts KA, Barrot-Feixat C, Hogrebe G, Gene Badia M (2016) Fingermark ridge drift. Forensic Sci Int 258:26–31. https://doi.org/10.1016/j.forsciint.2015.11.003
7. De Alcaraz-Fossoul J, Barrot-Feixat C, Zapico SC, Mancenido M, Broatch J, Roberts KA et al (2017) Ridge width correlations between inked prints and powdered latent fingerprints. J Forensic Sci 63(4):1085–1091. https://doi.org/10.1111/1556-4029.13656
8. De Alcaraz-Fossoul J, Mancenido M, Soignard E, Silverman N (2018) Application of 3D imaging technology to latent fingermark aging studies. J Forensic Sci 64(2):570–576. https://doi.org/10.1111/1556-4029.13891
9. Cole SA, Scheck B (2018) Fingerprints and miscarriages of justice: 'other' types of error and a post-conviction right to database searching. Albany Law Rev 81(3):807–850
10. Merkel R, Gruhn S, Dittmann J, Vielhauer C, Brautigam A (2012) On non-invasive 2D and 3D chromatic white light image sensors for age determination of latent fingerprints. Forensic Sci Int 222(1–3):52–70. https://doi.org/10.1016/j.forsciint.2012.05.001
11. Merkel R, Otte K, Clausing R, Dittmann J, Vielhauer C, Brautigam A (2013) First investigation of latent fingerprints long-term aging using chromatic white light sensors. In: Proceedings of the 1st ACM workshop on information hiding and multimedia security (IH&MMSec), pp 95–104. https://doi.org/10.1145/2482513.2482519
12. Bailey MJ, Bright NJ, Croxton RS, Francese S, Ferguson LS et al (2012) Chemical characterizations of latent fingerprints by matrix-assisted laser desorption ionization, time-of-flight secondary ion mass spectrometry, mega electron volt secondary mass spectrometry, gas chromatography/mass spectrometry, X-ray photoelectron spectroscopy, and attenuated total reflection Fourier transform infrared spectroscopic imaging: an intercomparison. Anal Chem 84(20):8514–8523. https://doi.org/10.1021/ac302441y

13. Bailey MJ, Ismail M, Bleay S, Bright N, Levin Elad M et al (2013) Enhanced imaging of developed fingerprints using mass spectrometry imaging. Analyst 138(21):6246–6250. https://doi.org/10.1039/c3an01204b
14. Sapstead RM, Corden N, Hillman AR (2015) Latent fingerprint enhancement via conducting electrochromic copolymer films of pyrrole and 3,4-ethylenedioxythiophene on stainless steel. Electrochimica Acta 162:119–128. https://doi.org/10.1016/j.electacta.2014.11.061
15. Dorakumbura BN, Becker T, Lewis SW (2016) Nanochemical mapping of latent fingermarks: a preliminary investigation into the changes in surface interactions and topography over time. Forensic Sci Int 267:16–24. https://doi.org/10.1016/j.forsciint.2016.07.024
16. O Hagan A, Green S (2018) Crime scene to court: a study on finger-mark aging. Forensic Res Criminol Int J 6(6):491–503. https://doi.org/10.15406/frcij.2018.06.00249
17. National Institute of Justice (U.S.) (2018) The fingerprint sourcebook. Department of Justice, Office of Justice Programs, National Institute of Justice
18. Hicklin RA, Buscaglia J, Roberts MA (2013) Assessing the clarity of friction ridge impression. Forensic Sci Int 226(1–3):106–117. https://doi.org/10.1016/j.forsciint.2012.12.015
19. Scientific Working Group on Friction Ridge Analysis, Study, and Technology (2011) Standards for examining friction ridge impressions and resulting conclusions, ver 1.0
20. Home Office Scientific Development Branch (2006) Fingerprint Development and Imaging Newsletter: Special Edition
21. Pulsifer DP, Muhlberger SA, Williams SF, Shaler RC (2013) An objective fingerprint quality-grading system. Forensic Sci Int 231(1–3):204–207. https://doi.org/10.1016/j.forsciint.2013.05.003
22. Ulery BT, Hicklin RA, Roberts MA, Buscaglia J (2014) Measuring what latent fingerprint examiners consider sufficient information for individualization determinations. PLoS One 9(11):e110179. https://doi.org/10.1371/journal.pone.0110179.
23. Reed H, Stanton A, Wheat J, Kelley J, David L, Rao W, Smith A, Owen D, Francese S (2016) The Reed-Stanton press rig for the general of reproducible fingermarks: towards a standardised methodology for fingermark research. Sci Justice 56(1):9–17. https://doi.org/10.1016/j.scijus.2015.10.001
24. Thomas GL, Reynoldson TE (1975) Some observations on fingerprint deposits. J Phys D Appl Phys 8(6):724–729. https://doi.org/10.1088/0022-3727/8/6/017
25. Fieldhouse SJ (2015) An investigation into the effects of force applied during deposition on latent fingermarks and inked fingerprints using a variable force fingerprint sampler. J Forensic Sci 60(2):422–427. https://doi.org/10.1111/1556-4029.12661
26. Richmond S (1997) Do fingerprint ridges and characteristics within ridges change with pressure? Australian Federal Police, Forensic Services. 2004. In: Roddy A, Stosz J (eds) Fingerprint features – statistical analysis and system performance estimates. Proceedings of the IEEE 1997, vol 85, no 9, pp 1390–1421
27. Serina ER, Mockensturm E, Mote CD Jr, Rempe DA (1998) Structural model of the forced compression of the fingertip pulp. J Biomech 31(7):639–646. https://doi.org/10.1016/s0021-9290(98)00067-0
28. Maceo AV (2009) Qualitative assessment of skin deformation: a pilot study. J Forensic Ident 59(4):390–440
29. Sheets HD, Torres A, Langenburg G, Bush PJ, Bush MA (2014) Distortion in fingerprints: a statistical investigation using shape measurement tools. J Forensic Sci 59(4):1113–1120. https://doi.org/10.1111/1556-4029.12503
30. Chadwick S, Moret S, Jayashanka N, Lennard C, Spindler X, Roux C (2018) Investigation of some of the factors influencing fingermark detection. Forensic Sci Int 289:381–389. https://doi.org/10.1016/j.forsciint.2018.06.014
31. De Alcaraz-Fossoul J, Barrot-Feixat C, Tasker J, McGarr L, Stow K, Carreras-Marin C, Turbany Oset J, Gene Badia M (2016) Latent fingermark aging patterns (part II): color contrast between ridges and furrows as one Indicator of degradation. J Forensic Sci 61(4):947–958. https://doi.org/10.1111/1556-4029.13099

32. De Alcaraz-Fossoul J, Barrot-Feixat C, Zapico SC, McGarr L, Carreras-Marin C, Tasker J, Gene Badia M (2019) Latent fingermark aging patterns (part IV): ridge width as one indicator of degradation. J Forensic Sci 64:1–10. https://doi.org/10.1111/1556-4029.14018
33. De Alcaraz-Fossoul J, Barrot Feixat C, Carreras-Marin C, Tasker J, Zapico S, Gene Badia M (2017) Latent fingermark aging patterns (part III): discontinuity index as one indicator of degradation. J Forensic Sci 64:1–9. https://doi.org/10.1111/1556-4029.13438
34. De Alcaraz-Fossoul J, Mestres Patris C, Barrot Feixat C, McGarr L, Brandelli D, Stow K, Gene Badia M (2016) Latent fingermark aging patterns (part I): minutiae count as one indicator of degradation. J Forensic Sci 61(2):322–333. https://doi.org/10.1111/1556-4029.13007
35. Dror I, Rosenthal R (2008) Meta-analytically quantifying the reliability and biasability of forensic experts. J Forensic Sci 53(4):900–903. https://doi.org/10.1111/j.1556-4029.2008.00762.x
36. Dror I, Hampikian G (2011) Subjectivity and bias in forensic DNA mixture interpretation. Sci Justice 51(4):204–208. https://doi.org/10.1016/j.scijus.2011.08.004
37. Bordas L, Bonsutto J (2020) Adermatoglyphia: the loss or lack of fingerprints and its causes. J Forensic Ident 70(2):154–162
38. Kahn HS, Ravindranath R, Valdez R, Venkat Narayan KM (2001) Fingerprint ridge – count difference between adjacent fingertips (dR45) predicts upper-body tissue distribution: evidence for early gestational programming. Am J Epidemiol 153(4):338–344. https://doi.org/10.1093/aje/153.4.338
39. Bradshaw R, Wolstenholme R, Ferguson LS, Sammon C, Mader K et al (2013) Spectroscopic imaging based approach for condom identification in condom contaminated fingermarks. Analyst 138:2546–2557. https://doi.org/10.1039/c3an00195d
40. Jasuja OP, Toofany MA, Singh G, Sodhi GS (2009) Dynamics of latent fingerprints: the effect of physical factors on quality of ninhydrin developed prints – a preliminary study. Sci Justice 49(1):8–11. https://doi.org/10.1016/j.scijus.2008.08.001
41. Arora SS, Liu E, Cao K, Jain AK (2014) Latent fingerprint matching: performance gain via feedback from exemplar prints. IEEE Trans Pattern Anal Machine Intel 36(12):2452–2465. https://doi.org/10.1109/TPAMI.2014.2330609
42. Meyers A (2017) The assessment of fingerprint quality for a more effective match score in minutiae-based matching performers. West Virginia University: The Research Repository – Graduate theses, dissertations, and problem reports. https://doi.org/10.33915/etd.6226
43. De Alcaraz-Fossoul J, Mestres Patris C, Balaciart Muntaner A, Barrot Feixat A, Gene Badia M (2013) Determination of latent fingerprint degradation patterns – a real fieldwork study. Int J Legal Med 127(4):857–870. https://doi.org/10.1007/s00414-012-0797-0
44. Czech A, Szabelak A, Sowinski A (2019) Changes in fingerprints depending on physiological factors. J Forensic Sci 64(3):711–716. https://doi.org/10.1111/1556-4029.13937
45. Ahmed AA, Osman S (2016) Topological variability and sex difference in fingerprint ridge density in a sample of the Sudanese population. J Forensic Legal Med 42:25–32. https://doi.org/10.1016/j.jflm.2016.05.005
46. Fournier NA, Ross AH (2016) Sex, ancestral, and pattern type variation of fingerprint minutiae: a forensic perspective on anthropological dermatoglyphics. Am J Phys Anthropol 160(4):625–632. https://doi.org/10.1002/ajpa.22869
47. Girod A, Ramotowski A, Lambrechts A, Misrielal P, Aalders M, Weyermann C (2016) Fingermark age determinations: legal considerations, review of the literature, and practical propositions. Forensic Sci Int 262:212–226. https://doi.org/10.1016/j.forsciint.2016.03.021
48. Czech A, Gryszczyk N, Szabelak A, Sowinski A (2020) Changes in fingerprints and the quantity of material forming the print depending on hand cleanliness, gender, and ambient conditions. J Forensic Sci 65(1):84–89. https://doi.org/10.1111/1556-4029.14164
49. Fainman Y, Lenz E, Shamir J (1982) Optical profilometer: a new method for high sensitivity and wide dynamic range. Appl Opt 21(17):3200–3208. https://doi.org/10.1364/AO.21.003200
50. Kent T (2016) Water content of latent fingerprints – dispelling the myth. Forensic Sci Int 266:134–138. https://doi.org/10.1016/j.forsciint.2016.05.016

51. Bumbrah GS, Sharma RM, Jasuja OM (2016) Emerging latent fingerprint technologies: a review. Res Rep Forensic Med Sci 6:39–50. https://doi.org/10.2147/RRFMS.S94192
52. Ametek Ultra Precision Technologies (2018) Zygo Mx surface texture parameters manual, pp 1–28
53. Kuwayama K, Tsujkawa K, Miyaguchi H, Kanamore T, Togawa Iwata Y, Inoue H (2013) Time-course measurements of caffeine and its metabolites extracted from fingertips after coffee intake: a preliminary study for the detection of drugs from fingerprints. Anal Bioanal Chem 405(12):3945–3952. https://doi.org/10.1007/s00216-012-6569-3
54. Lambrechts SA, van Dam A, de Vos J, van Weert A, Sijen T, Aalders MC (2012) On the autofluorescence of fingermarks. Forensic Sci Int 222(1–3):89–93. https://doi.org/10.1016/j.forsciint.2012.05.004
55. Barnett PD, Berger RA (1976) The effects of temperature and humidity on the permanency of latent fingerprints. J Forensic Sci Soc 16(3):249–254. https://doi.org/10.1016/s0015-7368(76)71068-5
56. Akiba N, Kuroki K, Kurosawa K, Tsuchiya K (2018) Visualization of aged fingerprints with an ultraviolet laser. J Forensic Sci 63(2):556–562. https://doi.org/10.1111/1556-4029.13588
57. Payne IC, McCarthy I, Almond M, Baum J, Bond J (2014) The effect of light exposure on the degradation of latent fingerprints on brass surfaces: the use of silver electroless deposition as a visualization technique. J Forensic Sci 59(5):1368–1371. https://doi.org/10.1111/1556-4029.12524
58. Fischer R, Vielhauer C (2012) Forensic ballistic analysis using a 3D sensor device. The 14th ACM Workshop on Multimedia and Security, pp 67–76. https://doi.org/10.1145/2361407.2361418
59. Bolton-King RS, Evans O, Smith CL, Painter JD, Allsop DF, Cranton WM (2010) What are the prospects of 3D profiling systems applied to firearms and toolmark identification? AFTE J 42(1):23–33
60. Bachrach B (2002) Development of a 3D-based automated firearms evidence comparison system. J Forensic Sci 47(6):1253–1264. https://doi.org/10.1520/JFS15557J
61. Senin N, Groppetti R, Garfano L, Fratini P, Pierni M (2006) Three-dimensional surface topography acquisition and analysis for firearm identification. J Forensic Sci 51(2):282–295. https://doi.org/10.1111/j.1556-4029.2006.00048.x
62. Dektak stylus profiler (2020) Brutak website. https://www.bruker.com/products/surface-and-dimensional-analysis/stylus-profilometers.html
63. Petraco NDK, Chan H, De Forest PH, Diaczuk P, Gambino C et al (2012) Application of machine learning to toolmarks: statistically based methods for impression pattern comparisons. NCJ Report 239048. pp 1–101
64. Optical profiler basics (2020) Zygo Corporation website. https://www.zygo.com/?/met/profilers/opticalprofilersabout.htm
65. Kosir AB, Divieto C, Pavsic J, Pavarelli S, Dobnik D, Dreo T, Bellotti R, Sassi MP, Zel J (2017) Droplet volume variability as a critical factor for accuracy of absolute quantification using droplet digital PCR. Anal Bioanal Chem 409(28):6689–6697. https://doi.org/10.1007/s00216-017-0625-y
66. Cadd S, Islam M, Manson P, Bleay S (2015) Fingerprint composition and aging: a literature review. Sci Justice 55(4):219–238. https://doi.org/10.1016/j.scijus.2015.02.004
67. Muramoto S, Sisco E (2015) Strategies for potential age dating of fingerprints through the diffusion of sebum molecules on a nonporous surface analyzed using time-of-flight secondary ion mass spectrometry. Anal Chem 87(16):8035–8038. https://doi.org/10.1021/acs.analchem.5b02018
68. O'Neill KC, Jin Lee Y (2018) Effect of aging and surface interactions on the diffusion of endogenous compounds in latent fingerprints studied by mass spectrometry imaging. J Forensic Sci 63(3):708–713. https://doi.org/10.1111/1556-4029.13591
69. Exline DL, Wallace C, Roux C, Lennard C, Nelson MP, Treado PJ (2003) Forensic applications of chemical imaging: latent fingerprint detection using visible absorption and luminescence. J Forensic Sci 48(5):1047–1053. https://doi.org/10.1520/JFS2002333

70. Andersson PO, Lejon C, Mikaelsson T, Landstrom L (2017) Towards fingermark dating: a Raman spectroscopy proof-of-concept study. ChemistryOpen 6(6):706–709. https://doi.org/10.1002/open.201700129
71. Dorakumbura BN, Busetti F, Lewis SW (2020) Analysis of squalene and its transformation by-products in latent fingermarks by ultrahigh-performance liquid chromatography-high resolution accurate mass Orbitrap mass spectrometry. Forensic Chem 17(100193). https://doi.org/10.1016/j.forc.2019.100193
72. Daluz HM (2014) Fundamentals of fingerprint analysis, 2nd edn. CRC Press, New York
73. Pinto M, Langer TM, Huffer T, Hofmann T, Herndl GJ (2019) The composition of bacterial communities associated with plastic biofilm differs between different polymers and stages of biofilm succession. PLoS One 14(6):e0217165. https://doi.org/10.1371/journal.pone.0217165
74. Oorschot RAH, Szkuta B, Meakin GE, Kokshoorn B, Goray M (2019) DNA transfer in forensic science: a review. Forensic Sci Int Genet 38:140–166. https://doi.org/10.1016/j.fsigen.2018.10.014
75. Kanokwongnuwut P, Kirkbride KP, Linacre A (2018) Detection of latent DNA. Forensic Sci Int Genet 37:95–101. https://doi.org/10.1016/j.fsigen.2018.08.004
76. Archer N, Charles Y, Elliott J, Jickells S (2005) Changes in the lipid composition of latent fingerprint residue with time after deposition on a surface. Forensic Sci Int 154(2–3):224–239. https://doi.org/10.1016/j.forsciint.2004.09.120
77. Croxton R, Baron M, Butler S, Kent T, Sears V (2010) Variation in amino acid and lipid composition in latent fingerprints. Forensic Sci Int 199(1–3):93–102. https://doi.org/10.1016/j.forsciint.2010.03.019
78. Mountfort KA, Bronstein H, Archer N, Jickells SA (2007) Identification of oxidation products of squalene in solution and in latent fingerprints by ESI-MS and LC/APCI-MS. Anal Chem 79(7):2650–2657. https://doi.org/10.1021/ac0623944
79. Raymond JJ, Oorschot RAH, Walsh SJ, Gunn PR (2009) Trace DNA and street robbery: a criminalistic approach to DNA evidence. Forensic Sci Int Genet Supp Ser 2(1):544–546. https://doi.org/10.1016/j.fsigss.2009.08.073
80. Matuszewski S (2015) Age-dependent changes of contrast in fingermarks of various lipid content. Prob Forensic Sci 101:5–13
81. Romano CG, Mangiaracina R, Donata L, D'Angelo R, Scimone C, Sidota A (2019) Aged fingerprints for DNA profile: first report of successful typing. Forensic Sci Int 302. https://doi.org/10.1016/j.forsciint.2019.109905
82. Matei A (2019) Chapter 4: end user commentary on novel technological applications for latent and blood-stained fingermark aging studies. In: Francese S (ed) Emerging technologies for the analysis of forensic traces. Springer, Cham, pp 67–74
83. Oonk S, Schuurmans T, Pabst M, de Smet LCPM, de Puit M (2018) Proteomics as a new tool to study fingermark ageing in forensics. Sci Rep 8(16425):1–11. https://doi.org/10.1038/s41598-018-34791-z
84. Christensen AM, Crowder CM, Ousley SD, Houck MM (2014) Error and its meaning in forensic science. J Forensic Sci 59(1):123–126. https://doi.org/10.1111/1556-4029.12275
85. Dror IE, Thompson WC, Meissner CA, Kornfield I, Krane D, Saks M, Risinger M (2015) Context management toolbox: a linear sequential unmasking (LSU) approach for minimizing cognitive bias in forensic decision making. J Forensic Sci 60(4):1111–1112. https://doi.org/10.1111/1556-4029.12805
86. De Alcaraz-Fossoul J, Barrot-Feixat C, Roberts KA (2015) The paradigm of fingerprint age determination. J Forensic Sci 1(1):006
87. Cole SA (2007) More than zero: accounting for error in latent fingerprint identification. J Crim Law Crim 95(3)

Chapter 7
Latent Fingermark Aging: Chemical Degradation Over Time

Amanda A. Frick, Aline Girod-Frais, Ana Moraleda, and Céline Weyermann

Abstract This chapter presents an overview of the current knowledge regarding the chemical composition of fingermarks and its modification as a function of time, focusing primarily on the compounds targeted in fingermark detection and/or age estimation purposes. As numerous studies have been published regarding these constituents in various contexts, this discussion will pertain mainly to fingermarks and human skin secretions, with some in vitro experiments where relevant additional information is provided.

Thus, chemical degradation of the main eccrine (e.g., water, amino acids, proteins, lactic acid) and sebaceous (e.g., fatty acids, triglycerides, wax esters) components will be detailed, including illustrations of molecular structures, degradation mechanisms, and products. Four main factors affecting the chemical degradation of fingermark components will also be described, namely, the initial chemical composition (i.e., variations due to interactions between fingermark components, donors' characteristics, and transfer conditions), the nature of the substrate, the environment (i.e., variations due to light, air flow, water and ambient humidity, temperature, and microbial activity), and time. Finally, key elements to be considered when conducting fingermark aging studies will be discussed.

Keywords Eccrine · Sebaceous · Amino acids · Lipids · Fatty acids · Squalene · Oxidation · Ozonolysis · Donor variation · Deposition · Environment · Fingerprints

---

A. A. Frick (✉)
École des Sciences Criminelles, University of Lausanne, Lausanne, Switzerland

Department of Applied Sciences, Northumbria University, Newcastle upon Tyne, UK
e-mail: amanda.frick@northumbria.ac.uk

A. Girod-Frais
Criminal Intelligence Service (Bundeskriminalamt), Crime Scene Unit, Vienna, Austria

Department of Criminal Law and Criminology, University of Vienna, Vienna, Austria
e-mail: aline.girod-frais@bmi.gv.at; aline.girod-frais@univie.ac.at

A. Moraleda · C. Weyermann
Ecole des Sciences Criminelles, University of Lausanne, Lausanne, Switzerland
e-mail: anabelen.moraledamerlo@unil.ch; celine.weyermann@unil.ch

J. De Alcaraz-Fossoul (ed.), *Technologies for Fingermark Age Estimations: A Step Forward*, https://doi.org/10.1007/978-3-030-69337-4_7

Acronyms

ChOOH	Hydroperoxycholesterol
FA	Fatty Acid
FTIR	Fourier Transform Infrared
GC-MS	Gas Chromatography-Mass Spectrometry
LC-MS	Liquid Chromatography-Mass Spectrometry
MALDI-MS	Matrix-Assisted Laser Desorption/Ionization-Mass Spectrometry
ORO	Oil Red O
SQ	Squalene
TG	Triglyceride
TOP	4,8,13,17,21-Tetramethyl-Octadeca-4,8,12,16,20-Pentaene-al
TOT	4,9,13,17-Tetramethyl-Octadeca-4,8,12,16-Tetraeneal
TTT	5,9,13-Trimethyl-Tetradeca-4,8,12-Triene-al
UHPLC-HRAM Orbitrap™ MS	Ultrahigh-Performance Liquid Chromatography-High-Resolution Accurate Mass Orbitrap™ Mass Spectrometry
UPLC-IMS-QToF-MS^E	Ultra-performance Liquid Chromatography-Ion Mobility Spectroscopy-Quadrupole Time-Of-Flight Mass Spectrometry

7.1 Introduction

After deposition on a surface, the residue that constitutes a latent fingermark is susceptible to a number of chemical (as well as physical) modifications. The nature and rates of these processes can have a significant impact on the persistence of target compounds, and hence detectability, but may also provide a basis for estimating the age of a latent fingermark. The idea of estimating fingermark age dates back at least to the early twentieth century [1], with the first reported fingermark dating approaches published in the 1960s (to the best of the authors' knowledge) [2, 3]. However, early propositions based on the study of chemical modifications did not show reproducible results [3–8] or were often based on facile observations such as the appearance of a fingermark after powdering [9–11]. In addition to being unreliable, methods based on such physical modifications of fingermarks have little to no basis in any understanding of the relevant aging processes involved, and the fact that excellent ridge quality may be observed from old fingermarks can be difficult to explain in an informed manner [12, 13]. With continuing advances in analytical technology, several investigations into the chemical changes in latent fingermark composition over time have been reported in recent decades, with a focus on several broad goals:

1. Understanding how aging and environmental factors impact the detection of latent fingermarks [14–17]
2. Identification of stable products as potential targets for fingermark detection [14, 16, 18]
3. Modeling of target compound aging, toward the development of a reliable method for latent fingermark age estimation [19–26]

This chapter provides an overview of the current state of the literature regarding chemical modifications undergone by latent fingermark constituents as a function of time, with a primary focus on compounds of particular relevance to fingermark detection and/or studies into the development of methods for age estimation. This is followed by a discussion of the factors that can influence fingermark residue degradation and so must be taken into account in any attempt to estimate fingermark age. Owing to the substantial number of compounds present in latent fingermarks and the large volume of information regarding their reactivity in various matrices, this overview will focus on studies pertaining to latent fingermarks and human skin secretions, although discussion of in vitro experiments is included where appropriate.

7.2 Chemical Degradation of Eccrine Components

The earliest reported works into the effects of aging on fingermark composition focused on the eccrine components targeted by detection methods, i.e., chlorides [3], as well as amino acids and other nitrogenous components such as urea [27]. The key aim of Cuthbertson and Morris' investigations was to investigate the effects of compositional variation on fingermark detection, but the possibility of estimating age was also highlighted, based on changes detected in the concentrations of these compounds as a function of time following deposition.

The aging of eccrine components is of particular relevance due to the fact that eccrine glands are present along the ridges of the fingertips, and so their secretions are the most likely to be present in latent fingermarks. Besides water, salts, and small organic compounds, eccrine glands secrete several large organic molecules (e.g., peptides and proteins), the behavior of which over time may provide interesting results for the development of fingermark dating methodologies, particularly over longer time periods (e.g., up to several months or even years). The following sections describe several of the main eccrine compounds found in fingermark residue and their degradation processes (where known).

7.2.1 Water

While not strictly a chemical process, the evaporation of water is one of the first processes that occurs after fingermark deposition, and its loss thereby alters the initial chemical composition of the latent residue, as well as its physical properties. Water comprises approximately 99% of eccrine sweat, but the actual water content of latent fingermarks can vary considerably (20–70%), depending on factors such as the amount of eccrine secretions present on the ridge skin at the time of deposition and the presence of other constituents such as sebaceous lipids or exogenous substances such as food [28, 29].

A rapid decrease in mass that occurs soon after deposition has been observed in several investigations into fingermark composition and aging and has often been attributed primarily to the volatilization of water, as well as low molecular weight organic constituents [16, 30, 31]. Recently, Keisar et al. utilized temperature-programmed desorption mass spectrometry in combination with mass measurements using a quartz crystal microbalance to confirm whether this short-term mass loss was due solely to water evaporation [29]. When fingermarks were kept at 37 °C, water evaporation occurred within minutes following deposition, with samples from unwashed hands (containing some lipid material) exhibiting a slower mass loss rate.

7.2.2 Amino Acids and Proteins

Free amino acids, which are the targets of several fingermark detection methods such as ninhydrin and 1,2-indanedione, are often considered to be relatively stable fingermark components. Fingermark detection with amino acid-sensitive reagents has been successfully carried out on porous substrates known to be several decades to over a century old [9, 32]. It is therefore widely assumed that as long as paper substrates are not exposed to elevated humidity or immersion in water, thereby dissolving the amino acids from the latent residue, the amino acids remain stably bonded to the cellulose fibers [32]. However, work by Cuthbertson and Morris indicates that (at least some) amino acids do undergo some kind of degradation process over time, as the total mass of amino acids decreased significantly over 236 days after deposition on aluminum [27]. The reaction pathways involved in this process have not yet been identified. This degradation process is considered a slow one, as the effectiveness of amino acid-sensitive reagents on older specimens has been proven several times [33].

Other studies into amino acid degradation in latent fingermarks have focused on the effects of temperature, in order to simulate the effects of extreme temperatures (400–500 °C) that would be produced during combustion, e.g., in cases of arson or explosions [18]. Under these conditions, alanine was observed to undergo condensation to produce dimethylpiperazine-2,5-dione, while aspartic acid decomposed to form maleimide and 2,5-furandione. Pyrolysis products of other abundant eccrine amino acids, such as serine, glycine, and lysine, were not observed. De Paoli et al. similarly observed that amino acids (aspartic acid, glutamic acid, glycine, histidine, ornithine, serine, and threonine) in solution underwent thermal degradation at 100–150 °C but were unable to detect any degradation products using liquid chromatography-mass spectrometry (LC-MS) [34].

van Dam et al. proposed a method of age estimation based on the autofluorescence of aged fingermarks, which was attributed to an unspecified complex formed between tryptophan-containing proteins and oxidized lipid material [35]. Later investigations utilizing thin layer chromatography indicated that protein-bound tryptophan in fact degrades to produce a number of fluorescent derivatives, including indoleacetic acid, β-carbolines, and xanthurenic acid (Fig. 7.1), possibly as a result of bacterial activity [36].

In an investigation into compositional variation, van Helmond et al. tentatively identified a series of endogenous amino acid degradation products in fingermarks, including urocanic acid, pyroglutamic acid, and 4-methylene-l-glutamine, based on accurate mass measurements [37]. Urocanic acid is produced by the deamination of histidine, while the other two are derived from glutamine. It was observed that the relative quantities of these compounds were greater in natural fingermarks (samples collected without prior preparation of the fingertip skin by cleaning and can include substances from incidental touching of body parts, objects, and other chemical/biological products) than eccrine samples. This may be indicative of the longer time the precursor amino acids had been present on the fingertips, compared to the eccrine samples, indicating that these possible degradation processes take place on the skin surface [37].

Until recently, fingermark proteins remained overlooked in comparison to free amino acids. Oonk et al. studied the aging of target proteins utilizing a proteomics-based approach [21]. From a pool of identified proteins, five, including several keratins and dermcidin, were selected as potential markers for fingermark age determination. Dermcidin was found to decrease in signal over 16 days at ambient temperature (20 °C). Conversely, the keratins showed a relative increase, which was

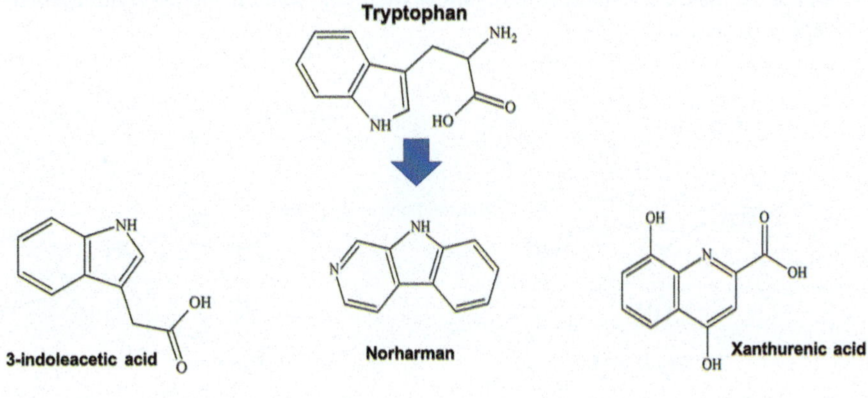

Fig. 7.1 Molecular structures of tryptophan and its fluorescent degradation products

attributed to the degradation of less stable protein constituents. It should be noted that the investigation of degradation pathways was beyond the scope of this study, and so no degradation products have yet been identified.

7.2.3 Other Eccrine Components

There are a number of other eccrine constituents that undergo degradation, though many of these are less relevant to fingermark detection than those described above. One exception is lactic acid, which is a target of cyanoacrylate fuming. De Paoli et al. reported that lactic acid photooxidizes to pyruvate at 50 °C with simulated sunlight exposure (Fig. 7.2) [34]. Both lactic acid and pyruvate were shown to decompose at 50 °C. Wargacki et al. inferred that this process is responsible for the poor performance of cyanoacrylate fuming on fingermarks that have been exposed to light and air for 1 day following deposition [38]. However, further studies on the degradation of lactic acid in fingermark residue and its effect on detection techniques have not been reported, and so this hypothesis is not yet confirmed.

Ammonia and urea were part of an investigation into fingermark composition and degradation in the 1970s, as it was thought that nitrogen-containing components might be chemically altered by bacteria and hence provide a means to estimate fingermark age [2]. Cuthbertson and Morris observed that both compounds underwent a quantitative decrease over 236 days [27]. However, this degradation did not appear to be reproducible, due to sample variation and donor-dependent degradation rates. More recently, De Paoli et al. reported that urea standards were thermally degraded at 150 °C but appeared unaffected by photodegradation [34].

Generally, studies on the degradation processes of the eccrine components, even of key target compounds for the development of detection or age estimation methods, are rather limited. Latent fingermarks contain a number of other eccrine constituents in addition to those discussed above, such as B-complex vitamins [6], phenol [39], uric acid, and creatinine [40]. Thus, further research focusing on these components may be of interest in regard to their degradation and potential markers for age estimation.

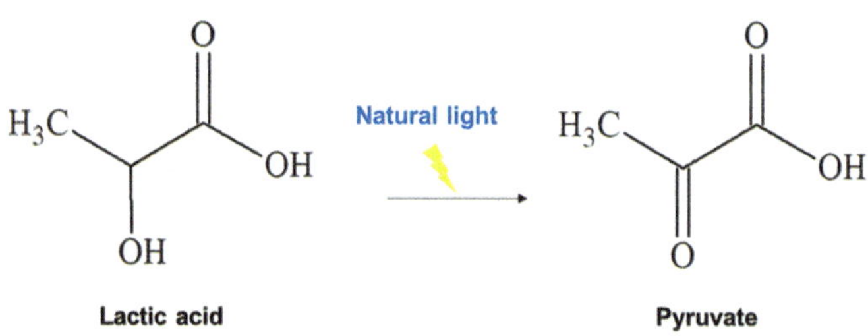

Fig. 7.2 Molecular structures of lactic acid and pyruvate

7.3 Chemical Degradation of Sebaceous Components

The aging processes of the water-insoluble fraction of latent fingermarks, of which lipids are a subgroup, are relevant to a variety of fingermark detection techniques. This is particularly true in instances where substrates have been wetted, and eccrine constituents may have been removed [41, 42]. Sebum is a major source of fingermark lipids and contains a number of unsaturated compounds that are susceptible to chemical modifications soon after deposition, making them attractive as potential targets for age estimation. Many studies into fingermark aging have focused on specific sebaceous lipids or lipid classes, such as squalene, fatty acids, and glycerides [15, 23, 24, 43, 44]. However, it must be noted that the sebaceous lipids in fingermark residue originate from transfer through physical contact of the fingertips with body parts such as the face or hair; thus, the amount present can vary to a greater extent than eccrine compounds. This may explain some of the (quantitative) variability observed among fresh fingermark residue from the same person (intraindividual variation) and is an important issue to overcome for age estimation purposes.

Concerning the general behavior over time of lipids found in fingermark residue, Mong et al. observed that older fingermark residue appears to solidify and therefore become less receptive to staining by histological dyes [45]. These changes are consistent with the increase in viscosity that occurs with the progressive oxidation of various unsaturated sebaceous lipids, including squalene and glycerides, as well as the evaporation of volatile organic compounds produced by these processes.

Outside of the forensic literature, indications as to the degradation processes undergone in latent fingermark residue may be found in fields such as dermatology and air quality research. Oxidation of skin surface lipids, including squalene, cholesterol, and unsaturated fatty acids, has been observed to take place in situ [46–50] and provides a complement to the more preliminary information gained so far based on fingermark samples. The following sections provide an overview of the current state of knowledge regarding the degradation of the sebaceous lipids present in fingermark residue.

7.3.1 *Free Fatty Acids (FAs) and Triglycerides (TGs)*

A study by Archer et al. reported that most free fatty acids in latent fingermarks appeared to undergo an initial increase in concentration, followed by a subsequent decrease over 33 days following sample collection [15, 51]. They hypothesized that there might be competing mechanisms of fatty acid generation and degradation, such as the hydrolysis of triglycerides; however, neither glycerides nor proposed degradation products could be identified using gas chromatography-mass spectrometry (GC-MS). Other researchers have reported the degradation of unsaturated free fatty acids with increasing fingermark age over comparable aging times [16, 23];

however, the observation described by Archer et al. has not been replicated. It should be noted that this trend was not significant in a majority of the samples [15]. Weyermann et al. found high variability in free fatty acid concentrations over 1 month, and thus, no significant trends could be determined [44], while other investigations have described that unsaturated fatty acids appear to be significantly affected by aging processes [16, 52].

Mong et al. proposed that "air oxidation" of unsaturated fatty acids involved cleavage at the double bond [16] after observing that aged fingermarks contained several short-chain fatty acids as well as dicarboxylic acids. As an example, they suggested nonanoic acid and nonanedioic acid as oxidation products of oleic acid (FA 18:1Δ9). Work by Pleik et al. identified similar degradation products derived from FA 16:1 and FA 18:1 over 14 days of aging using GC-MS but pointed out that cleavage at the double bond yielded different degradation products in fingermark samples compared to the Δ9 unsaturated fatty acid standards [23]. These results enabled them to definitively conclude that the fatty acids in latent fingermarks were sapienic acid (FA 16:1Δ6) and FA 18:1Δ8. Ozonolysis of these fatty acids results in the formation of decanal and oxohexanoic and oxooctanoic acid (from FA 16:1Δ6 and FA 18:1Δ8, respectively). These can be oxidized further to form decanoic acid, as well as dicarboxylic acids (i.e., hexanedioic acid and octanedioic acid) [23]. Figure 7.3 provides a summary of this process. Similar products will be formed from other unsaturated free fatty acids, depending on the total chain length and double bond position.

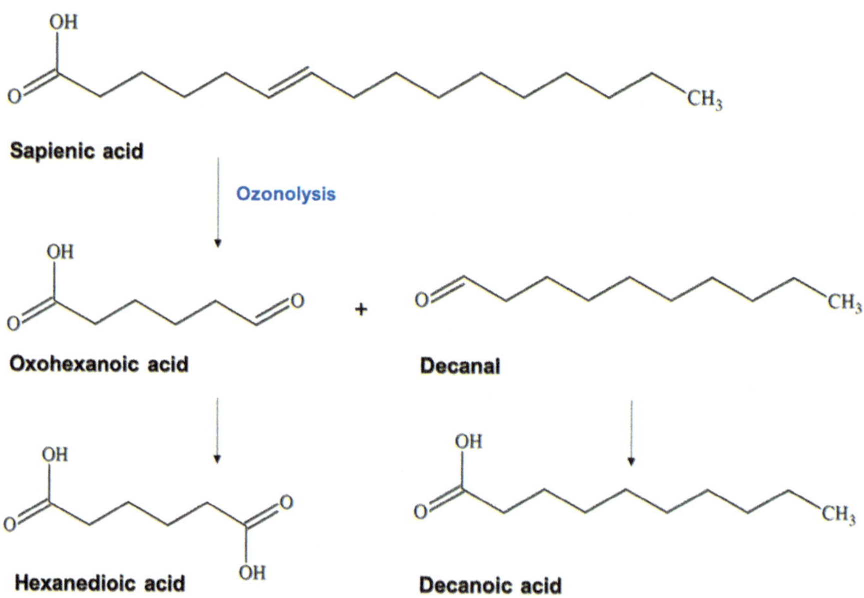

Fig. 7.3 Degradation of sapienic acid (FA 16:1Δ6)

It has been proposed that the oxidation of fingermark glycerides occurs via similar processes to drying oils such as linseed oil, which undergoes autoxidation to form hydroperoxides, as well as polymerization [16], though these processes have never been directly observed in fingermark residue. Johnston and Rogers inferred, based on Fourier transform infrared (FTIR) microspectroscopy measurements of fingermarks exposed to elevated temperatures, that increases in the OH stretch band may be due in part to the oxidation of unsaturated triglycerides, in addition to other lipids [52]. However, hydroperoxide formation is favored in polyunsaturated fatty acids (the high "drying" rate of linseed oil is due to its high content of linolenic acid (FA 18:3)) [53, 54]. In human sebum, monounsaturated and saturated fatty acids are predominant, with polyunsaturated fatty acids (primarily sebaleic acid and linoleic acid) being present in trace amounts [55]. Therefore, fatty acid hydroperoxides are unlikely to be major fingermark degradation products [52]. The peroxidation rate of triglycerides is reported to be significantly lower relative to that of other skin surface lipids such as squalene and cholesterol (see Sects. 7.3.2 and 7.3.3) [49]. Zhou et al. noted that triolein did not undergo degradation in vitro in the absence of ozone [56]. This is consistent with the fact that most abundant unsaturated fatty acids in human skin surface lipids are monounsaturated and so will be more readily oxidized by ozone [46, 56].

It is accepted that atmospheric ozone reacts with a double bond to ultimately produce an ozonide as described by the Criegee mechanism (Fig. 7.4) [57, 58]. In brief, ozone attacks the double bond to form a primary ozonide (1,2,3-trioxolane), which almost immediately decomposes to produce a carbonyl and a zwitterion known as a Criegee intermediate. In the absence of a protic solvent (such as alcohols or water), these two intermediates recombine to form a more stable secondary ozonide (1,2,4-trioxolane). Complete ozonolysis of a polyunsaturated species is progressive, with monoozonides being the first oxidation products formed, before eventual total conversion to the fully oxidized compound [24, 56, 58].

Unsurprisingly, the identified degradation processes of fingermark glycerides are largely similar to the free fatty acids. Unsaturated sebaceous triglycerides can contain up to nine double bonds, providing multiple target sites for oxidative processes [59–61]. While ozonolysis products of diglycerides have not been observed in fingermark samples, it has been demonstrated using standard solutions that they also undergo the same reaction [43]. Pleik et al. first identified ozonolysis as a major aging process for unsaturated fingermark triglycerides under ambient atmospheric conditions through monitoring the degradation of a single triglyceride (TG 48:1) to its ozonide [24]. Frick et al. were able to identify several monoozonides and diozonides of several triglycerides in fingermarks collected from a small sample population, using ultra-performance liquid chromatography-ion mobility spectroscopy-quadrupole time-of-flight mass spectrometry (UPLC-IMS-QToF-MS^E) [62]. Additionally, they were able to tentatively identify other ozonolysis products resulting from the cleavage of the primary triglyceride ozonides to form aldehyde and carboxylic acid derivatives (summarized in Fig. 7.5).

In the presence of protic solvents, the Criegee intermediates can react with these rather than recombining with the carbonyl intermediate [46, 57, 58, 63]. With

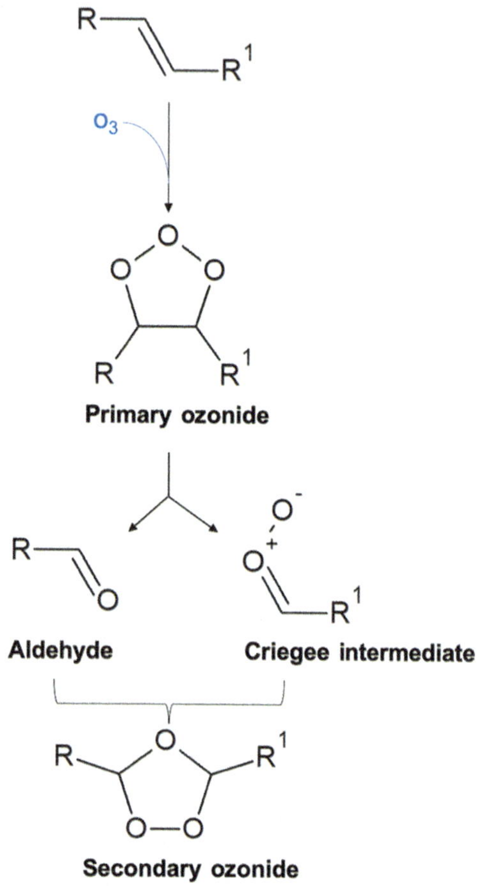

Fig. 7.4 Criegee mechanism of ozonolysis

increasing numbers of double bonds, there are therefore more instances where an unsaturated lipid may undergo side reactions before secondary ozonide formation is completed. Lower molecular weight degradation products are formed from the decomposition products of the primary ozonide through reactions with water [64–67]. Aldehyde and carboxylic acid derivatives of fingermark triglycerides were identified in addition to their ozonides via UPLC-IMS-QToF-MS^{E} [62]. Such structures were also reported by Hinners et al. via matrix-assisted laser desorption/ionization mass spectrometry (MALDI-MS) but were attributed to in-source decomposition of secondary ozonides rather than fingermark aging [43]. Sun et al. reported that similar decomposition products were formed as minor products of ozonolysis of a triolein standard [58]. Zhou et al. observed that ozonide yields were much lower when ozonolysis was performed at high humidity (50% RH) compared to low humidity conditions [56], which would be an important factor to consider in the age estimation of fingermarks based on the oxidation of unsaturated lipids.

In summary, with increasing fingermark age, the lipid fraction becomes predominantly saturated, as the unsaturated glycerides and free fatty acids are rapidly depleted by ozonolysis [24, 43, 62]. This change in lipid composition occurs soon

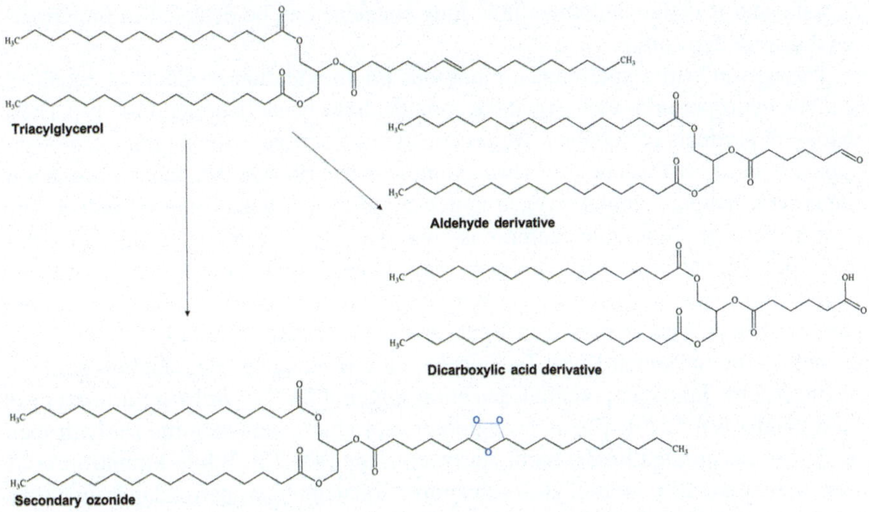

Fig. 7.5 Example molecular structures of degradation products of triglycerides

after deposition and has been reported to take place within the first few days following deposition on nonporous substrates [24, 43]. The resulting aged residue is therefore more viscous, consistent with the "waxy" residue described by Mong et al. [16]. Oxidized fingermark residue would therefore be less compatible with detection methods such as lipophilic dyes (which partition into lipid residue), powdering (which relies upon mechanical adhesion to the fingermark), and methods which target the double bond functional group, such as iodine fuming. Due to the high triglyceride content of human sebum and the high number of unsaturated constituents, this degradation process has potential as a means to estimate the age of a latent fingermark [24, 43].

7.3.2 *Squalene*

Due to its abundance in human sebum and its high degree of unsaturation, squalene is considered to be the most significant skin surface lipid in terms of chemical reactivity, acting as a major scavenger for both singlet oxygen and ozone. Once exposed to the environment of the skin surface, squalene is exposed to numerous other stressors, including UV radiation and skin flora, culminating in the formation of a wide variety of volatile degradation products and intermediates [46, 49]. The oxidation of squalene also occurs in latent fingermarks; the rapid degradation of this compound is one of the most prevalent chemical changes reported in investigations into fingermark aging [14–16, 44, 68]. The longevity of squalene in latent fingermarks has been reported in the range of 7 days to over a month, depending on factors such as substrate and storage conditions (discussed in greater details in Sect. 7.4). No further studies on squalene have been reported covering longer aging periods, and it is thus unknown how long squalene can be detected in fingermark residue after deposition.

Through in vitro experiments, numerous photooxidation products of squalene, namely, hydroperoxides and squalene epoxide, have been identified [14, 47]. These oxidation products of squalene are produced via reactions with singlet oxygen and appear to be short-lived both in vitro and in fingermarks. On the skin surface and in fingermark residue, singlet oxygen is generated by UVA exposure of endogenous photosensitizers. A notable example is coproporphyrin, which is produced by skin flora [69], though it is unknown how long these compounds would last in latent fingermarks. However, squalene peroxidation can also occur in the absence of a photosensitizer when exposed to light, albeit at a slower rate [14]. A study conducted by the Savannah River Technical Centre investigated degradation products of fingermark lipids as potential detection targets [70, 71], utilizing a mixture of lipid standards [45, 70]. The main degradation products were identified as hydroperoxides of unsaturated lipids, particularly squalene [45, 72]. While Mountfort et al. were able to identify more highly substituted squalene hydroperoxides from oxidation of a standard solution after 1 day, only the monohydroperoxides (SQ-[OOH]) were detected by Mountfort et al. and Dorakumbura et al. in fingermark samples up to 7 days old [14, 73]. Furthermore, it was observed in both studies that the formation of squalene oxidation products is very rapid, as SQ-[OOH] and SQ-epoxide can also be detected in freshly deposited fingermarks.

Squalene is also progressively degraded via ozonolysis, yielding a plethora of short-chain degradation products in both the gas and condensed phases [46, 74]. Ozonolysis products which themselves contain double bonds continue to oxidize further to secondary and tertiary degradation products, namely, volatile ketones and aldehydes [46, 75]. Four primary squalene ozonolysis products (6,10-dimethyl-5,9-undecadien-2-one (geranyl acetone), 5,9,13-trimethyl-tetradeca-4,8,12-triene-al (TTT), 4,8,13,17,21-tetramethyl-octadeca-4,8,12,16,20-pentaene-al (TOP), and 4,9,13,17-tetramethyl-octadeca-4,8,12,16-tetraeneal (TOT)) have been identified via GC-MS in latent fingermarks extracted 5–10 min following deposition [76]. These results are unsurprising given the reactivity of squalene, and that ozonolysis would also take place on the donors' skin prior to sample deposition. More recently, Dorakumbura et al. identified SQ-[OOH], SQ-epoxide, and TOT in fingermarks up to 7 days old using ultrahigh-performance liquid chromatography-high-resolution accurate mass Orbitrap™ mass spectrometry (UHPLC-HRAM Orbitrap™ MS), confirming that several squalene oxidation pathways occur simultaneously during aging [73]. Figure 7.6 provides a summary of squalene oxidation products that have been identified in latent fingermarks.

7.3.3 Cholesterol and Sterol Esters

Cholesterol and its esters comprise a very minor fraction of human sebum (approximately 5%), and their aging in fingermark residue has not been investigated in as much detail as the lipids described above. Weyermann et al. determined that

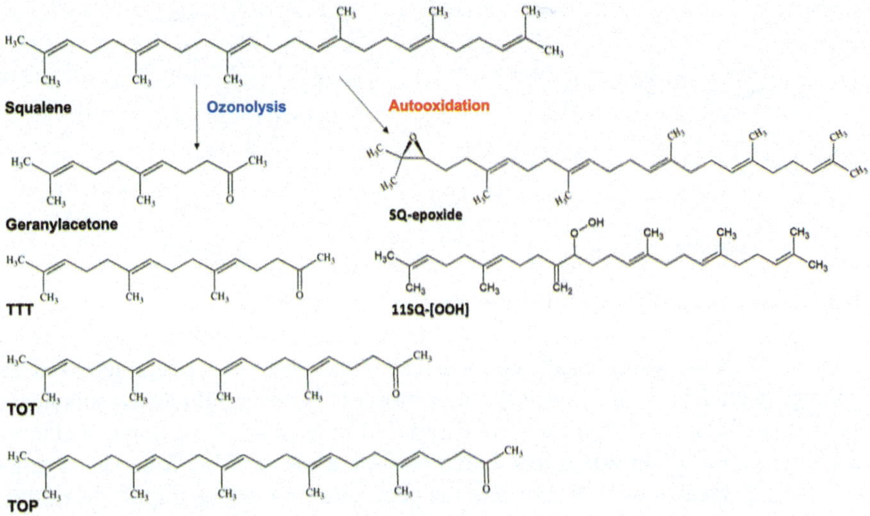

Fig. 7.6 Molecular structures of oxidation products of squalene identified in fingermarks

cholesterol appeared to be a relatively stable fingermark component, compared to the more readily oxidized squalene [44]. Being unsaturated, cholesterol and sterol esters (particularly those esterified to unsaturated fatty acids) are subject to similar oxidation processes to those described in previous sections of this chapter. Oxidation products of cholesterol in fingermark samples were tentatively identified by Dorakumbura et al. using UHPLC-HRAM Orbitrap™ MS, but characterization was not possible as isobaric compounds were unable to be distinguished [73].

Cholesterol is oxidized by both singlet oxygen and ozone to produce a series of hydroperoxycholesterols (ChOOH) (including 5α-, 6α-, 6β-, 7α-, and 7β-ChOOH) and 3β-hydroxy-5-oxo-5,6-secocholestan-6-al (together with its aldolization product 3β-hydroxy-5β-hydroxy-B-norcholestane-6β-carboxaldehyde), respectively [77, 78]. Picardo et al. observed that both free and esterified cholesterols were degraded by UV exposure in vitro and in vivo on human skin [49]. Similarly, Yamazaki et al. reported that higher amounts of 7α- and 7β-ChOOH of both free and esterified cholesterol were recovered on the skin following sunlight exposure (Fig. 7.7), indicating photooxidation as a major pathway for cholesterol degradation [78]. Thus, it seems reasonable to envisage that these degradation processes could also take place in fingermark residue. The research on this topic is currently lacking, and further studies on this aspect of fingermark aging are required.

7.3.4 Wax Esters

Wax esters have been reported as components of both fresh and aged fingermark residues [70]. However, there is little information currently available in the literature concerning the degradation products of these compounds due to age or other

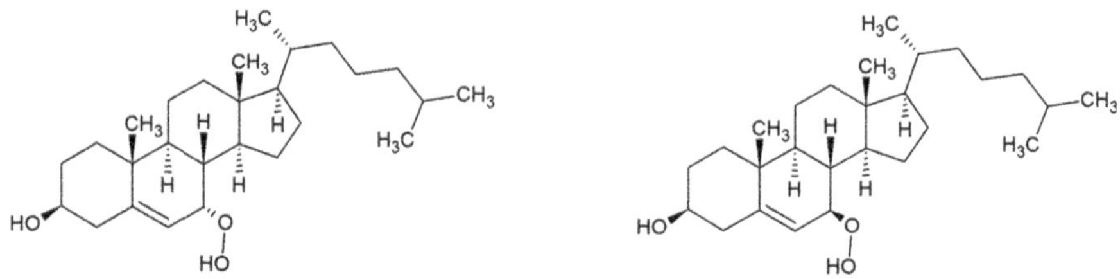

7α-ChOOH 7β -ChOOH

Fig. 7.7 Molecular structures of cholesterol 7α- and 7β-hydroperoxides

influence factors. Mong et al. reported that unsaturated wax esters appeared to undergo degradation but at a slower rate compared to other unsaturated components such as fatty acids and squalene [16]. Girod et al. studied the aging of targeted lipid compounds, including wax esters, based on peak areas resulting from GC separation [25]. Decreases were observed in the peak areas of the target lipids over time, indicating the occurrence of degradation processes, but specific mechanisms were not proposed. Given that a number of wax ester species are unsaturated, it follows that they would behave similarly to other unsaturated sebaceous components. For example, wax esters appear to peroxidize to a small extent under UV radiation but to a lesser extent to more reactive species such as squalene [49].

Recent investigations using wax ester standards shed more light on processes likely to occur in fingermarks. It has been demonstrated that as with the unsaturated lipids described above, unsaturated wax esters undergo ozonolysis [43]. However, saturated wax esters do not seem to react in any way [52]. These preliminary results show that the question of how wax esters degrade in fingermark residue still requires further investigation.

7.4 Factors Affecting Chemical Degradation of Fingermark Components

As described by Wertheim, a reliable age estimation method would ideally be based on a reaction that occurs mainly as a function of time, with minimal influence from external factors [9]. However, many different influence factors have been reported in the literature as having a significant impact on the initial and aged composition of latent fingermarks [15, 16, 25, 35, 68, 79, 80]. These factors encompass a number of variables that affect fingermark composition before (i.e., donor influence and presence of exogenous substances on the fingertips), during (e.g., substrate, time of contact, deposition pressure), and after deposition (interaction of residue with substrate and environment).

For both fingermark detection and construction of aging models, it must be borne in mind that different factors will affect the degradation of certain compounds to a

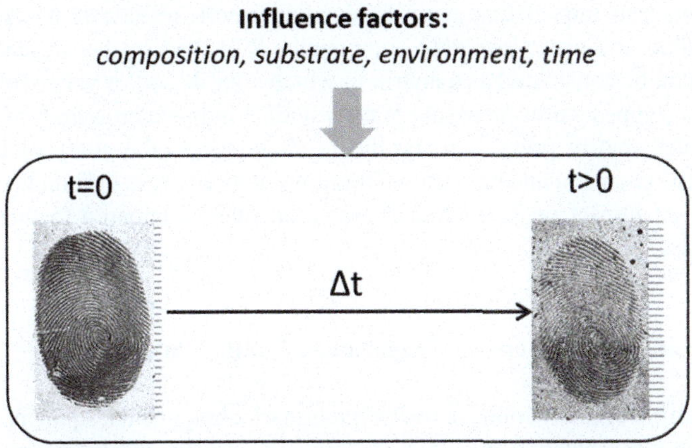

Fig. 7.8 Schematic representation of factors affecting fingermark composition over time, from the initial moment of deposition ($t = 0$) to the aged mark at the time of the crime scene investigation ($t > 0$)

greater extent than others [81]. As an example, the performance of oil red O (ORO) for the visualization of latent fingermarks is generally considered to be inferior to that of physical developer on specimens more than 4 weeks old, due to oxidation of the neutral lipids that ORO targets, as well as their diffusion into porous substrates [41]. Physical developer, on the other hand, continues to be successful on fingermarks up to several decades old [32], presumably as the targets of this detection method remain fairly stable. However, Beaudoin reported the successful application of ORO to develop a fingermark on a paper substrate recovered by police 21 years prior [82].

The first part of this section will discuss the particular impact of the initial composition ($t = 0$) on the aging of fingermarks and the factors that contribute to its variation across individuals over time. The remaining discussion will detail some of the main factors influencing the known chemical changes in fingermark composition that occur over time ($t > 0$) (Fig. 7.8).

7.4.1 Initial Chemical Composition (t = 0)

Latent fingermark residue can vary considerably in qualitative and quantitative composition, depending upon factors such as, but not limited to, an individual's age, the time since they last washed their hands, the substances they recently handled (e.g., sebum and skin products from touching the face or hair), or the transfer conditions applied (e.g., pressure or time of contact) when depositing he fingermark [25, 76, 79, 83]. The wide array of combinations has a subsequent effect upon the way a fingermark ages, in regard to the limits of detection of certain components and/or their degradation products, but also the interactions between certain fingermark components can impact longevity. Latent fingermark residue is an emulsion of water-soluble and water-insoluble components [84], and so the presence of both fractions can be considered as a matrix in which volatile and/or water-soluble compounds are trapped within a mixture of nonvolatile lipids and proteins [85]. However, this complex matrix presents a significant challenge to the study of fingermark aging, as the chemical and physical interactions of all of these different components will have an impact on individual degradation rates, compared to those of pure compounds.

7.4.1.1 Interactions Between Fingermark Components

As an example of interactions between fingermark components, squalene is considered to have a protective role in skin surface chemistry, by being preferentially oxidized in lieu of other unsaturated skin surface lipids upon exposure to UV radiation [86]. As squalene is present in unusually high concentrations in sebum, the onset of degradation of unsaturated fatty acids, for example, could be delayed in fingermark residue until its squalene content is sufficiently depleted. Furthermore, Mountfort et al. observed that the photooxidation of squalene was more rapid in fingermarks than in vitro and attributed this to the higher concentration of squalene in standard solutions. However, the fingermark degradation was compared to a squalene standard that had been oxidized in the absence of a photosensitizer. As stated in Sect. 7.3.2, the skin surface contains several endogenous photosensitizers including coproporphyrin and riboflavin. Thus, these observations could also be due to the influence of photosensitizers on squalene oxidation.

Another interesting point to highlight is that the bulk residue may physically protect some components from degradation or evaporation, slowing the rate of aging mechanisms compared to standard compounds stored under the same conditions [24]. For example, individual lipid standards deposited onto a surface exhibit faster reaction rates compared to analogous lipids within fingermark samples [24, 73]. Moreover, Kim et al. observed that concentrations of squalene and cholesterol standards decreased significantly at 60 °C, whereas in fingermark samples, these compounds showed greater stability at the same temperature [81]. Johnston and Rogers proposed that the chemical species that comprise human sebum might interact in such a way as to influence each other's degradation rates [87]. Possible interactions in fingermark residue include cholesterol with triglycerides and free fatty acids [88] and squalene and unsaturated fatty acids [89]. Johnston and Rogers investigated the influence of major sebaceous lipids on degradation of the bulk residue using a series of "fingermark analogues," mixtures of representative fingermark components that were each missing a component such as triglycerides or serine [87]. Examination of the analogues using FTIR microspectroscopy indicated that squalene and cholesterol play a role in interactions with other fingermark lipids and suggested that this could impact aging processes.

7.4.1.2 Donor Variation

The interindividual variability (differences between persons) of both fresh and aged fingermark composition is clearly illustrated by the differences in persistence between fingermarks deposited by adults and children. There have been several case examples where a child was known from eyewitness testimony to have been in physical contact with a surface (such as the interior of a car), but upon forensic examination, only the fingermarks of the adult suspect could be detected [80, 90]. Subsequently, this phenomenon became the subject of a series of investigations. Using GC-MS, Buchanan et al. found that the fingermarks of prepubescent children contained far fewer nonvolatile lipids, such as wax esters, which accounted for the observation that children's fingermarks could "disappear" within hours of deposition, while under the same conditions, adults' fingermarks could still be detected after several weeks [80, 90]. Similar results were obtained by Mong et al. [16]. The difference in composition can be attributed to the onset of adrenarche, which occurs at approximately 7–10 years of age, during which the production of androgens stimulates increased sebum production (before this stage of development, the sebaceous glands are largely inactive) [91]. Due to the inter-variability already present in fresh fingermarks, samples from adults and children will age differently, such that they can be clearly distinguished several weeks after deposition [17].

Other sources of interindividual variability in fingermark composition have been attributed to factors such as the biological sex of the donor and ethnicity; however, the specific impacts of such characteristics have never been determined in detail [79]. It has been shown that fingermark composition can vary to such an extent that individuals can be classified as poor or rich lipid donors [76]. Furthermore, the intraindividual variability of fingermark residue (variability among fingermarks from the same individual deposited at different moments) may also influence aging. Changes in lifestyle factors such as diet, certain metabolic disorders, consumption of medication, or drugs can affect the initial composition through qualitative or quantitative variation [22, 92]. As an example, fingermarks deposited by an individual who used acne cream were observed to contain significantly higher quantities of lipids when its use was ceased [22]. It has also been demonstrated that handwashing can influence the initial relative water content of fingermarks [29], through removal of the secretions and contaminants present on the fingertips. However, this effect may not be long-lived, as eccrine sweat is continuously secreted onto the fingertips. Lewis et al. demonstrated that when deposited 5 min after handwashing, fingermarks already contained enough eccrine components to be successfully developed using cyanoacrylate [38].

In summary, interindividual and intraindividual variation presents a significant complicating factor into understanding fingermark composition and its degradation processes [27, 68, 84]. Girod et al. have suggested that interindividual variation may be impossible to overcome, such that a means to estimate fingermark age necessitates the construction of an aging model for each individual [93].

7.4.1.3 Transfer Conditions

While intra-variability has been identified as one of the main factors influencing the initial fingermark composition, conditions surrounding the transfer of material from the fingertip to a surface also have an impact. For example, the effects of deposition pressure have been investigated in the transfer of fingermarks onto a porous substrate [25]. A greater mass of residue was transferred when fingermarks were deposited at a pressure of 1000 g compared to 100 g, though intra-donor variability of the normalized data was greater than the variability induced by differences in pressure. Subsequently, the aging of squalene contained in the fingermark deposited at 100 g pressure could no longer be monitored after 5 days following deposition using GC-MS, as the remaining quantities fell below the limits of detection (conversely, squalene aging could be observed up to 18 days in samples deposited at 1000 g). Fieldhouse et al. reported that increased deposition pressures created fingermarks with larger surface areas and hence a greater amount of deposited residue [83]. Further studies are required to determine the impact of other transfer conditions, such as contact time, and the temperature of the substrate.

7.4.2 Substrate

The characteristics of the substrate on which a fingermark is deposited can affect the initial composition and its aging. In fact, a porous substrate such as paper will absorb a greater quantity of residue from the fingertip ridge skin than a nonporous surface such as metal or glass [44]. Fingermark components such as amino acids will penetrate into such substrates, with the extent of residue absorption depending on factors such as the smoothness and porosity of the paper [94]. The porosity of substrates enables fingermark detection long after deposition, for example, with techniques such as amino acid-sensitive reagents and physical developer, as shown in a very recent study [32]. Similarly, the "fixing" of lipid-rich fingermarks on certain polymer substrates such as powder-coated aluminum PVC or painted surfaces has also been attributed to the absorption of fingermark residue into porous and semi-porous substrates [12]. On the contrary, smooth, nonporous substrates are less absorbent and so may leave deposited residue vulnerable to chemical degradation or mechanical abrasion. Wargacki et al. observed that eccrine fingermarks appeared to degrade due to physical removal of brittle salt residue from eccrine fingermarks by air currents, in tandem with chemical degradation [38]. Furthermore, it has also been observed that squalene and cholesterol degrade faster on a nonporous substrate [44].

Substrate properties other than porosity will also impact fingermark aging, due to the interactions of certain materials with fingermark constituents. For example, it has been observed that the photodegradation of lactic acid is faster on galvanized steel than Teflon, due to the iron content of the former substrate [34]. Concerning metal substrates, inorganic salts originating from eccrine sweat react with metals such as aluminum and brass, resulting in a durable image of the fingermark pattern via corrosion of the substrate [95]. This process occurs at temperatures up to 600 °C, preserving the ridge pattern even under circumstances that would destroy the organic fingermark constituents (e.g., arson and ballistics). The resulting quality of the fingermark pattern is affected by both the type of the metal substrate and the concentration of salt in the latent residue (dependent on donor factors) [96]. A study by Girod et al. found that fingermark samples deposited on glass and aluminum stored in the dark and analyzed with FTIR microspectroscopy could be differentiated by substrate, rather than by age [20]. While this was not observed in samples that had been stored in the light, it is possible that the acceleration of aging processes brought on by light exposure masked any substrate effects.

7.4.3 Environment

7.4.3.1 Light

Exposure to direct natural light, particularly UV radiation, has been linked to the increased degradation rate of several lipid components, particularly squalene [15, 19, 68, 73]. As described in Sect. 7.3.2, UVA exposure leads to the increased formation of singlet oxygen and therefore enhanced fingermark peroxidation. Thus, even fingermarks stored indoors behind a window can exhibit UV-induced degradation, as ordinary glass transmits most UVA radiation. Conversely, it has been suggested that exposure to UV light might preserve certain fingermark components by destroying microbial activity [97]. However, several analytical investigations have concluded the degradation rate of sebaceous components including squalene, cholesterol, and free fatty acids is accelerated when latent fingermarks are exposed to light over a period of days to weeks [15, 19, 24, 45].

It is interesting to note that even when fingermarks are stored in the dark, the degradation of some unsaturated lipid compounds (particularly squalene) may still be observed [15, 19, 44]. The quantity of low molecular weight saturated acids has been reported to increase over time in fingermark samples stored in the dark [16]. These components could originate from the oxidation of fatty acids, glycerides, and squalene as described in Sects. 7.3.1 and 7.3.2, and hence, such degradation processes would occur in fingermarks regardless of light exposure.

Information is rather limited in the literature concerning the effects of light on the degradation of eccrine components. As described in Sect. 7.2.3, lactic acid degrades to pyruvate when exposed to artificial sunlight [34]. This is consistent with observations that fingermarks exposed to UV light exhibit poorer ridge quality upon cyanoacrylate fuming [38] and that pyruvate is a less effective initiator of superglue polymerization than sodium lactate [98]. This was attributed to the lower pH value of the pyruvate standard. Other fingermark eccrine components, such as amino acids and proteins, appear to not be affected by photodegradation processes [19, 34]. Conversely, Girod et al. observed that IR peaks relating to esters and eccrine constituents decreased faster when fingermark samples were exposed to artificial and natural light (through a window) [20]. Further research on this topic should be conducted in order to obtain definitive results.

7.4.3.2 Air Flow, Water, and Ambient Humidity

The oxidation of fingermark lipids occurs rapidly, regardless of temperature and exposure to natural or artificial light [15, 52]. Ozone in particular is a highly reactive species and can be rapidly depleted in small, constricted spaces, such as drawers, if there is no circulation of air from ozone-rich sources to replenish ozone levels. As a result, lipid ozonolysis may be significantly slower [24]. Furthermore, human presence in an indoor setting has a significant effect on ozone concentration, due to the presence of sebaceous lipids on the skin, hair, and clothing, as well as handled surfaces, and the reactivity of surface lipids within the respiratory system [46, 65]. Thus, if fingermarks are deposited in a relatively contained environment with many people present (such as a shared office or home), the ozone content of the air will decrease, and the ozonolysis rate of fingermark residue will probably be slower compared to an outdoor setting [46, 74].

It should be noted that aging studies, particularly comparisons between degradation rates of fingermark compounds with and without exposure to light, normally involve wrapping or otherwise containing samples in order to protect them from natural and artificial light [15, 16, 24]. This may have the unintended consequence of also impeding degradation processes by restricting both the initial availability of ozone and other reactive gases and their replenishment by airflow. Both opaque and transparent containers have been demonstrated to significantly reduce the rate of ozonolysis in fingermark lipids compared to when samples were exposed to ambient air for the same amount of time [24, 43].

In instances where objects are submerged, it might be assumed that sebaceous lipids may be preserved, due to the lower availability of oxygen and ozone in water compared to air. However, Dorakumbura et al. observed that squalene degradation was increased when fingermarks were submerged and stored in covered containers compared to samples stored in dark and dry conditions [73]. The persistence of sebum-rich fingermarks is also affected by water quality. Sutton et al. found that fingermarks on nonporous substrates were of worse quality after immersion in water collected from natural water sources and stored in tanks than those submerged directly in water sources (e.g., lakes or rivers) [99]. This was attributed to microbial accumulation in the tanks used to contain samples in the laboratory, which in turn degraded the fingermark lipids.

As described in Sect. 7.2.2, elevated humidity can affect fingermark composition by dissolving water-soluble components, such as amino acids and salts, out of the residue [2]. Additionally, humid storage environments can affect the formation of lipid oxidation products, through interaction of water vapor with ozonolysis intermediates, as described in Sect. 7.3.1.

7.4.3.3 Temperature

Increased temperatures can accelerate the oxidation of fingermark lipids, which would subsequently lead to greater mass loss through evaporation of the volatile degradation products. Johnston and Rogers noted that at temperatures of 55 °C or more, peak signals corresponding to sebaceous material decreased more rapidly over a 5-h period compared to when samples were stored at 45 °C, which was attributed to an overall decrease in sample mass and to chemical modifications of some functional groups [52]. Girod et al. investigated the effect of environmental temperature (15 °C, 20 °C, and 25 °C) on lipid degradation and demonstrated that squalene degradation was accelerated by increased temperatures [25]. Similarly, Kim et al. reported that short-term storage (up to 8 h) at 100 °C increased the degradation rates of both squalene and cholesterol in fingermark samples [81]. Wolstenholme et al. investigated the effects of increasing temperatures on fatty acid aging, reporting that oleic acid degraded at faster rates at 37 °C and 60 °C, compared to 4 °C [100]. Degradation products were identified using MALDI-MS imaging as dehydrated and didehydrated oleic acid, indicating that temperature may influence the specific degradation reactions that occur, as well as degradation rates. Lower signals from the dehydrated forms of oleic acid were detected at 60 °C compared to 37 °C, indicating further decomposition of the fatty acid.

Amide groups appear to be less affected by storage temperature [52], which is consistent with studies indicating that fingermark amino acids do not decompose at temperatures below 100 °C [18, 34]. This would suggest that amino acid degradation due to temperature is not of particular concern from a forensic perspective, except for cases involving extreme environmental conditions, such as arson [18, 34]. As described in Sect. 7.2.3, some additional eccrine constituents such as urea also undergo thermally induced degradation, while FTIR analyses have demonstrated that carboxylic acid salts (i.e., sodium lactate) may still be detected in fingermarks residue after storage at 70 °C for 72 h [101].

7.4.3.4 Microbial Activity

The free fatty acids in human skin surface lipids are derived from digestion of sebaceous triglycerides by facultative anaerobes within the secretion ducts of the sebaceous glands [102–104]. *Cutibacterium acnes* (formerly referred to as *Propionibacterium* or *Corynebacterium acnes*) has been identified as one of the main species responsible for hydrolyzing triglycerides to produce a variety of diglycerides and monoglycerides, as well as free fatty acids [105–107]. It has been proposed that given appropriate environmental conditions, such bacteria could survive on a substrate such that this or other enzymatic process would continue to modify the composition of the latent fingermark residue [23, 33]. Preferential bacterial growth in latent fingermarks has been demonstrated, under controlled conditions, as a novel visualization method [108]. Skin flora from multiple genera may be deposited onto a surface through contact with friction ridge skin; however, their subsequent longevity has not been investigated [109]. A more in-depth discussion on the potential effects of microbes on fingermark degradation is discussed in Chap. 11.

7.4.4 Time

The passage of time following deposition will also have an influence on the aging of fingermarks, and the investigated timeframe is thus an important factor to take into consideration when studying aged composition. As described in Sects. 7.2 and 7.3, some aging processes, such as the evaporation of water, will occur within minutes [29]. Other compounds, such as fatty acids, squalene, and glycerides, will degrade more slowly over a period of several days to weeks [15, 16, 23, 24]. Squalene and cholesterol have been shown to decrease exponentially in the first few hours after deposition [44], and the remaining quantities of these compounds may fall below the limits of detection within a month, depending on other factors such as light exposure, poor lipid donors, and low deposition pressures [15, 25]. Saturated lipids appear to be relatively unaffected by degradation processes [16, 52] and may persist for longer periods than have currently been explored. While few investigations have been carried out over timeframes greater than 1 or 2 months, there are a limited number of studies which indicate that some compounds (such as amino acids [27, 32]) can still be detected years after deposition. The fact that detection methods such as powdering, ORO, or physical developer may still be effective on years old fingermarks [12, 26, 32, 82] indicates the presence of target compounds in the aged latent residue. It is, however, currently unclear if target compounds which comprise the initial composition can still persist in the aged latent residue or whether some degradation products may also act as targets for detection methods. Further, long-term studies would be required to determine such processes and whether degradation products may act as detection or aging targets.

7.5 Considerations for Aging Studies

As stated in the introduction of this chapter, research concerning the aging processes of fingermarks based on chemical degradation serves several purposes. It is essential to acquire fundamental knowledge of the composition of fingermarks discovered at crime scenes (i.e., at time $t > 0$) to target appropriate compounds for detection. Thus, such knowledge will help to understand, develop, or improve detection techniques. While a detection method should ideally work on all fingermarks, regardless of age, it is theoretically possible that a method that targets compounds more common to aged fingermarks could be employed in instances where the approximate age of a fingermark is known and thereby "filter out" other fingermarks less relevant to the investigation. The use of detection techniques for such purposes is, however, rather difficult to implement in practice. For the development of methods for age estimation, one proposed approach to place fingermarks in time is to measure their chemical degradation. For both lines of enquiry (i.e., detection and age estimation), it is necessary to consider fingermarks as complex biological matrices. While experiments with standard solutions can be performed to confirm observations noted in fingermark samples [14, 18, 23, 24, 52], these cannot be reliably used to infer fingermark aging processes (particularly degradation rates) on their own. To ensure that such investigations are relevant in practice, some parameters should be observed as discussed below.

First of all, it must be noted that many aging studies, particularly those which focus upon lipid constituents, utilize groomed or charged samples, which are typically prepared by having fingermark donors rub their fingertips on their nose and forehead, where sebaceous glands are found in greatest density [15, 16, 25, 43, 68]. While this ensures that sebaceous compounds are present at detectable levels, it also risks misrepresenting the composition and degradation processes of latent fingermarks recovered in criminal investigations [79]. As per recommendations from the International Fingerprint Research Group [110], research needs to move away from groomed fingermarks as the sole type of sample investigated. Several studies in more recent years acknowledge the importance of investigating "natural" fingermarks deposited without prior preparative steps on the part of the donor, as well as eccrine fingermarks prepared by cleaning the hands of any sebum and exogenous materials (using water and soap or solvent) [37, 43, 73].

Because of high intraindividual and interindividual variability in initial fingermark composition, several approaches have been reported in order to obtain more reproducible samples, particularly in studies utilizing destructive methods such as GC- and LC-MS. A homogenization step, consisting of rubbing the fingertips together before depositing fingermarks, is sometimes employed in order to minimize intraindividual variation [15, 62, 111]. The use of synthetic mixtures intended to mimic fingermark composition has also been proposed to minimize sample variation [112]. While such approaches may be used to gain initial insight into fingermark aging, the use of more realistic samples would be required in subsequent stages of research, particularly when developing methods aimed to be used in practice. As significant variability of fingermark composition has been observed by many researchers, it is important to understand the extent of this variation by using more realistic, natural (i.e., "non-groomed") fingermarks. This will enable the identification of the (groups of) components present in most or all fingermarks, whatever their age, as these "common factors" will serve as optimal targets for the development of more broadly applicable detection and/or age estimation approaches.

Substrates should also be chosen in accordance with the purpose of the conducted study. Clean substrates that are most compatible with the analytical technique(s) employed are again useful to gain fundamental knowledge about fingermark composition and aging in the initial stages of research. Fingermark studies often utilize brand new, thoroughly cleaned laboratory materials such as filter paper or microscope slides [15, 21, 43, 62, 76]. This is necessary to minimize complications such as contamination and low signal-to-noise ratios, as well as for more prosaic reasons such as ease of handling by the researcher, or limitations of certain analytical instruments. However, for validation and implementation of age estimation approaches, eventually the methodology must comply with substrates which more closely resemble those encountered at crime scenes (e.g., white copy paper rather than filter paper or patterned aluminum cans rather than pre-cleaned sheets).

In the first stages of research, it is vital to know the conditions of the sample storage environment to evaluate their specific influence on aging processes and the resulting aged fingermark composition [112]. At a minimum, monitoring of storage conditions should be conducted. If these are not directly controlled, the absence of information of aging conditions in experiments can make it difficult to compare results, as well as infer variables that may cause differences in results, in a rational and informed manner. In particular, conditions such as the type and duration of light exposure, temperature, and relative humidity should be disclosed. However, uncontrolled (or realistic) conditions should also be explored in the final stages of the development of age estimation approaches, as these might have a nonnegligible impact on the targeted chemical degradation process(es) (as explained in Sect. 7.4).

Crime scene investigations rarely commence within the first few hours after fingermark deposition, and thus, the timeframe of aging experiments should be designed appropriately [110]. For example, if the goal is to monitor certain target compounds for the purpose of fundamental data, it may be appropriate to conduct the study for as long as the target compounds can be detected, weeks to months for sebaceous compounds and possibly several years for some eccrine components, starting from immediately after sample deposition. However, for the development of an age estimation methodology, more relevant timeframes should be selected with consideration to the reality of forensic investigations. The ability to differentiate between fingermarks that are a few days old and a few weeks or few months old would already represent a significant gain to forensic practice [26]. Thus, analyses should be conducted up to at least 3–4 months (if target compounds are still detectable). Sebaceous lipids would be appropriate targets for such timeframes. If the focus is set on longer timeframes of up to several years, eccrine compounds may be preferable analytes.

The choice of the analytical technique(s) will also be dependent on the pursued aims. In a fundamental research context, high-cost, state-of-the-art instruments may be preferred when available for their advantages such as improved sensitivity and high mass resolution. Such approaches might also be applied in high-profile cases, but complete validation of the methodology in realistic conditions should not be underestimated. However, for the development of an age estimation model intended for eventual routine practical use, analytical techniques more readily available to forensic laboratories should be prioritized. Moreover, nondestructive approaches (e.g., FTIR on nonporous substrates or headspace extraction approaches) that can be applied in sequence with detection techniques present advantages over methods that require solvent extraction, such as GC- or LC-MS. Following data acquisition, reliable data processing is of uttermost importance. Preprocessing techniques may be used to increase reproducibility of selected aging parameters, and chemometrics should be considered to enable the evaluation of univariate and multivariate modeling. However, it should be stressed that while data processing can be employed to reduce variability, it will never be entirely controlled, as a forensic trace is by its very nature produced in an uncontrolled manner, and is not always representative of its source [113].

7.6 Conclusions

Fingermarks are complex biological matrices whose initial composition differs from person to person, as well as over time from a single person, and is influenced by many factors that come into play before, during, and after deposition. The identification and understanding of aging processes are therefore a considerable challenge. Indeed, fingermarks are comprised of a diverse range of components mainly originating from the skin, but also external contaminants such as cosmetics, foods, and other handled substances. Several of these components, depending on the timeframe and reproducibility of their aging processes, represent potential targets for fingermark detection or age estimation.

In order to respond to the practice-oriented needs expressed by the forensic community, a detailed understanding of the chemical processes involved in latent fingermark degradation is essential. While some early investigations have inferred chemical degradation through the poor (or good) performance of certain detection methods with increasing sample age, more knowledge of the specific mechanisms involved is needed to truly understand and improve current detection capabilities and to reliably place latent fingermarks in time. While there is some information already available concerning the degradation mechanisms of some components as reported in this chapter, many have yet to be fully elucidated. Furthermore, investigations into chemical degradation of fingermark components in the context of the latent residue are required, which can produce results that diverge from more simple in vitro experiments. In particular, fundamental information is needed on how the initial composition, the substrate, and the environment to which the latent residue is exposed interact to influence the nature and rates of the degradation mechanisms that occur between deposition and detection.

To date, the sebaceous secretions have been the main focus of aging studies, presumably because their chemical modification is known to occur in a relatively short timeframe (hours to months). The degradation of compounds such as squalene, cholesterol, fatty acids, and glycerides has been studied as a function of different donors (e.g., poor to good lipid donors), on different substrates (e.g., paper, glass, and aluminum), and under different storage conditions (e.g., light exposure, ambient humidity, and temperature). Many investigations indicate significant variability that may represent an insurmountable issue in the development of age estimation approaches. Currently, the implementation of an age estimation methodology applicable to all fingermarks seems unrealistic. It has been suggested that in practice, samples from the person who left a questioned fingermark (provided that an identification is possible) would need to be used in order to build an age estimation model that would then only be valid for a particular individual and case conditions [26]. Further research needs to be conducted, particularly concerning eccrine compounds and other components that remain relatively overlooked, and might provide interesting aging parameters over longer timeframes. The possibility of utilizing other, complementary methods (such as optical or physical characterization, as well as detection techniques) in conjunction with chemical characterization to better understand aging processes might also be considered as a complex but promising approach to tackle the issue of estimating fingermark age.

While future research could focus on identifying degradation processes that are less impacted by the factors discussed in this chapter, the variability of fingermark properties (chemical and physical) over time will never be completely overcome, and thus, research efforts need to coordinate and focus on the practical forensic context, rather than only on the technological aspects. Combined multivariate approaches might be more suited to better deal with this complexity, as different aging parameters will be influenced differently by composition, substrate, environment, and time. Thus, further research on the chemical composition of natural fingermarks, to characterize degradation processes and factors influencing composition at the time of deposition, should be a primary aspect of future endeavors.

Acknowledgments The authors would like to thank the Swiss National Science Foundation (SNF no. PP00P1_123358/1 and 205121_169677) for funding the research carried out by the authors in the field of fingermark composition and aging.

References

1. Heindl R (1927) System und Praxis der Daktyloskopie und der sonstigen technischen Methoden der Kriminalpolizei, 3rd edn. De Gruyter, Berlin
2. Cuthbertson F (1969) The chemistry of fingerprints. Atomic Weapons Research Establishment, Aldermaston
3. Angst E (1961) Untersuchungen zur Bestimmung des Alters von daktyloskopischen Spuren auf Papier. Kriminalistik 15:1–12
4. Dikshitulu YS, Prasad L, Pal JN et al (1986) Aging studies on fingerprint residues using thin-layer chromatography and high performance liquid chromatography. Forensic Sci Int 31(4):261–266
5. Dalrymple B, Duff J, Menzel E (1977) Inherent fingerprint luminescence – detection by laser. J For Sci 22:106–115
6. Duff JM, Menzel ER (1978) Laser-assisted thin-layer chromatography and luminescence of fingerprints: an approach to fingerprint age determination. J For Sci 23(1):129–134
7. Menzel ER (1992) Letters to the editor: fingerprint age determination by fluorescence. J For Sci 37(5):1212–1213
8. Olsen RD (1987) Chemical dating techniques for latent fingerprints: a preliminary report. Identif News 37(2):10–12
9. Wertheim K (2003) Fingerprint age determination: is there any hope? J Forensic Identif 53(1):42–49
10. De Alcaraz-Fossoul J, Mestres PC, Balaciart MA et al (2013) Determination of latent fingerprint degradation patterns – a real fieldwork study. Int J Legal Med 127(4):857–870

11. Midkiff CR (1993) Lifetime of a latent print: how long? Can you tell? J Forensic Identif 43(4):386–392
12. Cohen Y, Rozen E, Azoury M et al (2012) Survivability of latent fingerprints part 1: adhesion of latent fingerprints to smooth surfaces. J Forensic Identif 62(1):47–53
13. Barnett PD, Berger RA (1977) The effects of temperature and humidity on the permanency of latent fingerprints. J For Sci 16(3):249–254
14. Mountfort KA, Bronstein H, Archer N (2007) Identification of oxidation products of squalene in solution and in latent fingerprints by ESI-MS and LC/APCI-MS. Anal Chem 79(7):2650–2657
15. Archer NE, Charles Y, Elliott JA et al (2005) Changes in the lipid composition of latent fingerprint residue with time after deposition on a surface. Forensic Sci Int 154(2–3):224–239
16. Mong GM, Petersen CE, Clauss TRW (1999) Advanced fingerprint analysis project: fingerprint constituents. Pacific Northwest National Laboratory, Richland
17. Antoine KM, Mortazavi S, Miller AD et al (2010) Chemical differences are observed in children's versus adults' latent fingerprints as a function of time. J For Sci 55(2):513–518
18. Richmond-Aylor A, Bell S, Callery P et al (2007) Thermal degradation analysis of amino acids in fingerprint residue by pyrolysis GC-MS to develop new latent fingerprint developing reagents. J For Sci 52(2):380–382
19. Andersson PO, Lejon C, Mikaelsson T et al (2017) Towards fingermark dating: a Raman spectroscopy proof-of-concept study. ChemistryOpen 6(6):706–709
20. Girod A, Xiao L, Reedy B et al (2015) Fingermark initial composition and aging using Fourier transform infrared microscopy (u-FTIR). Forensic Sci Int 254:185–196
21. Oonk S, Schuurmans T, Pabst M et al (2018) Proteomics as a new tool to study fingermark ageing in forensics. Sci Rep 8:16425
22. Koenig A, Girod A, Weyermann C (2011) Identification of wax esters in latent print residues by gas chromatography-mass spectrometry and their potential use as ageing parameters. J Forensic Identif 61(6):652–676
23. Pleik S, Spengler B, Schäfer T et al (2016) Fatty acid structure and degradation analysis in fingerprint residues. J Am Soc Mass Spectrom 27(9):1565–1574
24. Pleik S, Spengler B, Bandan DR et al (2018) Ambient-air ozonolysis of triglycerides in aged fingerprint residues. Analyst 143(5):1197–1209
25. Girod A, Spyratou A, Holmes D et al (2016) Aging of target lipid parameters in fingermark residue using GC/MS: effects of influence factors and perspectives for dating purposes. Sci Justice 56(3):165–180
26. Girod A, Ramotowski R, Lambrechts S et al (2016) Fingermark age determinations: legal considerations, review of the literature and practical propositions. Forensic Sci Int 262:212–226
27. Cuthbertson F, Morris JR (1972) The chemistry of fingerprints. Atomic Weapons Research Establishment, Aldermaston
28. Kent T (2016) Water content of latent fingerprints – dispelling the myth. Forensic Sci Int 266:134–138
29. Keisar O, Cohen Y, Finkelstein Y et al (2019) Measuring the water content in freshly-deposited fingermarks. Forensic Sci Int 294:204–210
30. Scruton B, Robins BW, Blott BH (1975) The deposition of fingerprint films. J Phys D 8:714–723
31. Bright NJ, Willson TR, Driscoll DJ et al (2013) Chemical changes exhibited by latent fingerprints after exposure to vacuum conditions. Forensic Sci Int 230(1–3):81–86
32. Bleay S, Fitzgerald L, Sears V et al (2019) Visualising the past – an evaluation of processes and sequences for fingermark recovery from old documents. Sci Justice 59(2):125–137
33. Cadd S, Islam M, Manson P et al (2015) Fingerprint composition and aging: a literature review. Sci Justice 55(4):219–238
34. De Paoli G, Lewis SA, Schuette EL et al (2010) Photo- and thermal-degradation studies of selected eccrine fingerprint constituents. J For Sci 55(4):962–969

35. van Dam A, Schwarz JCV, de Vos J et al (2014) Oxidation monitoring by fluorescence spectroscopy reveals the age of fingermarks. Angew Chem Int Ed 53(24):6272–6275
36. van Dam A, Aalders MCG, Todorovski T et al (2016) On the autofluorescence of aged fingermarks. Forensic Sci Int 258:19–25
37. van Helmond W, van Herwijnen AW, van Riemsdijk JJH et al (2019) Chemical profiling of fingerprints using mass spectrometry. Forensic Chem 16:100183
38. Wargacki SP, Lewis LA, Dadmun MD (2008) Enhancing the quality of aged latent fingerprints developed by superglue fuming: loss and replenishment of initiator. J For Sci 53(5):1138–1144
39. Cuthbertson F (1965) The chemistry of fingerprints. Atomic Weapons Research Establishment, Aldermaston
40. Connatser RM, Prokes SM, Glembocki OJ et al (2010) Toward surface-enhanced Raman imaging of latent fingerprints. J For Sci 55(6):1462–1470
41. Salama J, Aumeer-Donovan S, Lennard C et al (2008) Evaluation of the fingermark reagent oil red O as a possible replacement for physical developer. J Forensic Identif 58(2):203–237
42. de la Hunty M, Moret S, Chadwick S et al (2015) Understanding physical developer (PD): part I – is PD targeting lipids? Forensic Sci Int 257:481–487
43. Hinners P, Thomas M, Lee YJ (2020) Determining fingerprint age with mass spectrometry imaging via ozonolysis of triacylglycerols. Anal Chem 92(4):3125–3132
44. Weyermann C, Roux C, Champod C (2011) Initial results on the composition of fingerprints and its evolution as a function of time by GC/MS analysis. J For Sci 56(1):102–108
45. Mong G, Walter S, Ramotoski R et al (1999) The chemistry of latent prints from children and adults. Chesapeake Examiner 37(2):4–6
46. Wisthaler A, Weschler CJ (2010) Reactions of ozone with human skin lipids: sources of carbonyls, dicarbonyls, and hydroxycarbonyls in indoor air. Proc Natl Acad Sci 107(15):6568–6575
47. Ekanayake Mudiyanselage S, Hamburger M, Elsner P (2003) Ultraviolet A induces generation of squalene monohydroperoxide isomers in human sebum and skin surface lipids in vitro and in vivo. J Invest Dermatol 120(6):915–922
48. Hayashi N, Togawa K, Yanagiswa M et al (2003) Effect of sunlight exposure and aging on skin surface lipids and urate. Exp Dermatol 12(s2):13–17
49. Picardo M, Zompetta C, De Luca C et al (1991) Role of skin surface lipids in UV-induced epidermal cell changes. Arch Dermatol Res 283(3):191–197
50. Gallagher M, Wysocki CJ, Leyden JJ et al (2008) Analyses of volatile organic compounds from human skin. Br J Dermatol 159:780–791
51. Baniuk K (1990) Determination of age of fingerprints. Forensic Sci Int 46(1–2):133–137
52. Johnston A, Rogers K (2017) The effect of moderate temperatures on latent fingerprint chemistry. Appl Spectrosc 71(9):2102–2110
53. Gomez M, Reggio D, Lazzari M (2019) Linseed oil as a model system for surface enhanced Raman spectroscopy detection of degradation products in artworks. J Raman Spectrosc 50(2):242–249
54. Lazzari M, Chiantore O (1999) Drying and oxidative degradation of linseed oil. Polym Degrad Stab 65(2):303–313
55. Nicolaides N (1974) Skin lipids: their biochemical uniqueness. Science 186(4158):19–26
56. Zhou Z, Zhou S, Abbatt J (2019) Kinetics and condensed-phase products in multiphase ozonolysis of an unsaturated triglyceride. Environ Sci Technol 53(21):12467–12475
57. Criegee R (1975) Mechanism of ozonolysis. Angew Chem Int Ed Eng 14(11):745–752
58. Sun C, Zhao Y, Curtis JM (2012) A study of the ozonolysis of model lipids by electrospray ionization mass spectrometry. Rapid Commun Mass Spectrom 26(8):921–930
59. Camera E, Ludovici M, Galante M et al (2010) Comprehensive analysis of the major lipid classes in sebum by rapid resolution high-performance liquid chromatography and electrospray mass spectrometry. J Lipid Res 51:3377–3388

60. Michael-Jubeli R, Bleton J, Baillet-Gufroy A (2011) High-temperature gas chromatography-mass spectrometry for skin surface lipids profiling. J Lipid Res 52:143–155
61. Sadowski T, Klose C, Gerl MJ et al (2017) Large-scale human skin lipidomics by quantitative, high-throughput shotgun mass spectrometry. Sci Rep 7:43761
62. Frick AA, Kummer N, Moraleda A et al (2020) Changes in latent fingermark glyceride composition as a function of sample age using UPLC-IMS-QToF-MS^E. Analyst 145(12):4212–4223
63. Zhou S, Forbes MW, Katrib Y et al (2016) Rapid oxidation of skin oil by ozone. Environ Sci Technol Lett 3(4):170–174
64. Soriano NU, Migo VP, Matsumura M (2003) Ozonation of sunflower oil: spectroscopic monitoring of the degree of unsaturation. J Am Oil Chem Soc 80(10):997–1001
65. Pryor WA, Squadrito GL, Friedman M (1995) The cascade mechanism to explain ozone toxicity: the role of lipid ozonation products. Free Radic Biol Med 19(6):935–941
66. Chu Y, Cheng TW, Gen M et al (2019) Effect of ozone concentration and relative humidity on the heterogeneous oxidation of linoleic acid particles by ozone: an insight into the interchangeability of ozone concentration and time. ACS Earth Space Chem 3(5):779–788
67. Zhou S, Forbes MW, Abbatt JPD (2016) Kinetics and products from heterogeneous oxidation of squalene with ozone. Environ Sci Technol 50(21):11688–11697
68. Frick AA, Chidlow G, Goodpaster JV et al (2016) Monitoring compositional changes of the lipid fraction of fingermark residues deposited on paper during storage. Forensic Chem 2:29–36
69. Ryu A, Arakane K, Koide C et al (2009) Squalene as a target molecule in skin hyperpigmentation caused by singlet oxygen. Biol Pharm Bull 32(9):1504–1509
70. Girod A, Ramotowski R, Weyermann C (2012) Composition of fingermark residue: a qualitative and quantitative review. Forensic Sci Int 223(1):10–24
71. Ramotowski R (2001) Composition of latent print residue. In: Lee H, Gaensslen RE (eds) Advances in fingerprint technology. CRC Press, Boca Raton
72. Yamashita B, French M (2011) Latent print development. In: The fingerprint sourcebook. National Institute of Justice, Washington, DC
73. Dorakumbura BN, Busetti F, Lewis SW (2020) Analysis of squalene and its transformation by-products in latent fingermarks by ultrahigh-performance liquid chromatography-high resolution accurate mass Orbitrap™ mass spectrometry. Forensic Chem 17:100193
74. Weschler CJ (2016) Roles of the human occupant in indoor chemistry. Indoor Air 26(1):6–24
75. Yeo HCH, Shibamoto T (1992) Formation of formaldehyde and malonaldehyde by photooxidation of squalene. Lipids 27(1):50–53
76. Girod A, Weyermann C (2014) Lipid composition of fingermark residue and donor classification using GC/MS. Forensic Sci Int 238:68–82
77. Niki E (2015) Lipid oxidation in the skin. Free Radic Res 49(7):827–834
78. Yamazaki S, Ozawa N, Hiratsuka A et al (1999) Increases in cholesterol 7-hydroperoxides in lipids of human skin by sunlight exposure. Free Radic Biol Med 26(9–10):1126–1133
79. Croxton RS, Baron MG, Butler D et al (2010) Variation in amino acid and lipid composition of latent fingerprints. Forensic Sci Int 199(1–3):93–102
80. Buchanan MV, Asano K, Bohanon A (1997) Chemical characterisation of fingerprints from adults and children. In: Forensic evidence analysis and crime scene investigation. SPIE (International Society for Optical Engineering), Boston
81. Kim Y, Choi W, Jeon B et al (2020) The effect of temperature and exposure time on stability of cholesterol and squalene in latent fingermarks deposited on PVDF membrane. J For Sci 65(2):458–464
82. Beaudoin A (2011) Oil red O: fingerprint development on a 21-year-old cold case. J Forensic Identif 61(1):50–59
83. Fieldhouse SJ (2015) An investigation into the effects of force applied during deposition on latent fingermarks and inked fingerprints using a variable force fingerprint sampler. J For Sci 50(2):422–427

84. Dorakumbura BN, Boseley RE, Becker T et al (2018) Revealing the spatial distribution of chemical species within latent fingermarks using vibrational spectroscopy. Analyst 143(17):4027–4039
85. de la Hunty M, Moret S, Chadwick S et al (2015) Understanding Physical Developer (PD): part II – is PD targeting eccrine constituents? Forensic Sci Int 257:488–495
86. Shimizu N, Ito J, Kato S et al (2018) Oxidation of squalene by singlet oxygen and free radicals results in different compositions of squalene monohydroperoxide isomers. Sci Rep 8:9116
87. Johnston A, Rogers K (2018) A study of the intermolecular interactions of lipid components from analogue fingerprint residues. Sci Justice 58(2):121–127
88. Kim SK, Nawar WW (1991) Oxidative interactions of cholesterol with triacylglycerols. J Am Oil Chem Soc 68(12):931–934
89. Dessì MA, Meiana M, Day BW et al (2002) Oxidative stability of polyunsaturated fatty acids: effect of squalene. Eur J Lipid Sci Technol 104(8):506–512
90. Bohanon AM (1998) Latents from pre-pubescent children versus latents from adults. J Forensic Identif 48(5):570–573
91. Zouboulis CC (2004) Acne and sebaceous gland function. Clin Dermatol 22:360–366
92. Jones NE, Davies LM, Russell CAL et al (2001) A systematic approach to latent fingerprint sample preparation for comparative chemical studies. J Forensic Identif 52(5):504–515
93. Girod A, Roux C, Weyermann C (2014) La datation des traces digitales (partie II): proposition d'une approche formelle. Revue Internationale de Criminologie et Police Technique et Scientifique 67(2):226–249
94. Almog J, Azoury M, Elmaliah Y et al (2004) Fingerprints' third dimension: the depth and shape of fingerprints penetration into paper – cross section examination by fluorescence microscopy. J For Sci 49(5):1–5
95. Bobev K (1995) Fingerprints and factors affecting their condition. J Forensic Identif 45(2):176–183
96. Bond JW (2008) Visualization of latent fingerprint corrosion of metallic surfaces. J For Sci 53(4):812–822
97. Braasch K, de la Hunty M, Deppe J et al (2013) Nile red: alternative to physical developer for the detection of latent fingermarks on wet porous surfaces? Forensic Sci Int 230(1–3):74–80
98. Dadmun MD (2010) Cultivating methods to enhance the quality of aged fingerprints developed by cyanoacrylate fuming. University of Tennessee, Knoxville
99. Sutton R, Grenci C, Hrubesova L (2014) A comparison on the longevity of submerged marks in field and laboratory conditions. J Forensic Identif 64(2):143–156
100. Wolstenholme R, Bradshaw R, Clench MR et al (2009) Study of latent fingermarks by matrix-assisted laser desorption/ionisation mass spectrometry imaging of endogenous lipids. Rapid Commun Mass Spectrom 23(19):3031–3039
101. Williams DK, Brown CJ, Bruker J (2011) Characterization of children's latent fingerprint residues by infrared microspectroscopy: forensic implications. Forensic Sci Int 206(1–3):161–165
102. Marples RR, Kligman AM, Lantis LR et al (1970) The role of the aerobic microflora in the genesis of fatty acids in human surface lipids. J Invest Dermatol 55(3):173–178
103. Shalita AR (1974) Genesis of free fatty acids. J Invest Dermatol 62(3):332–335
104. Downing DT (1970) Lipolysis by human skin surface debris in organic solvents. J Invest Dermatol 54:395–398
105. Byrd AL, Belkaid Y, Segre JA (2018) The human skin microbiome. Nat Rev Microbiol 16:143–155
106. Grice EA, Segre JA (2011) The skin microbiome. Nat Rev Microbiol 9:244–253
107. Goetz N, Burgaud H, Berrebi C et al (1984) Analysis of the lipid content of single hair bulbs. Comparison with the content of the sebaceous glands and with surface lipids. Int J Cosmet Sci 4(2):39–52
108. Harper DR, Clare CM, Heaps CD et al (1987) A bacteriological technique for the development of latent fingerprints. Forensic Sci Int 33(3):209–214

109. Phan K, Barash M, Spindler X et al (2019) Retrieving forensic information about the donor through bacterial profiling. Int J Legal Med 131(1):21–29
110. IFRG (2014) Guidelines for the assessment of fingermark detection techniques. J Forensic Identif 64(2):174–200
111. Dorakumbura BN, Busetti F, Lewis SW (2019) Investigations into sampling approaches for chemical analysis of latent fingermark residue. Forensic Chem 14:100166
112. Steiner R, Roux C, Moret S (2019) Controlling fingermark variability for research purposes: a review. WIREs Forensic Sci 1(4):e1338
113. Margot P (2011) Forensic science on trial – what is the law of the land? Aus J Forensic Sci 43(2–3):89–103

Chapter 8
New Chemical Imaging Approaches to Fingermark Dating by Mass Spectrometry

Simona Francese and Robert Bradshaw

Abstract Time since deposition is a type of intelligence enabling to address activity level propositions. In relation to a fingermark, this information would provide an additional evidential value by proving/disproving the presence of a suspect at the time of the crime. Whilst fundamentally important, this type of intelligence remains, to date, elusive to the forensic community. This is due to the significant number and interplay of factors that can affect the determination of age. Mass spectrometry (MS) has greatly contributed to the understanding of the changes occurring within the molecular composition of a fingermark over time, especially with regard to lipids. In particular, the hyphenation of chromatography with mass spectrometry has enabled to analyse and monitor over time the complexity of the molecular lipid composition for both volatile and non-volatile species. However, these approaches require extraction of the fingermark molecular components, and as such they are completely destructive. Conversely, chemical imaging methods allow to keep the fingermark intact by enabling in situ analysis. In addition, due to the very nature of mass spectrometry imaging, these methods offer contextual intelligence, that is, time since deposition (potentially) and biometric information for suspect identification; this is particularly desirable when conventional fingermark enhancement methods yield partial, scarce or no ridge detail.

In this chapter a survey of prior approaches to fingermark dating will be briefly covered as a point of comparing and contrasting with mass spectrometric imaging. In particular, the use of matrix-assisted laser desorption/ionization mass spectrometry imaging (MALDI MSI) will be illustrated as it is, to date, amongst the most applied techniques to the analysis of fingermarks. The authors provide a personal perspective, also supported by their own studies, on the state of the art of mass spectrometric imaging for fingermark dating. Although exciting and promising, these approaches are still considered to be in their infancy.

S. Francese (✉) · R. Bradshaw
Biomolecular Science Research Centre, Centre for Mass Spectrometry Imaging, Sheffield Hallam University, Sheffield, WB, UK
e-mail: s.francese@shu.ac.uk

J. De Alcaraz-Fossoul (ed.), *Technologies for Fingermark Age Estimations: A Step Forward*, https://doi.org/10.1007/978-3-030-69337-4_8

Keywords Aging · MALDI · Imaging · Fingermarks

Acronyms

ATR-FTIR	Attenuated total reflectance-Fourier transform infrared
CLSM	Confocal laser scanning microscope
CSI	Crime scene investigators
CWLS	Chromatic white light sensor
FET	Fingermark enhancement techniques
GC-MS	Gas chromatography-mass spectrometry
LC-MS	Liquid chromatography-mass spectrometry
MALDI MSI	Matrix-assisted laser desorption/ionization mass spectrometry imaging
MALDI MSP	Matrix-assisted laser desorption/ionization mass spectrometry profiling
m/z	Mass-to-charge
PCA	Principal component analysis
PLS	Partial least squares
OA	Oleic acid
OP	Optical profilometry
SIMS	Secondary ion mass spectrometry
TG	Triacylglycerides

8.1 Introduction

Evidence dating remains an unsolved issue across different types of traces that could be recovered at a crime scene; fingermarks, gunshot residues and biofluids are examined to provide information at a source level but not at activity levels. The age of inks (related to forensic document examination) also remains a challenge. In this case, a recent pan-European round robin study (Fisher et al, submitted) conducted within the EU-funded COST Action CA16101 – MULTI-modal Imaging of FOREnsic SciEnce Evidence – tools for Forensic Science (https://multiforesee.com/), none of the techniques employed, amongst which Raman, FTIR, Vis/NIR, multispectral imaging, secondary ion mass spectrometry and matrix-assisted laser desorption/ionization mass spectrometry, could address the question of "How old is this document?".

As for other types of evidence, the age of a fingermark would confirm or refute a suspect's statement/alibi and establish whether or not that fingermark is relevant for the investigation and, therefore, would address activity level propositions. The determination of time since deposition of a fingermark (or fingermark dating) has been debated since the 1970s with the first-ever report published in 1927 [1] and experimentation initially taking place in isolation and in a discontinuous fashion. The work of Archer et al. [2], Croxton et al. [3] and Girod et al. [4] gave a new impulse to this research with the so-called kinetic studies [5], providing meaningful insights into degradation of fingermark compounds over time.

However, while the entire forensic community certainly agrees on the importance of conquering this type of intelligence, occasional reports have been published questioning the possibility that this will ever be possible [6–8]. Girod has been consistently addressing this important gap in knowledge since 2012 [9]. In 2019, she published a review offering insights into the challenges presented by what is considered the "holy grail" in forensic science, from a practice-oriented point of view [10].

The severity of the challenges is likely due to the large range of cross-talking factors that would affect the interpretation of the age of a fingermark. The nature of the surface of deposition, environmental conditions, donor's sweat characteristics and possibly also the prior application of fingermark enhancement techniques could affect measurements and yield an apparent age of the fingermark that is different from its real age [9]. Therefore, whilst kinetic studies under a range of conditions and combination of factors are not technically impossible, they are certainly extremely challenging from a practical point of view. Effective predictive models can only be built upon informed accounting of the range of factors that can potentially affect the determination of age. The practical challenges of this enormous data gathering exercise is, in the authors' view, one of the biggest barriers undermining the recovery of this type of intelligence in the future. It is also important to consider that even if this exercise can be successfully performed with one or more analytical techniques (complementing information), it will not be possible to anticipate how "good" the protocol will be in operational terms. This is because the "ground truth" on the age, deposition and survival history of a given fingermark can rarely be fully known.

That being said, in science, however severe, challenges have never stopped research from being pursued; today, a great deal of information around fingermark molecular aging and the factors affecting age determination is known (as presented in Chaps. 3 and 7) compared to when this topic started to be debated and researched in the 1970s.

In the context of fingermark aging, analytical techniques can be classified mainly in two groups: (i) physical methods and (ii) analytical chemistry methods.

Physical methods have involved inspection of visual appearance and monitoring of the changes of physical characteristics over time. However, it has been shown that general appearance of a mark (with respect to quality), whilst sometimes used in informal practice, is unreliable. There have been documented cases of older marks being developed with "very good" quality (in itself poorly described in quantitative terms) or of similar visual quality as "younger marks" [10].

Much less empirical techniques, such as contactless chromatic white light sensors (CWLS) and confocal laser scanning microscopes (CLSM) [11, 12], have been employed for the analysis of marks on the only amenable type of deposition surface (smooth) to make quantitative determinations on "the appearance of a mark" and associations with its age [13]. However, even if an aging model was built and successfully applied to smooth surfaces, its reproducibility should be re-evaluated when factors affecting aging (other than surfaces) are considered (as discussed in Chap. 4).

Exciting quantitative investigations have also been undertaken with optical profilometry (OP) [14–17] aiming to determine the width, the area, the volume and the height of ridges, the number of *minutiae* and number and shape of pores as, collectively, a measure of time since deposition. OP has clearly shown interesting potential in the determination of these parameters. It is able to detect ridge metrics with a surface topography repeatability of up to 0.15 nm without the need of any pretreatment for visualization, yielding data in seconds (as discussed in Chap. 6).

However, the ability of "physical techniques" to estimate the age of a mark in an operational setting (crime scene) remains to be demonstrated. In fact, only in few cases an assessment was made and reported post application of crime scene investigations (CSI) or crime lab enhancement methods. These techniques have not been tested in conjunction with tape lifting either and this would be important as lifting, by its very nature, could alter some of the parameters described above (e.g. width and height of ridges). However, in the latter case, there is a clear trend to move away from tape lifting and to adopt digital portable devices (e.g. cameras and scanners) for the acquisition and transmission of photographic evidence. Some of these scientific publications have not addressed the influence of external factors [18, 19]; this again is precisely one of the major challenges in developing an implementable method that could account for all the possible intertwined factors that have an impact on fingermark age determination.

Molecular techniques have further and greatly contributed to advance our understanding of fingermark aging and will be briefly covered in the next section.

8.1.1 Spectroscopic Methods for Fingermark Age Determination

Analytical chemistry methods, such as spectroscopic approaches, have been investigated in the effort to provide both basic knowledge of a fingermark chemical degradation over time and a method that could pinpoint age on the basis of these molecular changes. However, as for physical methods, spectroscopic methods also have been largely constrained by lack of knowledge of the initial quantity of components present in a mark, which, understandably, varies significantly, both intra- and inter-donor. The vast majority of these approaches have targeted lipids and studied their degradation.

On reviewing available literature, Raman spectroscopy appears to be behind in capabilities, with the most recent study only showing applications to monitor molecular degradation over time but with no concurrent assessment of age impacting factors [20]. In any case, tape lifted fingermarks may still be problematic due to the interference of the tape chemical components. Portability may overcome this issue as the analysis would be performed *in situ*. However, there are no comprehensive studies showing compatibility of Raman with other CSI or crime labs enhancement techniques. Furthermore, Raman notoriously suffers from fluorescence back scattering from certain surfaces of deposition; this occurrence, together with possible sensitivity issues,[1] may explain why only one study employing Raman for fingermark dating could be retrieved from the literature.

Conversely, the literature shows that Fourier Transform Infrared (FT IR) spectroscopy has been a much more pursued and developed analytical approach to fingermark dating. In a one of a kind study (preceded by a pioneering one by Hemmila et al. in 2008 [21]), Antoine et al. investigated lipid degradation in children versus adults using FT IR in 2010 [22]. The results of their research, however, had pointed out a method to pinpoint age of an individual rather than indicating the time since deposition, thanks to the relative abundance of classes of compounds normally present in sebum.

To the authors' knowledge, Ricci et al. were the first to utilize an initial explorative attenuated total reflectance (ATR)-FTIR approach for kinetic studies in the context of fingermark aging [23]. In particular, lipid degradation and other compounds such as urea were monitored in undeveloped fingermarks over 44 days and at different temperatures ranging from room temperature (1 time point at 15 days) to 80 °C. Molecular images of the relevant vibrational bands were generated showing gradual degradation with increasing temperature and time. In 2015, Girod et al. delved deeper into the use of ATR-FTIR to investigate aging marks and the impact of substrates and storage conditions, using a chemometric approach and one donor [4]. Interestingly, they discovered that marks of the same age on different surfaces could be grouped together on the basis of their infrared spectral patterns (reflecting molecular composition), when they were stored in an open box kept in an air-conditioned laboratory at around 20 °C next to a window. However, when stored in the dark, only fingermarks deposited on the same substrate could be grouped by age, thus showing the importance to study aging within cross-talking conditions where two or more variables at any one time are tested for their impact on fingermark age classification. The study allowed the determination of fingermark age (for 1-day-old, 3-day-old, 7-day-old, 9-day-old, 20-day-old and 34-day-old fingermarks) within a 3-day error (Fig. 8.1).

Very importantly, whilst in the study of Ricci et al. [22], fingermarks were laid directly on a ZnSe ATR crystal, in the study by Girod et al., the tip of the germanium crystal was contacted with the deposited fingermarks making the last study potentially more "operational" in the field.

Although spectroscopic methods are characterized by higher specificity than "physical methods", as changes are measured at a molecular level, the degree of

[1] Due to the fact that approximately only one photon out of a million undergoes Raman scattering, the Raman signal is often very weak for analytes present at low concentrations, such as within fingermarks

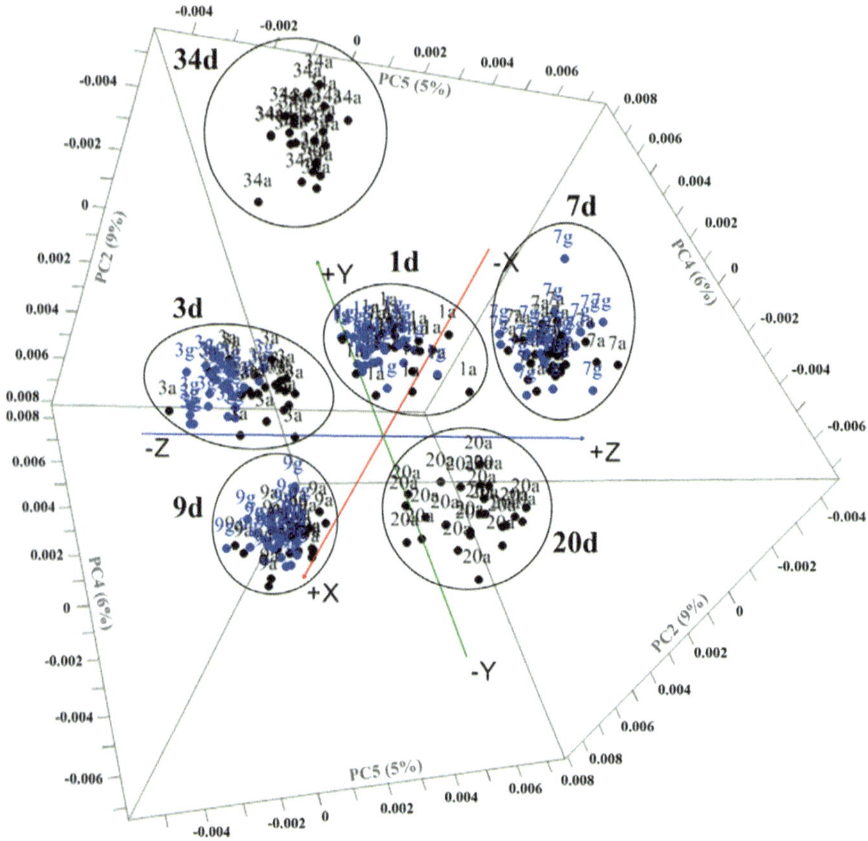

Fig. 8.1 PCA score plot of ATR-FTIR spectra of fingermarks aging in the light over 34 days. Fingermarks aged over 9 days on glass are in blue, and those aged over 34 days on aluminium are in black. Three principal components enabled separation of all ages into six groups (1, 3, 7, 9, 20 and 34 days). Reprinted with permission from Ref. [4], 2015 Elsevier Ireland Ltd

specificity is still limited. In fact, all of these techniques do not specifically detect a potential molecular marker but chemical groups, and this information is used to infer the presence of particular classes of compounds and sometimes of a specific compound when spectral libraries are available. Therefore, these techniques are likely more suitable for the provision of spectral patterns changing over time which can then be statistically modelled. In a different spectroscopic approach employing fluorescence spectroscopy, van Dam et al. reported the opportunity to estimate the age of 55% of the male fingermarks, up to 3 weeks old with an uncertainty of 1.9 days. The statistical model was built on the measurement of the fluorescence emitted by the oxidation products of proteins and lipids in time. The model accounted for both the relative amount of fluorescent oxidation products and the rate at which these products are formed [24].

An approach that stands out in the literature is the application of electrochemical impedance spectroscopy. Reported by Rosa et al. in 2017, this approach, only applicable to conductive surfaces, exploited the morphological changes to which marks undergo (clearly dictated by compositional changes) to inform spectroscopic measures [25]. Indeed, the authors found that physical modifications of the mark in time (stored in laboratory conditions) significantly affect the impedance spectra of the system (the mark and the deposition surface). This approach is extensively covered in Chap. 9. Whilst very encouraging, similarly to the studies of van Dam et al. [24] and Girod et al. [4], robustness and versatility of the technique (including compatibility with prior application of fingermark enhancement techniques) must be investigated prior to the methods providing an indication of their potential to be operationally implementable.

It is important to note that some studies do not report how fingermarks were prepared [22], and in most of the other studies, groomed marks were investigated, with the exception of the study by Rosa et al. [25]. Here, donors were asked not to wash their hands for at least 60 min before deposition, as well as to rub their hands together before deposition. As the authors of this chapter consider "natural marks" those that are generated without any restrictions and recommendation, those employed by Rosa et al. do not fit this definition. However, it is important that Rosa et al.'s marks still move away from sebum-loaded fingermarks and are closer to the concept of natural marks as described and applied in a few of our group's studies on fingermarks [24, 26, 27].

If spectroscopic methods are more specific than physical methods, mass spectrometric approaches offer an even greater level of specificity in the investigation of potential biomarkers of fingermark aging. This is due to mass spectrometry (MS) measuring an intrinsic property of molecules, namely, their molecular weight through their mass-to-charge (*m/z*) ratio; therefore, detection of individual compounds, or classes of compounds, is not inferred from library matches or from "characteristic" vibrational bands such as in FTIR, but it is ascertained through the measurement of the *m/z* of the compounds with the option of molecular fragmentation for structural characterization and confirmation. These methods have provided spectral profiles of the molecular content of a fingermark in various age determination studies and will be covered in Sect. 8.1.2.

8.1.2 Hyphenated Mass Spectrometric Techniques for Fingermark Dating

Chromatography coupled to mass spectrometry (hyphenated techniques) has proven to be an invaluable approach in deciphering the chemical composition of latent fingermarks for more than two decades. In fingermark aging studies, these sensitive and informative analytical techniques have been predominantly employed for two main reasons: (a) to determine changes in the chemistry of fingermark residue over time, informing on the choices and/or development of fingermark enhancement techniques and (b) to investigate changes in the composition of residue constituents over time as a means to perform fingermark age estimation (kinetic studies). An overview of the chemical changes that occur in fingermarks, including the informative results obtained from the use of hyphenated techniques in aging studies, has been covered extensively in Chap. 7. Therefore, this section will reintroduce the topic but will discuss it in the light of the potential pitfalls associated with these methodologies.

Gas chromatography-mass spectrometry (GC-MS) is the "gold standard" for chemical profiling in forensic science, and, historically, it has been the most extensively employed of the hyphenated techniques to specifically pinpoint chemical differences in fingermark residue as a result of aging. Its first documented use was in 1996 when Buchanan et al. [28] indicated that adult fingermarks persist for longer when deposited onto a surface due to a greater abundance of less volatile components which increase following puberty. Although this paper did not specifically focus on aged marks, it highlighted how the initial composition of residue can have an effect on the composition of fingermarks as they age. A number of GC-MS-based studies have since been undertaken in order to determine fingermark composition as a function of time [2, 29–33]. However, the findings of Buchanan et al. remain important as a reminder that the accuracy of predictive age models may be negatively affected by the fact that they cannot take into account the initial fingermark chemical composition (as it is not known). Archer et al. (2005) [2] acknowledged that storage conditions can affect changes in fingermark composition. This raised an important issue for the potential age determination of fingermarks, based on the inevitability that they can be exposed to a variety of different environmental conditions at crime scenes. In 2011, Weyermann et al. highlighted additional pitfalls in exploiting chemical changes as an indication of fingermark age [30]. They discovered that differences were observed in the composition of residue deposited between donors (intervariability) and by the same donor (intravariability), the amount of material transferred depending on the substrate (porous/non-porous) and also compositional changes dependent on the substrate.

Importantly, the application of the aforementioned techniques in a forensic context is subject to locating any fingermarks prior to analysis. Koenig et al. trialled three different enhancement techniques, i.e., magnetic powder, 1,2-indandione and cyanoacrylate fuming, prior to residue extraction [31]. Issues that were encountered included instrumentation contamination resulting in co-elution with analytes of interest (in the case of magnetic powder and 1,2-indandione), a reduction in target lipids and potential cross contamination due to dissolution in the immersion solvent (observed with 1,2-indandione). No issues were encountered using cyanoacrylate fuming in that particular sequential workflow. Ultimately, the authors were able to demonstrate that an optimized methodology for determining analyte abundance (employing ratios of peak areas derived from common fingermark constituents) proved to be relatively consistent between samples and tackled issues with inter- and intravariability of fingermarks. This provided a positive indication for the potential application of this methodology for fingermark age determination.

However, it is important to note that analysis of aged marks was not conducted and, therefore, its feasibility in this context was not proven.

In a later study, Girod et al. applied chemometric techniques, including principal component analysis (PCA) and univariate exponential regression to determine the effects of aluminium powder development on fingermark age determination [33]. Initially, PCA showed that there were clear differences in the chemical composition of enhanced versus non-enhanced fingermarks deposited in the same conditions. Subsequent application of an exponential linear regression model showed that fingermark enhancement hindered age estimation as this factor provided greater variation than the aging effects on fingermark composition. Most certainly, these studies demonstrated that fingermark age determination is hampered by a complex system of cross-talking factors.

From an analytical perspective, one of the major issues with the use of GC-MS is the inability to analyse non-volatile components of the residue, which often requires derivatization to improve analyte volatility [2, 28, 29, 32]. One potential means to overcome this is to use analytical methodologies which do not depend upon the volatile nature of compounds, such as liquid chromatography-mass spectrometry (LC-MS). Indeed, such an approach has been reported in the literature covering the analysis of fingermark constituents such as lipids [34, 35].

In an entirely different approach to what has previously been encountered in the literature, Oonk et al. targeted endogenous peptides and proteins using LC coupled to high-resolution MS [36]. Employing a bottom-up proteomics methodology, the authors were able to identify ~50 proteins/peptides within the fingermark proteome, 5 of which showed measurable responses as a result of aging and were not significantly affected by the age or sex of the donors. Specifically, the authors focused on the abundance of some keratin proteins and an antimicrobial peptide (dermcidin) and also the extent of amino acid residue modifications following aging over a 12-day period. The authors showed that the analytes of interest could still be detected when fingermarks had been contaminated with other biological traces (such as semen and saliva), although the total extracted proteome was reduced to 30–50% of that obtained from uncontaminated marks. Importantly, "natural" marks were employed, making this study relevant to realistic forensic applications; however, this resulted in a significant reduction in total protein content compared to previous studies that used "eccrine" and "groomed" fingermarks followed by extraction of residue directly from the fingertip [37]. This suggests that sensitivity issues may be encountered if recovered fingermarks are particularly depleted (donors with extremely low sweat secretion levels, extremely old marks, extreme environmental conditions). It must also be noted that no fingermark development was performed prior to fingermark extraction, only one type of substrate (glass slides) was used, and the same environmental conditions were maintained throughout; therefore, the effects of these variables on the fingermark proteome over time are still unknown.

When collectively considering hyphenated techniques, they are restricted by the same prerequisite, a necessity to extract fingermark constituents into an appropriate solvent prior to analysis. Inevitably, this will result in the destruction of the fingermark sample. A potential solution would be the use of alternative mass spectrometry techniques, such as those that can analyse directly from the surface of deposition or from lifting tape that has been used to recover marks from a surface. In Sect. 8.2, these authors will cover the application of mass spectrometry imaging approaches as a potential alternative for latent fingermark dating.

8.2 Mass Spectrometric Chemical Imaging Approaches for Fingermark Dating

The first example of non-hyphenated, *in situ* mass spectrometric method to investigate fingermark dating was offered by Wolstenholme et al. in 2009 using matrix-assisted laser desorption/ionization mass spectrometry (MALDI MS) [38]. In MALDI MS, ion generation is assisted by a chemical compound named the "matrix" and is triggered by a laser having a wavelength in the UV range (for UV-MALDI). The matrix is generally a low-molecular-weight organic compound with low volatility that protects the analytes from the direct impact of the laser, absorbing its energy due to an extended conjugation of double bonds and then returning to the ground state by protonating or deprotonating the analyte. Therefore, the technique is considered soft as it generates very little or no fragmentation, thus greatly simplifying spectral interpretation. Two ionization models are reported, and these have been briefly summarized by Francese et al. [39]. MALDI MS can be used in profiling mode (MALDI MSP), where discrete spectra are acquired within seconds, and imaging mode (MALDI MSI), where analytes are both detected and visualized within the specimen being analysed. The development of this technique has been pioneered by Francese's group, specifically for application to fingermark analysis, and has demonstrated the ability to perform chemical criminal profiling whilst generating images of the fingermark ridge pattern and, thus, biometric information for suspect identification [39]. Here, intelligence can be gathered on a suspect around their lifestyle and/or action prior to or during committing the crime. Importantly, the technology can be used compatibly after the application of many CSI and crime lab enhancement techniques. The latter capability has allowed for MALDI MS-based methods to be used operationally [40].

In unprecedented research, Francese's group employed MALDI MS and MALDI MSI to follow the degradation of standard oleic acid (OA) at *m/z* 283.2 which is a fatty acid commonly found in fingermarks [38]. MALDI MS spectra of this standard were obtained over 7 days in the dark, at different temperatures (4, 37 and 60 °C) showing degradation into dehydrated OA (*m/z* 265.2) and di-dehydrated OA (*m/z* 247.2). As expected, mass spectral profiles collected over 7 days did not show significant degradation of OA when it was kept at 4 °C. At higher temperatures, the OA signal decreased although with a higher rate in the first day. This can probably be explained by the degradation of higher-order lipids making it appear that OA is more slowly degrading after the first day. Dehydrated OA shows degradation trends that seem accelerated at higher temperatures. Whilst this approach uniquely provided insights into a single species degrading over time at different temperatures, it was clearly still insufficient to even suggest it as a potential method to pinpoint the age of a mark.

Chemical imaging methods are generally of interest to both the analytical and forensic community because of the reduced or no sample preparation, the ability to visualize the localization of molecules (with or without an a priori knowledge) and the much reduced invasiveness of analysis, which is important in terms of the preservation of the evidence. These techniques can analyse the specimen of interest *in situ*, and Francese's group was the first to explore the possibility to use a mass spectrometric chemical imaging method for dating a mark; here, MALDI MS Imaging (MALDI MSI) was employed to image groomed marks at zero hour and at 7-day time points at different temperatures (Fig. 8.2).

Whilst thousands of molecules could be collected in one single analysis, data processing focused again on OA and its degradation products. The density maps generated enabled a number of observations. First and foremost, for the 0 hour time point and the aging at 4 °C, the visualization of dehydrated OA and di-dehydrated OA yielded similar density maps, and it was consistent with the speculation of their formation in the mass spectrometer. Indeed, Bailey et al. reported on compound degradation by mass spectrometry techniques operated under vacuum, like in the case of typical MALDI [41]. While no study can disprove what was reported by Bailey et al., possible compound degradation due to vacuum conditions would not prevent the use of MALDI for dating studies, as once molecular degradation in vacuum has been observed and understood, appropriate normalization can be applied to the dataset to account for it.

As the temperature increased from 4 °C to 60 °C, the OA density map decreased, in agreement with the preceding OA profiling studies and with the spectra of OA obtained by extracting the region of interest from images. With respect to dehydrated OA, the density maps appeared different across the three temperatures investigated with an increase when raising the temperature from 4 °C to 37 °C (in line with the interpretation of OA degradation) and almost no visualization at 60 °C, presumably due to degradation of its carbonyl structure. The di-dehydrated OA followed a similar trend, but all density maps were different after signal normalization against the MALDI matrix at *m/z* 190.02. This is a significant point as it enabled to speculate on a different approach to fingermark dating. Rather than simply following lipid degradation ("disappearance") over time, a relative approach could be used where multiple molecular maps of intertwined species (such as a molecule and its degradation products) could be used to inform a statistical model enabling to predict age. In 2010, this was further tested in unpublished research by Bradshaw and Francese where the humidity and temperature were kept constant at 60% and 25 °C, respectively, using cycles of light and dark; groomed marks were analysed at 0, 1, 4, 10 and 14 days post-deposition (Fig. 8.3).

In this case, a trend was observed for OA image intensity and distribution which was similar between the fresh and 1-day-old marks and decreased over time until it was no longer useful for suspect identification following aging for 14 days. These preliminarily data suggested that for groomed marks, it was in principle possible to

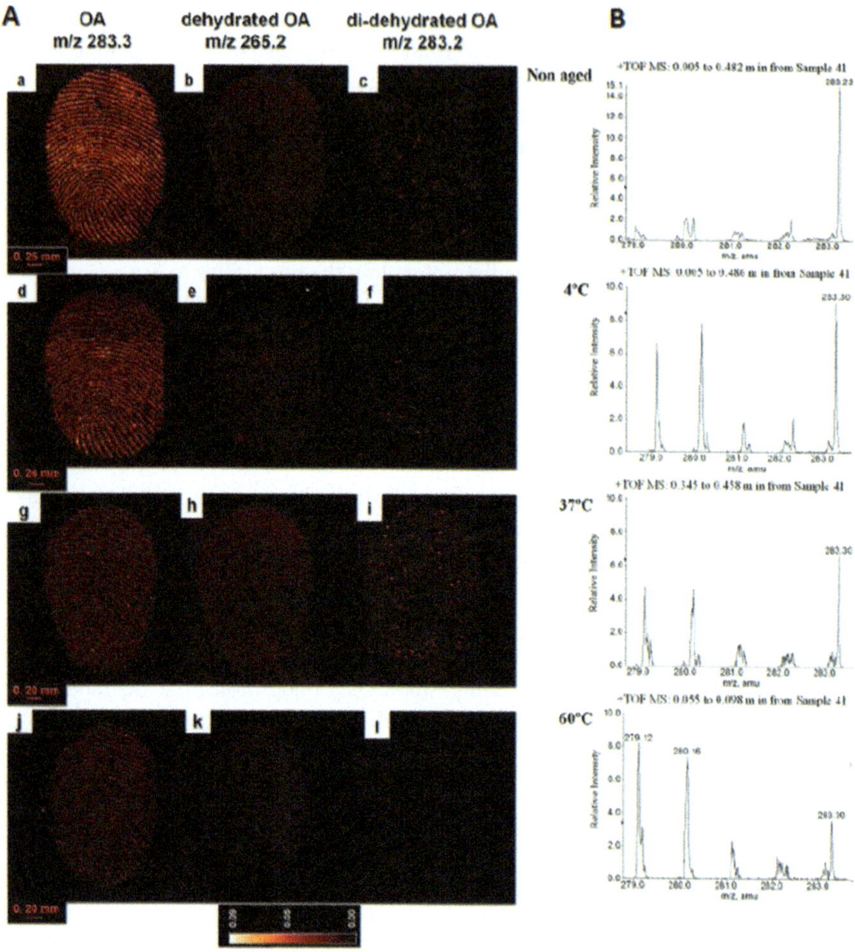

Fig. 8.2 MALDI-MSI analysis of aging groomed marks. Panel A, insets a–c show the MS images of OA (*m/z* 283.27), dehydrated OA (*m/z* 265.25) and di-dehydrated OA (*m/z* 247.24) in a non-aged groomed fingermark. The same set of MS images for marks stored at 4 °C, 37 °C and 60 °C are shown in panels d–f, g–i and j–l, respectively. Images indicate lipid degradation, as the temperature increases, with distribution maps for the species of interest that significantly change between 4 °C, 37 °C and 60 °C. Panel B reports MALDI-MS extracted spectra of OA from the corresponding fresh and aged fingermark images, showing the decrease in OA signal intensity upon storage temperature increase (Reprinted with permission from Ref. [38], Copyright © 2009 John Wiley & Sons, Ltd.)

predict age over a 14-day period. The images at 10- and 14-days post-deposition for the three molecular species showed a particularly different distribution and intensity. However, lack of detection of di-dehydrated OA at 10 and 14 days suggested that using this approach for dating much older marks than 10 days may be unsuitable. A natural follow-up was a quantitative densitometric assessment of the images for all three species and the development of a model pinpointing the age of a mark

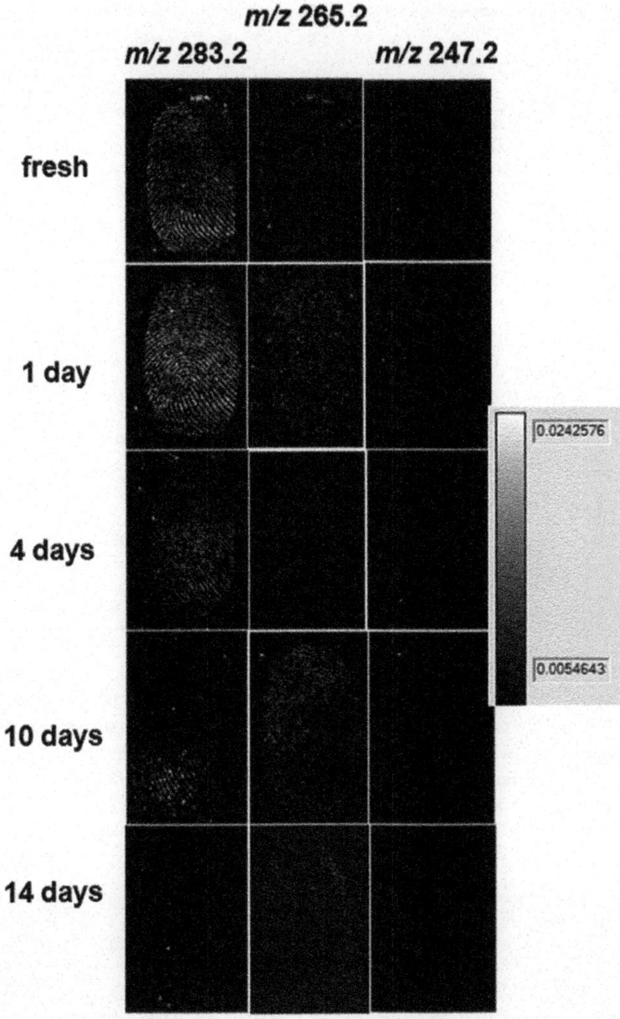

Fig. 8.3 MALDI-MSI analysis of aging groomed marks analysed at 25 °C and over a 14-day time course. Unpublished data, Bradshaw and Francese. MS images of OA (*m/z* 283.27), dehydrated OA (*m/z* 265.25) and di-dehydrated OA (*m/z* 247.24) are shown for the different time points

based on the relative intensity of the three molecular images. Initially, an existing software called Image-Pro was evaluated for its performance in providing such information from an image. Figure 8.4 shows unpublished data by Francese demonstrating the possibility of a (crude) quantitative determination of the "image pixel density" from molecular images of OA at different fingermark aging points.

Subsequently, in 2011, a fingermark image processing-dedicated in-house software (.NET Framework program, written in C#) was conceptualized whereby an "intensity score" was assigned. Figure 8.5 shows an initial proof of concept on

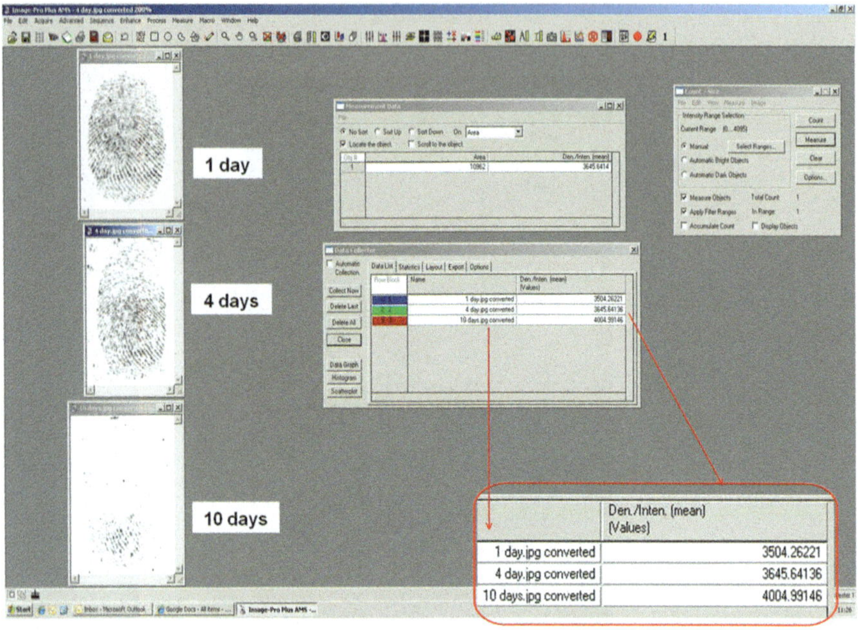

Fig. 8.4 Fingermark images density maps and quantitative assessment obtained using Image-Pro (Media Cybernetics). Panel A shows the fingermark images that were submitted for analysis. Panel B shows the Image-Pro processing with measures of the density map (highlighted in the red frame). Although the software has operated counter intuitively (one should expect the most visual intensity/density of distribution the higher the pixel count), it does show the possibility to measure changes quantitatively for future handling by statistical modelling

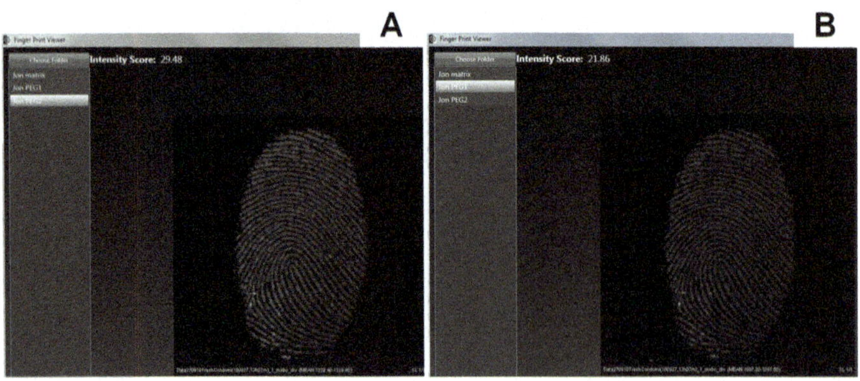

Fig. 8.5 Fingermark images at *m/z* 1229 and 1007 derived from polyethylene glycol (PEG) in a condom lubricant contaminated fingermark. The density map value for the *m/z* 1007 is lower (29.48) than that for the ion at *m/z* 1229 (21.86), as was visually expected. Unpublished data, courtesy of J. Wheat, Sheffield Hallam University

existing data published by Bradshaw et al. in 2011 [42] derived from "condom lubricant contaminated" fingermarks.

Due to lack of funding, these studies could no longer be pursued, though the natural progression would have been the collection of images of fingermarks aging at different times (and different conditions) and using the intensity score ratios (offered by the images of the species being targeted) to build a statistical model in both supervised and unsupervised approaches. In an alternative approach reported by Reed et al. in 2016, Francese's group investigated the entire detectable lipid content, as present in a lipid pad, within the *m/z* range 100–1000, rather than focusing only on one species and its degradation products [43]. Silicon fingers were used, and fingertips were enriched with the content of a lipid pad to generate marks in triplicates upon contact with an aluminium slide. The marks were aged at 0, 1, 4, and 8 days at constant 37 °C with 5% CO_2 in a cell incubator. Marks were then imaged by MALDI MSI, and regions of interest (the entire mark) were extracted to yield the averaged spectral profiles. This dataset was preliminarily processed by principle component analysis (PCA) which yielded discrimination of the different aging times by using three principal components. The data were subsequently analysed via partial least squares (PLS) regression where the age information was used as a part of the model. Here, clustering of aging points could still be observed (Fig. 8.6) within a linear model with $R^2 = 0.989$. This value of R^2 indicates a fairly good representation of the data.

The opportunity to cluster fingermarks of the same age, effectively determining age, further moved this study away from kinetic studies where compound

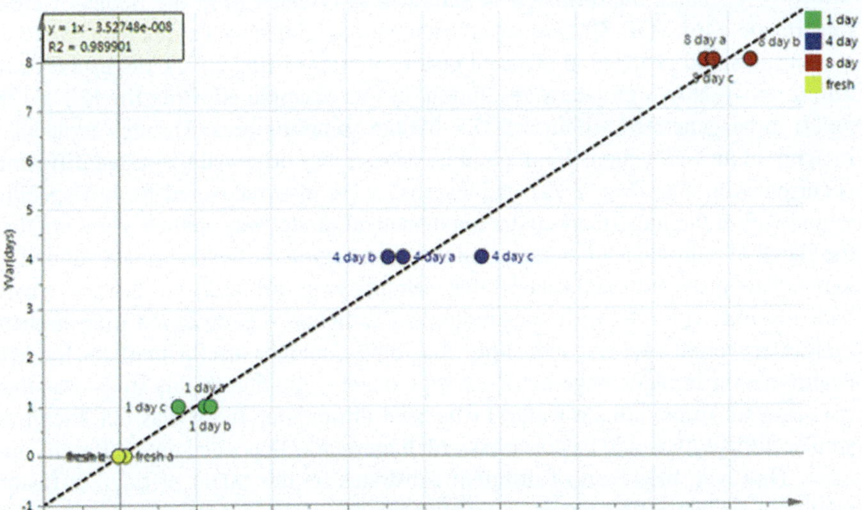

Fig. 8.6 PLS regression analysis of the MALDI MSI data for 0-, 1-, 4- and 8-day-old lipid spiked-silicon model originated fingermarks. (Reprinted with Elsevier permission from Ref. [43], under the C BY-NC-ND license (http://creativecommons.org/licenses/by-nc-nd/4.0/)

degradation is monitored over time. Whilst promising, this approach used lipid pads, which, despite containing fatty acids, diacylglycerols and triacylglycerols present commonly in fingermarks, it is not representative of the complexity of a real fingermark. As for many other studies, this study was also very simple by keeping the temperature and humidity constant. Finally, marks were not previously enhanced by any fingermark enhancement technique (FET) though many publications from our group suggest compatibility of MALDI MS based methods with several FET. For these reasons, this study was also limited in its significance with respect to the potential for fingermark dating. However, it was linked to the development of a contactless method to dispense fingerprint powders; as such, the study mainly aimed to prove that the matrix could be homogeneously deposited within the application of the *dry-wet* method, a method that enables visualization of the mark, recovery and analysis by MALDI MSI. Extensive details of the *dry-wet* method can be found in previous publications [44, 45].

On a general note, mass spectrometric chemical imaging is very interesting as it shows potential, with further developments, to not only date a fingermark but also to provide contextual suspect identification due to the reconstruction of the fingermark ridge pattern (Fig. 8.3).

Aside the studies from Francese's group, only four more scientific articles had been published on fingermark dating. When considering MSI studies for fingermark age determination overall, they fall into two main categories: (a) molecular imaging to display physical changes in the ridges through diffusion of selected analytes and (b) molecular imaging to show changes in the chemical composition of the residue. Muramoto and Sisco utilized time-of-flight secondary ion mass spectrometry (TOF-SIMS) to visualize the diffusion of saturated fatty acids from the ridges of aged fingermarks [46]. TOF-SIMS is an advantageous analytical technique with regard to measuring small changes in physical characteristics of fingermark ridges as it can afford remarkably high lateral resolution, in this example allowing for 128 × 128 pixels to be generated within a 500 × 500μm sampling area. Through molecular imaging of an endogenous lipid, palmitic acid, it was observed that ridge diffusion occurred over time (Fig. 8.7a). "Linescans" were also recovered from a specific region within the fingermark and a mathematical model was applied, showing that the signal of palmitic acid increased in the outer region of the fingermark following aging. This provided confirmation that palmitic acid diffused onto the previously bare substrate (Fig. 8.7b). Interestingly, less diffusion was observed from higher-molecular-weight species, indicating that they could be useful markers for age determination in older fingermarks. A significant limitation in this study was that the same substrate (silicon wafers) was used throughout and it was not known if similar diffusivity would be observed with fingermarks deposited onto different surfaces. However, the effect of different substrates on the extent of ridge diffusion would be an interesting factor to account for due to the variability of deposition surfaces that can be found at crime scenes. Indeed, this has since been addressed by O'Neill et al. who used MALDI MSI to visualize diffusion of endogenous molecules over time, specifically fatty acids and triacylglycerides (TGs), in fingermarks deposited onto four different substrates [47]. These authors showed that the

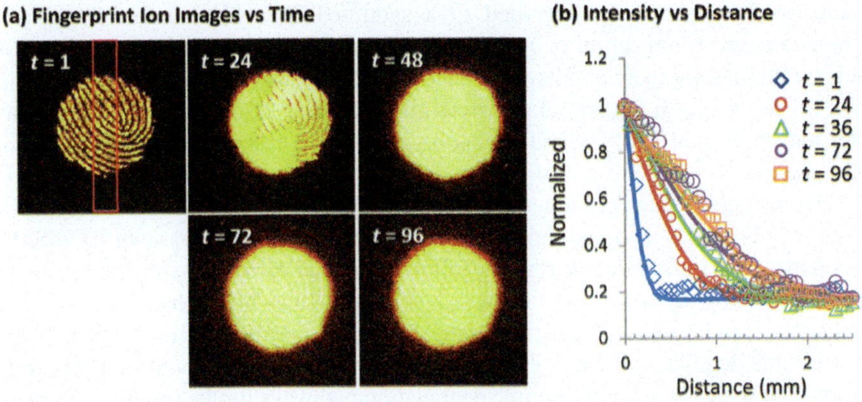

Fig. 8.7 (**a**) TOF-SIMS images of palmitic acid over different time points (t = 1–96 h) following deposition. The red rectangle shows a region of interest (ROI) from which "linescans" were obtained. (**b**) Linescans of palmitic acid signal intensity across a fingermark sample, with measurements taken at different time points (Reprinted with permission from Ref. [46]. Copyright (2015) American Chemical Society)

deposition surface had a much greater effect on diffusion of the endogenous compounds (due to the polarity of the substrate and surface interactions) than the molecular weight of the fingermark constituents which had previously been suggested as the prominent factor [46]. Either way, it would be very interesting to see an integrated approach combining the molecular changes, visualized by imaging techniques with the morphological ones observed, for example, by using optical profilometry. These results further emphasize complications in the use of diffusivity as a measure of fingermark age, and additional research is required to truly understand the effects of different and cross-talking variables encountered in forensic investigations.

Notable limitations were observed regarding the sample preparation employed in this study. Firstly, fingermarks were subjected to unrealistic, accelerated aging conditions in an oven at 55 °C for 24 h. This would have an impact on the observations made from fingermarks deposited onto polyethylene terephthalate (PET) more than other surfaces, though the authors were not able to determine what surface property of PET was changing at high temperatures. Due to this artificial aging, it would be challenging to directly translate the findings from this study to fingermarks typically found at crime scenes. Secondly, fingermarks were prepared differently for mass spectrometry analysis depending on the target analyte. For example, for fatty acids, sample preparation was performed by silver sputter coating, whereas for TG's, sublimation of the MALDI matrix (2,5-dihydroxybenzoic acid) was performed. The extent of lateral diffusion of analytes based on the sample preparation methodology is an unknown variable in this instance, thus making the results incomparable.

The predominant amount of research in fingermark age determination using MSI has been based on observing changes in the chemical composition of residues. In a study by Pleik et al., LC-MS was used to indicate that a prominent (TG) (48:1) detected in fresh marks degraded to a saturated TG (48:0) and a TG(48:0)-monoozonide as a result of ozonolysis [34]. MALDI MSI was used as a confirmatory technique for these results. It enabled the visualization of the TG (48:1) within the ridges of a fresh fingermark, whereas after aging for 26 days, the TG (48:1) was no longer present and TG(48:0)-monoozonide could instead be observed within the ridges.

These authors later generated LC-MS-based aging curves that showed a reduction in TG (48:1) and formation curves that showed an increase in TG(48:0)-monoozonide in fingermark residue following aging. However, similarly to prior studies employing the use of hyphenated techniques, they highlighted that variation was observed between donors and storage conditions. Interestingly, it was hypothesized that degradation of endogenous fingermark components could be affected by influencing factors such as multiple degradation pathways for the same endogenous marker (increasing degradation) or the presence of protective substances such as skin care products (slowing degradation). The high variability of exogenous contaminants on fingermarks in realistic conditions makes the latter point important for age determination based on changes in abundance of endogenous markers.

Hinners et al. provided additional evidence to the degradation of unsaturated TGs by ozonolysis [48]. Using MALDI MSI (Fig. 8.8), the authors showed that unsaturated TGs (in blue), contained within the fingermark ridges, decrease in abundance over a 7-day period. Other oxygen-containing hydrocarbons (of the O7 series) (sodiated $(C_{41}H_{76})_{7,}$ in purple), thought to derive as a degradation product of the TGs, increased over time. A greater extent of unsaturation in the molecule was shown to increase the rate of degradation which is further emphasized by molecular images of consistent quality/intensity obtained for a saturated TG across each time point (sodiated TG 48:0, in red).

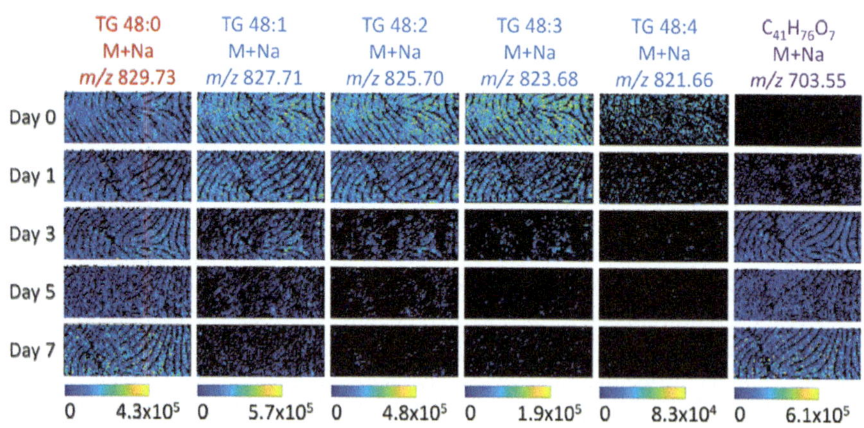

Fig. 8.8 MALDI MS images of fingermarks aged over a 7-day period. The images include a saturated TG (red), unsaturated TGs (48:1–4) (blue) and an ozonolysis product of the TGs (purple) (Adapted with permission from Ref. [48]. Copyright (2020) American Chemical Society)

In an additional set of experiments, a selection of the fingermarks was also analysed following carbon powder enhancement. Similar trends to the non-enhanced marks were observed, with an increase in ozonolysis peaks and decrease in unsaturated TGs within the MS images following fingermark aging. Importantly, this provided a positive indication for the forensic applicability of the proposed methodology and is a necessary development from previous studies [34, 38, 43, 46, 47].

8.3 Conclusions

From a strictly mass spectrometric point of view, hyphenated techniques such as GC-MS and LC-MS have greatly contributed to the understanding of the molecular changes occurring in a fingermark with aging. The findings obtained by these methods were probably responsible for triggering and inspiring further research into one of the most challenging and unresolved gaps in knowledge in forensic science, that is, the determination of the time since deposition of a fingermark. However, these methods require extraction of the fingermark residue and as such destruction of the evidence. Whilst, in some cases, this may be allowed (e.g. if a good photographic image of the mark has been acquired and the chain of custody has been maintained), partially destructive or non-destructive techniques will always be more desirable.

Mass spectrometric chemical imaging techniques may represent an interesting solution to fingermark dating as they can map chemical changes within a mark by keeping the ridge pattern intact as well as generating "molecular" images of the fingermark. This also means that, in principle, these methods may provide the age of a mark together with the provision of biometric information for suspect identification. However, the first-ever publication on a MSI method (MALDI MSI) to investigate fingermark aging was published by our group in 2009. Since then, only four more reports in this niche area of forensic science research have been published.

MALDI MS-based methods are the most published and investigated for chemical profiling and imaging of fingermarks (for forensic purposes). It is not surprising that five out of the overall six papers published in this area of fingermark dating have employed MALDI MSI as chemical imaging technique. Across the different studies, MALDI MSI approaches have followed either a particular molecular target or a class of molecules to measure molecular degradation which was then modelled to discriminate marks on the basis of their ages. Triacylglycerols' degradation by ozonolysis and molecular mapping by MALDI MSI appear to be an especially interesting system to be exploited for fingermark dating. This approach is similar to that published by Wolstenholme et al. in 2009 and resonates with our unpublished data in 2010 which, in the modelled estimation of fingermarks' age, consider the interplay between a marker and its degradation products over time rather than the degradation curve of a single species. Also, the ozonolysis MALDI MSI study has successfully assessed the compatibility with the prior application of carbon powder as fingermark developer, which is not surprising given the previous work of our group on MALDI compatibility with fingermark enhancement techniques, but it is pleasing nonetheless. Ultimately, forensic applicability of any method/techniques needs to consider and address compatibility with crime lab and crime scene enhancement techniques and/or establish at which point of the forensic workflow they can be employed. All the MALDI MS-based aging studies have considered lipids as potential aging biomarkers. Whilst amino acids, peptides and proteins have also been investigated as potential molecular targets using other techniques (although to a much lesser extent), they would be less suitable than lipids for MALDI MS-based investigations. These species can certainly be detected by MALDI MS and MSI, but lipid ionization efficiency is typically much greater which explains this target's selection for MALDI-based aging studies. Furthermore, whilst MALDI has a good dynamic mass range and can detect some of the above molecular species together (e.g. lipids and small peptides, peptides and small proteins, lipids and amino acids), lipids typically suppress ionization of other species. The ionization suppression is so significant that, for example, even if only the protein range is analysed in fingermarks, sample preparation is employed to eliminate the lipid content as much as possible [49, 50].

MSI studies have highlighted, once more, the complexity of the system investigated. This is reflected by the evident simplification of the system in the individual studies. Similar to studies employing hyphenated techniques, all the reports covering the use of MALDI MSI for fingermark dating have used so-called "groomed" fingermarks. The grooming procedure requires enrichment of the fingertips with sebaceous material from the face of the donor. This generated a fingermark abundant in certain classes of molecules, such as lipids, which were the target analytes in each of the studies. In one case only, fingermarks were deposited using the same pressure and contact time (Reed et al.) [43]; this design of experiments enabled to assess the approach, in its proof-of-concept phase, by excluding the possible complication given by the variable pressure and contact time. The downside of this approach is that no observation can be made on the robustness of the age classification model. However, in this study, the matrix was deployed with a contactless device that is being developed to possibly replace the use of brush and powders, known to be prone to crime scene cross-contamination. If the fingermark molecular composition is or will be the exploitable means to determine a fingermark age, then powder-based enhancement techniques will have to be deployed in a contactless manner. The use of groomed marks may be an unrealistic representation of crime scene fingermarks, and further work is most certainly required to determine whether the results are consistent with those that could be obtained from natural marks. Having said that, the ground truth on the initial molecular composition of the fingermark at the time of deposition will never be known. Therefore, the assessment of the robustness of a model would require the analysis of marks in the order of thousands of specimens to account for the variability of the fingermark composition.

Additionally, any method aiming to be employed operationally would also need the exploration of a number of cross-talking factors covering the wide range of environmental conditions (including extreme conditions such as immersion in water for long and brief times or exposure to very high temperature such as in a fire), various surfaces of deposition and the application of diverse enhancement techniques.

Whilst, initially, these variables can be investigated individually, it is also important to investigate them in a cross-talking fashion to assess response and robustness of the model.

The authors envisage a platform where chemical images are assessed in a quantitative and relative manner with respect to a marker and its degradation products. The platform should integrate reference libraries built from marks subjected to different and cross-talking conditions with statistical modelling assigning matching score to a probable age. This would require enormous collaborative efforts breaking the isolation of individual studies and scaling up the financial resources to be proportional to the level of the challenge presented by fingermark dating.

Ultimately, despite significant progress in the understanding of the molecular basis of fingermark aging as well as the use of sophisticated MSI technology, the authors feel that without global collaborative efforts, the forensic implementation of mass spectrometry-based approaches for fingermark age determination is still in its infancy.

Acknowledgements Prof. Jon Wheat (Sheffield Hallam University) is gratefully acknowledged for generating in 2011 unpublished data shown in Fig. 8.5 as part of the EPSRC Engineering for Life research grant. Dr. Leesa Ferguson is also gratefully thanked for generating unpublished data in 2010 during PhD research supervised by Prof. Francese and shown in Fig. 8.3.

References

1. Heindl R (1927) System und Praxis der Daktyloskopie und der sonstigen technischen Methoden der Kriminalpolizei, 3rd edn. De Gruyter, Berlin
2. Archer A, Charles Y, Elliott J, Jickells S (2005) Changes in the lipid composition of latent fingerprint residue with time after deposition on a surface. Forensic Sci Int 154(2–3):224–239
3. Croxton RS, Baron MG, Butler D, Kent T, Sears VG (2010) Variation in amino acid and lipid composition of latent fingerprints. Forensic Sci Int 199(1–3):93–102
4. Girod A, Xiao L, Reedy B, Roux C, Weyermann W (2015) Fingermark initial composition and aging using Fourier transform infrared microscopy (m-FTIR). Forensic Sci Int 254:185–196
5. Weyermann C, Ribaux O (2012) Situating forensic traces in time. Sci Justice 52(2):68–75
6. Cadd S, Islam M, Manson P, Bleay S (2015) Fingerprint composition and aging: a literature review. Sci Justice 55:219–238
7. Girod A, Ramotowski R, Lambrechts S, Misrielal P, Aalders M, Weyermann C (2016) Fingermark age determinations: legal considerations, review of the literature and practical propositions. Forensic Sci Int 262:212–226
8. Girod A, Roux C, Weyermann C (2014) La datation des traces digitales (partie II): proposition d'une approche formelle. Rev Int Criminol Police Tech 68(2):226–249
9. Girod A, Ramotowski R, Weyermann C (2012) Composition of fingermark residue: a qualitative and quantitative review. Forensic Sci Int 223(1–3):10–24
10. Girod-Frais A (2019) Dating of fingermarks: fiction or reality? SIAK J – J Police Sci Pract (Int Ed/2019): 21–34
11. Merkel R, Gruhn S, Dittmann J, Vielhauer C, Bräutigam A (2012) On noninvasive 2D and 3D chromatic white light image sensors for age determination of latent fingerprints. Forensic Sci Int 222(1–3):52–70

12. Merkel R, Dittmann J, Vielhauer C (2017) A first public research collection of high resolution latent fingerprint time series for short and long term print age estimation. IEEE Trans Inf Forensics Secur 12(10):2276–2291
13. Schwabenland JF (1992) Case report – determining the evaporation rate of latent impressions on the exterior surfaces of aluminium beverage cans. J Forensic Identif 42(2):84–90
14. De Alcaraz-Fossoul J, Mestres Patris C, Balaciart Muntaner A, Barrot Feixat C, Gené Badia M (2013) Determination of latent fingerprint degradation patterns – a real fieldwork study. Int J Legal Med 127(4):857–870
15. De Alcaraz-Fossoul J, Mestres Patris C, Barrot Feixat C, McGarr L, Brandelli D, Stow K et al (2016) Latent fingermark aging patterns (part I): minutiae count as one indicator of degradation. J Forensic Sci 61(2):322–333
16. De Alcaraz-Fossoul J, Barrot Feixat C, Tasker J, McGarr L, Stow K, Carreras-Marin C et al (2016) Latent fingermark aging patterns (part II): color contrast between ridges and furrows as one indicator of degradation. J Forensic Sci 61(4):947–958
17. Barros RM, Faria BE, Kuckelhaus SA (2013) Morphometry of latent palmprints as a function of time. Sci Justice 53(4):402–408
18. Midkiff CR (1992) Fingerprints – determination of time of placement. Fingerprint Whorld 18(70):125–128
19. Wertheim K (2003) Fingerprint age determination: is there any hope? J Forensic Identif 53(1):42–49
20. Andersson PO, Lejon C, Mikaelsson T, Landstrom L (2017) Towards fingermark dating: a Raman spectroscopy proof of concept study. ChemistryOpen 6(6):706–709
21. Hemmila A, McGill J, Ritter D (2008) Fourier transform infrared reflectance spectra of latent fingerprints: a biometric gauge for the age of an individual. J Forensic Sci 53(2):369–376
22. Antoine KM, Mortazavi S, Miller AD, Miller LM (2010) Chemical differences are observed in children's versus adults' latent fingerprints as a function of time. J Forensic Sci 55(2):513–518
23. Ricci C, Phiriyavityopas P, Curum N, Chan KL, Jickells S, Kazarian SG (2007) Chemical imaging of latent fingerprint residues. Appl Spectrosc 61(5):514–522
24. Van Dam A, Schwarz JC, de Vos J, Siebes M, Sijen T, van Leeuwen TG et al (2014) Oxidation monitoring by fluorescence spectroscopy reveals the age of fingermarks. Angew Chem Int Ed 53:6272–6275
25. Rosa R, Giovanardi R, Bozza A, Veronesi P, Leonelli C (2017) Electrochemical impedance spectroscopy: a deeper and quantitative insight into the fingermarks physical modifications over time. Forensic Sci Int 273:144–152
26. Bradshaw R, Rao W, Wolstenholme R, Clench MR, Bleay S, Francese S (2012) Separation of overlapping fingermarks by matrix assisted laser desorption ionisation mass spectrometry imaging. Forensic Sci Int 222:318–326
27. Francese S, Russo C (2019) Fingermarks as a new proteomic specimen: state of the art and perspective of in situ proteomics. In: Merkley (ed) Applications in forensic proteomics: protein identification and profiling, ACS symposium series. American Chemical Society, Washington, DC
28. Buchanan MV, Asano K, Bohanon A (1997) Chemical characterisation of fingerprints from adults and children. In: Forensic evidence analysis and crime scene investigation. SPIE (International Society for Optical Engineering), Boston, pp 89–95
29. Mong GM, Petersen CE, Clauss TRW (1999) Advanced fingerprint analysis project: fingerprint constituents, Report no.: PNNL-13019. Pacific Northwest National Laboratory, Richland
30. Weyermann C, Roux C, Champod C (2011) Initial results on the composition of fingerprints and its evolution as a function of time by GC/MS analysis. J Forensic Sci 56(1):102–108
31. Koenig A, Girod A, Weyermann C (2011) Identification of wax esters in fingermark residues by GC/MS and their potential use as aging parameters. J Forensic Identif 61(6):652–676
32. Pleik S, Spengler B, Schäfer T, Urbach D, Luhn S, Kirsch D (2016) Fatty acid structure and degradation analysis in fingerprint residues. J Am Soc Mass Spectrom 27(9):1565–1574

33. Girod A, Spyratou A, Holmes D, Weyermann C (2016) Aging of target lipid parameters in fingermark residue using GC/MS: effects of influence factors and perspectives for dating purposes. Sci Justice 56(3):165–180
34. Pleik S, Spengler B, Bhandari DR, Luhn S, Schäfer T, Urbach D et al (2018) Ambient-air ozonolysis of triglycerides in aged fingerprint residues. Analyst 143:1197–1209
35. Mountfort K, Bronstein H, Archer N, Jickells S (2007) Identification of oxidation products of squalene in solution and in latent fingerprints by ESI-MS and LC/APCI-MS. Anal Chem 79:2650–2657
36. Oonk S, Schuurmans T, Pabst M, de Smet L, de Puit M (2018) Proteomics as a new tool to study fingermark ageing in forensics. Sci Rep 8:16425
37. Drapel V, Becue A, Champod C, Margot P (2009) Identification of promising antigenic components in latent fingermark residues. Forensic Sci Int 184:47–53
38. Wolstenholme R, Bradshaw R, Clench MR, Francese S (2009) Study of latent fingermarks by matrix-assisted laser desorption/ionisation mass spectrometry imaging of endogenous lipids. Rapid Commun Mass Spectrom 23:3031–3039
39. Francese S, Bradshaw R, Denison N (2017) An update on MALDI mass spectrometry based technology for the analysis of fingermarks – stepping into operational deployment. Analyst 142(14):2518–2546
40. Bradshaw R, Denison N, Francese S (2017) Implementation of MALDI MS profiling and imaging methods for the analysis of real crime scene fingermarks. Analyst 142(9):1581–1590
41. Bright NJ, Willson TR, Driscoll DJ, Reddy SM, Webb RP, Bleay S et al (2013) Chemical changes exhibited by latent fingerprints after exposure to vacuum conditions. Forensic Sci Int 230(1–3):81–86
42. Bradshaw R, Wolstenholme R, Blackledge RD, Clench MR, Ferguson LS, Francese S (2011) A novel MALDI MSI based methodology for the identification of sexual assault suspects. Rapid Commun Mass Spectrom 25:415–422
43. Reed H, Stanton A, Wheat J, Kelley J, Davis L, Rao W et al (2016) The Reed-Stanton press rig for the generation of reproducible fingermarks: towards a standardised methodology for fingermark research. Sci Justice 56(1):9–17
44. Ferguson L, Bradshaw R, Wolstenholme R, Clench MR, Francese S (2011) A novel two step matrix application for the enhancement and imaging of latent fingermarks. Anal Chem 83(14):5585–5591
45. Ferguson L, Wolstenholme R, Francese S. GB2489215 (01/05/2013) Improvements to MALDI MSI (Dry-Wet Matrix Deposition). Granted Hong Kong patent (HK1169854A1) (15/03/2011) Granted US patent (US9261438B) (16/02/2016)
46. Muramoto S, Sisco E (2015) Strategies for potential age dating of fingerprints through the diffusion of sebum molecules on a nonporous surface analyzed using time-of-flight secondary ion mass spectrometry. Anal Chem 87:8035–8038
47. O'Neill K, Lee Y (2017) Effect of aging and surface interactions on the diffusion of endogenous compounds in latent fingerprints studied by mass spectrometry imaging. J Forensic Sci 63(3):708–713
48. Hinners P, Thomas M, Lee Y (2019) Determining fingerprint age with mass spectrometry imaging via ozonolysis of triacylglycerols. Anal Chem 92(4):3125–3132
49. Ferguson LS, Wulfert F, Wolstenholme R, Fonville JM, Clench MR, Carolan VA, Francese S (2012) Direct detection of peptides and small proteins in fingermarks and determination of sex by MALDI mass spectrometry profiling. Analyst 137(20):4686–4692
50. Francese S (2016) Techniques for fingermark analysis using MALDI MS – a practical overview. In: Kramer R (ed) Advances in MALDI and laser induced soft ionisation mass spectrometry. Springer, New York, pp 93–128. 978-3-319-04818-5

Chapter 9
Latent Fingermarks and Electrochemistry: Possibilities for Development and Aging Studies

Roberto Rosa, Consuelo Mugoni, Massimiliano Bononi, and Roberto Giovanardi

Abstract The aim of this chapter is to comprehensively review the application of the principles of reduction-oxidation reactions, electrochemistry and electrochemical-based techniques to the visualization of latent fingermarks and to study their physical and chemical modifications over time.

After a brief introduction, the first part discusses redox reactions, which are at the basis of both well-established latent fingermarks detection techniques as well as innovative and very promising procedures that need yet to be fully explored for their application in casework. These redox reactions are typically exploited to in situ synthesise the enhancement reactant. Generally, this reactant selectively adheres to the print residue, leading to the positive development of the mark. Examples of these procedures include the physical developer (PD) and the processes based on colloidal gold, i.e. multimetal deposition (MMD) and its subsequently modified versions like MMD II and single-metal deposition (SMD and SMD II). The selective deposition of the reactant to the fingermark ridges can be then exploited by scanning electrochemistry microscopy (SECM) for imaging purposes. Another option for the enhancement reactant is to be directed towards the regions of the substrate free from the fingermark, thus enhancing the furrows rather than the ridges and leading to the so-called inverse development. Examples of this approach include electrodeposition and electrophoretic deposition. The ability of the fingermark residue to act as a mask towards an underlying metal substrate is also at the basis of differential corrosion approaches.

The sensitivity of electrochemical techniques in monitoring the physical changes that a fingermark undergoes over time will be highlighted in the second part of the

R. Rosa (✉)
Department of Sciences and Methods for Engineering,
University of Modena and Reggio Emilia, Reggio Emilia, Italy
e-mail: roberto.rosa@unimore.it

C. Mugoni · M. Bononi · R. Giovanardi
Department of Engineering "Enzo Ferrari", University of Modena and Reggio Emilia, Modena, Italy

J. De Alcaraz-Fossoul (ed.), *Technologies for Fingermark Age Estimations: A Step Forward*, https://doi.org/10.1007/978-3-030-69337-4_9

chapter. Indeed, the possibility to concurrently monitor physical modifications and the variation in the chemical composition represents a highly desirable added value in order to reach trustworthy information of time since deposition.

Most techniques that record physical modifications of fingermarks over time are based on microscopy observations, thus significantly limiting the sensitivity of detectable changes and easily exposing the results to certain bias from the observer. However, electrochemical-based techniques possess the potential to overcome the limits inherent to microscopy observations and allows a deeper understanding of the complex mechanisms involved in the aging of fingermarks. The application of electrochemical impedance spectroscopy (EIS) and the use of an electric potential sensor (EPS) for imaging surface charges and monitoring their decay will be illustrated in this section.

Finally, concluding remarks will highlight the yet unexplored potentialities of electrochemistry to the discipline of latent fingermarks research, including both visualization and age determination possibilities.

Keywords Latent fingerprints · Electrochemistry · Visualization · Redox reactions · Time since deposition

Acronyms

PD	Physical developer
MMD	Multimetal deposition
SMD	Single-metal deposition
SECM	Scanning electrochemistry microscopy
EIS	Electrochemical impedance spectroscopy
EPS	Electric potential sensor
GSRs	Gunshot residues
ELD	Electrolytic deposition
EPD	Electrophoretic deposition
CNLS	Complex non-linear least squares
CPE	Constant phase element
GC	Glassy carbon
PTFE	Polytetrafluoroethylene

9.1 Introduction

Electrochemistry is the chemistry discipline dedicated to the study of the reciprocal relations existing between electrical and chemical phenomena [1]. Despite the utilisation of spontaneous reduction-oxidation (redox) reactions to produce electrical energy and the application of electrical energy to allow non-spontaneous redox reactions to occur, electrochemistry encompasses several phenomena and methods, which can be proficiently applied for the study of chemical systems. Electrochemical experiments are conducted in conveniently designed electrochemical cells, consisting of a collection of different electrode/electrolyte interfaces among which the processes and factors affecting the charge transport are measured. Indeed, charge transport involving a particular analyte, or occurring at a particular electrode, induces changes in electrical parameters like voltage, current and charge. These changes can be detected and monitored in order to gain qualitative and quantitative information on the target analyte or the electrochemical process under investigation. According to the parameter monitored, electrochemical experiments can be, for example, grouped as potentiometric (potential = E is measured), amperometric (current = I is measured) and conductimetric (resistance = R is measured). Some of these electrochemical methods were proficiently exploited in recent years in the field of forensic sciences, particularly for the detection of explosives [2], gunshot residues (GSRs) [3] and drugs of abuse [4, 5].

In the field of latent fingermarks, redox reactions and electrochemistry principles are at the basis of a wide number of detection technique. Redox reactions are mainly investigated in order to synthesise in situ visualization reagents.

The application of more complex electrochemical methods to study latent fingermarks (for development as well as aging purposes) still remains surprisingly scarce, especially considering the widely recognized advantages of electrochemical methods in terms of sensitivity, simplicity, low cost, flexibility and miniaturization possibilities [6, 7], with respect, for example, to the conventionally employed spectroscopic techniques.

9.2 Overview of Electrochemical Techniques for Latent Fingermarks Enhancement

9.2.1 Techniques Leading to Positive Development of Latent Fingermarks

Redox reactions are the basis of several techniques that have been optimized during the years for the visualization of latent fingermarks on different types of substrates, which were classified by Ramotowski as metal deposition methods [8]. The first example was the use of silver nitrate in solution that led to silver chloride (and of course sodium nitrate) by reacting with sodium chloride salt present in the latent fingermark residue. Subsequently, the reduction of Ag^+ to metallic Ag was realized by exposure to shortwave ultraviolet radiation (9.1), leading to the development of brown-greyish-black fingermark ridges [8, 9].

$$AgNO_{3(aq)} + NaCl_{(s)} \rightarrow AgCl_{(s)} + NaNO_{3(aq)} \xrightarrow{h\nu} Ag_{(s)} \quad (9.1)$$

Reduction of silver ions to metallic silver and its selective and spontaneous deposition onto the fingermark ridges is also the basis of one of the most successful techniques for latent fingermark development on porous, even wet, substrates known as physical developer (PD). It was patented by Morris and Wells in 1979 [10]. The complete redox reaction is shown in Eq. 9.2, while a depiction of the development mechanism is summarized in Fig. 9.1. In this case, the Ag physical developer exploits the in situ reduction of silver ions to metallic silver by means of iron (II) ions oxidation to iron (III). Despite the simple nature of the redox reaction in Eq. 9.2, the PD working solution consists of an extremely complex mixture of chemicals, requiring a very careful preparation procedure [11, 12]. Indeed, the PD working solution also contains citric acid, with the aim to subtract Fe^{3+} ions to the redox equilibrium, by their complexation, thus pushing the reaction forward towards the production of metallic silver. Moreover, in order to avoid aggregation of the synthesised colloidal silver particles, n-dodecylamine acetate is added to the solution.

$$Ag^{+}_{(aq)} + Fe^{2+}_{(aq)} \leftrightarrow Ag_{(s)} + Fe^{3+}_{(aq)} \tag{9.2}$$

The mechanism of selective interaction of colloidal silver with fingermark ridges was originally thought to be the result of the presence of organic constituents of the fingermark but still remains largely debated and the object of intense investigation [13, 14].

Similar electrochemistry approaches are being explored for the in situ preparation of colloidal gold [15]. The first use of colloidal gold to latent fingermark visualization purposes dates back to the work of Saunders [16]. He developed a

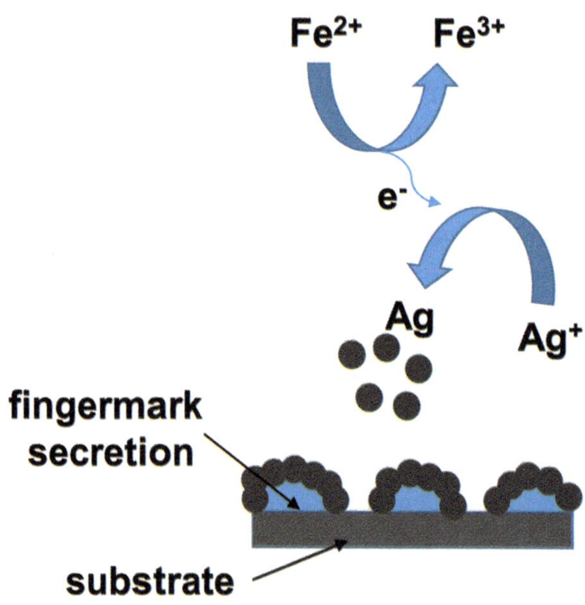

Fig. 9.1 Mechanism for latent fingermark development by Ag physical developer

technique called multimetal deposition (MMD) that proved effective on a wide variety of porous, non-porous and semi-porous substrates. MMD is a two-step procedure: in the first step, colloidal gold nanoparticles selectively adhere to the fingermark ridges, while in the second step the developed fingermark is enhanced by the use of a physical developer. Colloidal gold was originally produced by reduction of Au^{3+} ions inserted in solution as tetrachloroauric acid, by the action of sodium citrate, according to the well-known Turkevich synthesis [17], leading to gold nanoparticles with average diameter of approximately 30 nm. Although MMD was effective in the visualization of latent fingermarks on substrates with different porosities, the method was abandoned by the majority of forensic laboratories. This was in all likelihood a consequence of its tremendously tedious experimental procedure based on the application of a complex series of rinsing and immersion baths (of the object bearing the latent residue) in several solutions. These were characterized by having strict experimental parameters, like, for example, the pH of the Au nanoparticles suspension needed to be rigorously fixed to pH = 3.

The aforementioned issues contributed to push worldwide research efforts towards improvements in the MMD technique, leading to a great number of modifications to the original technique, some of which are summarized in Fig. 9.2 [18].

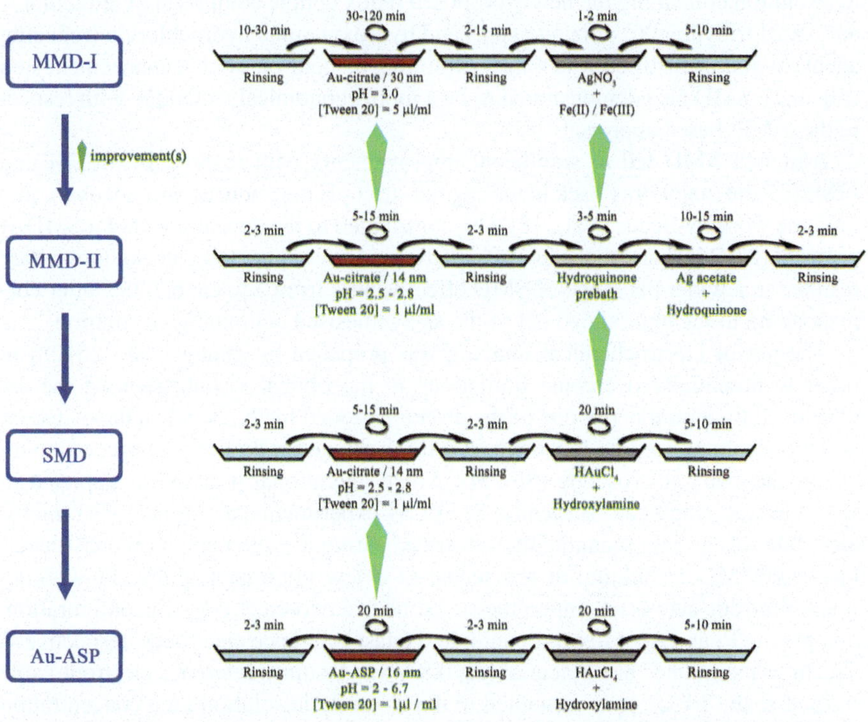

Fig. 9.2 Evolution of latent fingermark enhancement techniques based on colloidal gold and subsequent application of a physical developer. The green arrows highlight the major modification from the previous technique. (Reprinted from [18] Copyright (2011), with permission from Elsevier)

The first improvement to the original MMD (now referred as MMD-I) was developed by Schnetz and Margot that proposed the MMD-II methodology [19]. First, smaller (14 nm of average diameter) and more homogeneously distributed Au nanoparticles were prepared according to the procedure of Slot and Geuze [20]. This allowed to increase the sensitivity in the latent fingermark detection, together with the selectivity in targeting Au nanoparticles by means of the silver particles from the PD. In addition, the PD was modified, with the pair silver acetate – hydroquinone, resulting in a more effective and sensitive approach. The redox reaction at the basis of this innovative Ag-PD formulation is shown in Eq. 9.3.

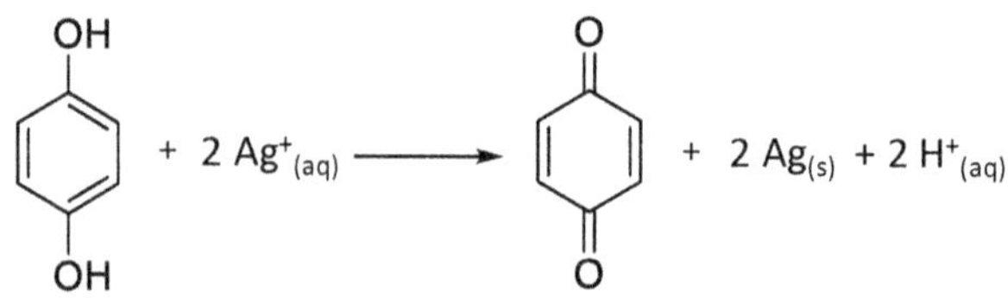

(9.3)

A less labour-intensive alternative, named single-metal deposition (SMD, Fig. 9.2), was proposed by Stauffer et al. [21, 22]. Substantially, it replaced both the hydroquinone and silver acetate/hydroquinone baths with a single immersion bath in a solution containing the newly proposed redox couple comprised of tetrachloroauric acid trihydrate (oxidizing agent) and hydroxylamine hydrochloride (reducing agent) in double distilled water, thus exploiting a "gold-on-gold" enhancement step (Fig. 9.2). SMD also demonstrated to be a more economical technique with respect to the MMD-based approach.

Although SMD led to significant improvements with respect to MMD-I and MMD-II, the major drawback involving the strict pH dependence was not overcome until the work by Becue et al. [18] who contributed to the development of a SMD-II protocol [23–25]. Indeed, the use of aspartic acid in conjunction with sodium citrate resulted in a wider pH range of SMD effectiveness from 2 to ca. 6.7, the latter corresponding to the unmodified pH of the as-synthesised colloidal gold solution.

The use of electrochemical imaging was proposed by Zhang et al. [26–28] in order to potentially overcome limitations in the optical visualization of the as-developed fingermarks (by one of the aforementioned methods) when deposited on surfaces characterized by low contrast with the fingermark. For instance, scanning electrochemistry microscopy (SECM), which operational principle is depicted in Fig. 9.3, was employed. Basically, in SECM experiments, a platinum (Pt) microelectrode acts as a tip scanning the substrate bearing the latent mark (already developed by MMD technique) at a distance of a few micrometres. The Pt working microelectrode acts as an amperometric probe in a two-electrode cell configuration, comprising Pt and Ag wires as counter and quasi-reference electrodes, respectively. The tip is maintained at a potential (ca. 0.8 V vs. Ag quasi reference electrode) able to oxidize the $IrCl_6^{3-}$ ions contained in the electrolytic solution (at a concentration of ca. 2 mM), and the generated steady state oxidation current is continuously monitored through amperometric measurements. Once the tip is in proximity of the

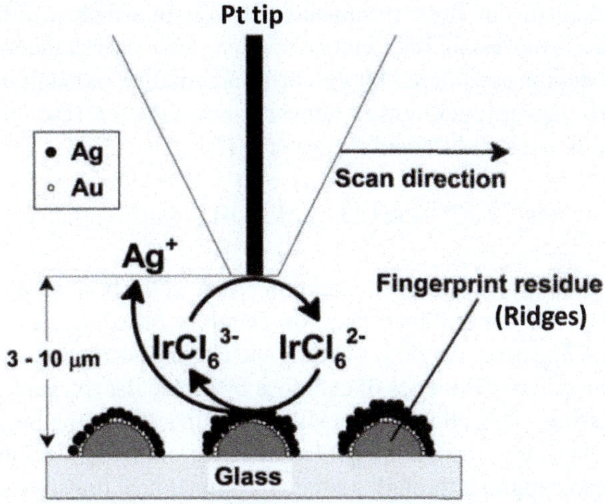

Fig. 9.3 Operating principles of SECM for imaging of fingermarks developed by MMD. (Reproduced from [27] with permission from The Royal Society of Chemistry)

ridges of the fingermark, previously covered by the Ag-PD, and in the presence of $IrCl_6^{2-}$, the redox reaction shown in Eq. 9.4 occurs, leading to an increase in the tip current. A decrease in the tip current is manifested as the tip approaches a furrow of the fingermark, or the substrate in general.

$$Ag_{(s)} + IrCl_{6(aq)}^{2-} \rightarrow Ag_{(aq)}^{+} + IrCl_{6(aq)}^{3-} \tag{9.4}$$

Reduction-oxidation reactions are not only employed for the direct synthesis of the enhancement reactant or for imaging purposes. Indeed, some inorganic salts comprised in the fingermark residue have demonstrated to significantly contribute to the corrosion of metal substrates on which they were deposited. Particularly, when a fingermark is deposited on a metal substrate, it can contribute to the wet corrosion of the metal itself, through the redox reaction shown in Eq. 9.5 [29].

$$2Me + nH_2O + \frac{n}{2}O_2 \rightarrow 2Me^{n+} + 2nOH^- \tag{9.5}$$

The metal ions formed at the anode can also attract negatively charged ions, like Cl^- present in the fingermark residue, leading to the formation of the corresponding metal chloride. The latter undergoes a hydrolysis reaction with water present in the fingermark residue, thus forming hydrochloric acid according to the overall reaction shown in Eq. 9.6 [29, 30], showing a localized pH reduction. This localized corrosion spreads down into the metal rather than laterally, creating a pit in the metal substrate and allowing fingermark visualization. The concurrent application of heat has demonstrated to have a strong effect on the developed fingermarks. Then, a possible mechanism has been postulated [31, 32] in which a differential oxide growth produces contrasting interference colours. This steered the improvement of further visualization procedures by exploring alternative oxidants towards which the fingermark deposit could either enhance the oxidation reaction or inhibit it, resulting in both cases to differential corrosion [33].

$$Me^{n+} + nCl^- + nH_2O \rightarrow Me(OH)_n + n(H^+ + Cl^-) \quad (9.6)$$

A process similar to differential oxidation is also at the basis of thermal development techniques for latent fingermarks on cellulose-based substrates. In fact, the presence of the fingermark residue (mainly the eccrine secretion constituents) produces different oxidation kinetics of cellulose by molecular oxygen, with respect to the naked substrate. This allows the positive visualization of the fingermark first at a fluorescent stage and, then, at a visible one (by prolonging heating). These principles have been exploited through both conventional heat transfer-based strategies [34] and an innovative direct electromagnetic energy transfer mechanism [35].

9.2.2 *Techniques Leading to Reversed Development of Latent Fingermarks*

A significant number of latent fingermarks still goes undetected, despite the astonishing advancements reached during the past years in the field of fingermark enhancement. They include not only the aforementioned techniques using electrochemistry principles but also further and very promising approaches based, for example, on nanotechnology [36–39] and immunological detection [40–42]. These techniques are not fully able to cover the tremendous variability in the secretion composition existing among persons [43, 44], possibly due to the not yet optimal sensitivity of most of the proposed procedures. A possible, and very promising, solution to this latter issue has been recently recognized by pursuing enhancement methods that would develop the substrate rather than the ridges of the fingermark residue, thus leading to the so-called reversed development [44].

In this latter framework, electrochemistry principles and electrochemical techniques have already demonstrated their potential, especially for metallic surfaces. Indeed, spatially selective deposition of the enhancement reactant into the furrows of latent fingermarks can be obtained by electrolytic deposition (ELD, or more simply electrodeposition). This uses the insulation abilities of the fingermark residue to behave as a mask on the underlying metal substrate. As summarized in Fig. 9.4, the conductive substrate bearing the latent fingermark was employed as the working electrode in a three-electrode cell configuration (with a Pt foil as counter electrode and an $Ag/AgCl/KCl_{(sat)}$ reference electrode) with the electrolytic solution containing silver nitrate or tetrachloroauric acid, in order to deposit Ag and Au, respectively [45], under potentiostatic conditions. The potential for Au electrodeposition was

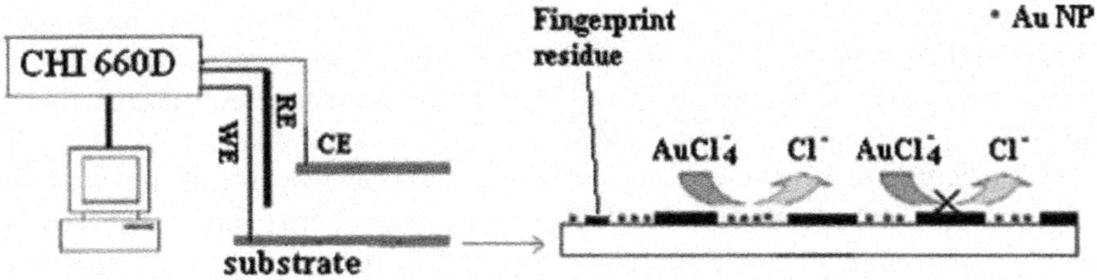

Fig. 9.4 Diagram of the experimental set-up used for the selective electrodeposition of metal (Au in the depicted case). (Reprinted from [45], Copyright (2013), with permission from Elsevier)

fixed at −0.5 V (vs. $Ag/AgCl/KCl_{(sat)}$), while in the case of Ag deposition, the voltage was set at −0.15 V.

A similar approach was investigated by the same authors in order to selectively deposit Prussian blue thin film on the furrows of latent fingermarks deposited on different conductive substrates [46]. Despite the similar configuration of the experimental set-up, electrodeposition was performed in an aqueous solution containing Fe^{3+} and $[Fe(CN)_6]^{3-}$ ions, under potentiodynamic conditions. In this case, cyclic voltammetry experiments were conducted between −0.1 V and 1.2 V and allowed to repeatedly reduce and oxidize the species in solution at the working electrode interfaces, leading to the growth of $Fe_4[Fe(CN)_6]_3$ films.

Potentiodynamic experiments performed in electrolytic solutions containing the appropriate monomers were also used to selectively induce electropolymerization of pyrrole and substituted porphyrins in the furrows of fingermarks deposited on various metallic objects [47].

However, all of the above approaches generated a visually unalterable reversed image of the fingermark, thus limiting the potential improvement of the visual contrast between the enhancement reactant and the metal substrate. This situation was overcome by Beresford and Hillman [48] that proposed a selective electrodeposition of an electrochromic material whose optical properties could be easily manipulated by the application of selected voltages. In that particular case [48], a polyaniline film was deposited potentiostatically (0.9 V) from 0.1 M aniline/aqueous 1 M H_2SO_4, while the fingermark visualization was subsequently performed potentiodynamically ($-0.2\ V < E < 0.9\ V$) in an aniline-free aqueous 1 M H_2SO_4 solution, with more positive potential values that significantly increased the visual contrast with the fingermark. Similar results were also obtained by selective electropolymerization of poly(3,4-ethylenedioxythiophene) and subsequent electrochromic enhancement [49]. Post-deposition functionalization strategies for pyrrole-based systems were also considered in order to significantly increase the plethora of colours obtained in order to further optimize the fingermark contrast against the substrate [50]. The greatest potential of these techniques relies on the reversibility of the variation in the redox states of the film. The optimum contrast can independently be reached in different regions of the fingermark, becoming extremely appealing for fingermark residues of variable thickness/definition and for substrates of non-uniform coloration.

More recently, another electrochemical-based technique, namely, electrophoretic deposition (EPD), has been proposed to easily develop the negative image of a latent fingermark, potentially with no limitations in contrast enhancement as it could be applied in the form of a fine powder to any solid available [51]. EPD fundamentally differs from the previously reported ELD techniques. EPD is a two-step electrochemical material processing technique carried out in a two-electrode cell. In the first step, an electric field is applied between two electrodes. The charged particles suspended in an appropriate liquid medium are forced to migrate towards the oppositely charged electrode (electrophoresis). During the practically simultaneous second step (i.e. deposition), the particles accumulate at the deposition electrode creating a relatively compact and homogeneous film [52]. Then, when the conductive substrate bearing the latent fingermark is used as the deposition electrode (according with the experimental apparatus depicted in Fig. 9.5), the fingermark behaves again as a mask, thus allowing the particles to be deposited only into the regions of the substrate free from the residue. In the particular scenario described in Fig. 9.5, an anodic EPD is used, i.e. negatively charged particles are forced to migrate towards the positively charged (i.e. the anode) electrode.

The stability of the particle suspension plays a crucial role among the several factors that can affect the EPD process. It is well-known that an electrical double layer (constituted by the Stern and by the diffuse layers) surrounds each charged particle in a liquid medium [53]. When the repulsive forces (generated by the overlapping of these electrical double layers) are able to counterbalance and overcome the attractive forces existing between approaching particles (i.e. Van der Waals attractions), a repulsion barrier appears. In order for the particles to aggregate, this repulsion barrier needs to be overcome. Zeta potential, ζ, which is the electrostatic

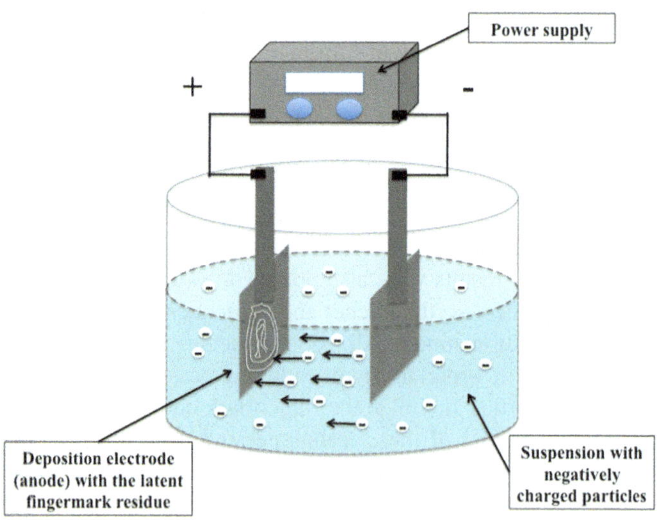

Fig. 9.5 Experimental set-up designed for the detection of latent fingermarks on conductive substrates by anodic EPD technique

potential at the slipping plane (i.e. the boundary, within the diffuse layer, inside which the ions are able to move together with the particle) is generally used as an index of the magnitude of the repulsive forces manifested by the particles and, thus, as an index of the stability of the suspension. High absolute values of ζ (generally $|\zeta| > 30$ mV) are required for a stable suspension. One of the main factors affecting the value of ζ, and the successful outcome of the EPD development of latent fingermarks, is the pH of the particle suspension in which the substrate bearing the fingermark is immersed [54]. Indeed, the charge of the residue is well-known to be strictly dependent upon pH (see Fig. 9.6a) as result of the complex series of protonation/deprotonation reactions involving both the carboxyl and amino groups in the proteins comprised in the fingermark residue. Concurrently, pH significantly affects the electrophoretic mobility (μ) of the suspended particles, through its dependence on zeta potential (ζ) expressed by the Henry Eq. (9.7), where ε_0 is the permittivity of the free space and ε_r and η are the permittivity and the viscosity of the liquid medium, respectively. $f(kr)$ represents the Henry coefficient, which relates the thickness of the electrical double layer $1/k$ (where k is the Debye-Huckle parameter, expressed in m^{-1}) to the radius of the suspended particles (r). The simple linear relationship between velocity (v, in cm s^{-1}) and electrophoretic mobility (resulting expressed in $cm^2\ s^{-1}\ V^{-1}$) is explained by Eq. 9.8, where E is the applied electric field between the electrodes (V cm^{-1}):

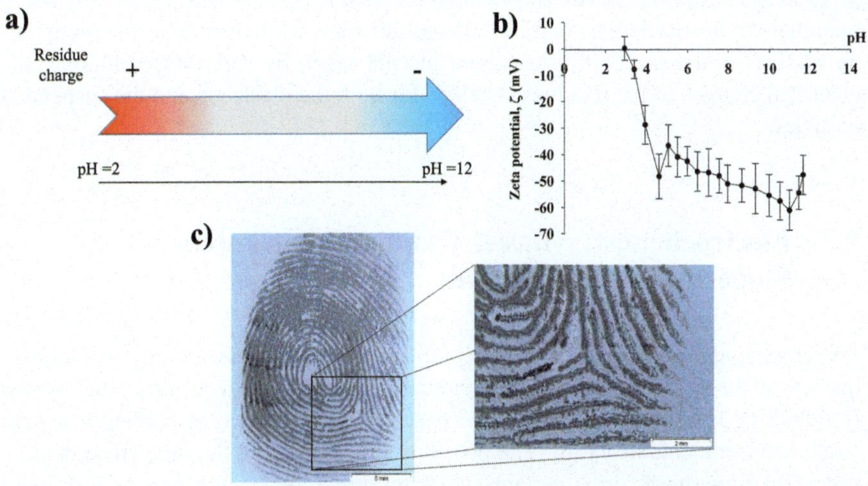

Fig. 9.6 (**a**) Charge of a fingermark residue as a function of the pH (Adapted from [39]). (**b**) Zeta potential variation for a TiO_2 aqueous suspension as a function of the pH. (**c**) Result of anodic EPD reversed development, on AISI 316 stainless steel substrate, employing a 1 wt.% TiO_2 nanoparticles suspension (pH = 7, 20 V of applied potential, 10 s deposition time and 10 mm of distance between the electrodes)

$$\mu = \frac{2}{3}\frac{\varepsilon_0 \varepsilon_r \zeta}{\eta} f(kr) \tag{9.7}$$

$$\mu = \frac{v}{E} \tag{9.8}$$

Zeta potential variation of a titanium dioxide (TiO_2) (anatase, >99% purity from Merck Chemicals) nanoparticle suspension (0.05 wt% in water) as a function of the pH of the suspension itself is shown in Fig. 9.6b. In this case, nanoparticles are negatively charged almost over the entire pH range investigated, except at the so-called isoelectric point, which is approximately at pH = 3. At this pH value, suspended particles globally carry (almost) no electrostatic charges; thus, their electrophoretic mobility and consequently their velocity while approaching the opposite charged electrode are extremely low, according to Eqs. 9.7 and 9.8, respectively.

At pH values close to the neutrality, the suspended particles possess zeta potential values of approximately −50 mV, thus providing them a good stability and electrophoretic mobility. This, together with the neutrality of the charge of the residue at this pH (see Fig. 9.6a), allows the use of anodic EPD for the development of latent fingermarks. In fact, by employing the metal substrate bearing the latent fingermark as the anode (+ pole) of the electrochemical cell (as depicted in Fig. 9.5), the approaching particles selectively adhere to the regions of the substrate free from the residue, leading to the reversed development of the mark as shown in the demonstrative example of Fig. 9.6c (referred to pH = 7).

EPD can be applied to any solid available as a fine powder, thus potentially being boundless in the reachable contrast with the substrate. Moreover, a further potentiality of EPD is the possibility to extend the pH range by carefully combining the electrical charges of the residue as well as for both the electrode and the suspended particles.

9.3 Electrochemistry-Based Techniques for Latent Fingermark Aging Studies

Fingermark aging is known to be a complex natural process, mainly as a consequence of the extremely complex chemical nature of the fingermarks themselves [55] and the large number of variables potentially affecting degradation (e.g. substrate, environment, etc.) [56]. The possibility of estimating the time since deposition of a fingermark on a certain substrate surely represents one of the grand challenges for future innovations in the field of forensic science research [57]. A successful technique/model will contribute to significantly advancing the probative potential of latent fingermarks found at crime scenes. In fact, a reliable methodology to estimate time since deposition would allow crime scene investigators and forensic scientist to discern between crime-related and not-crime-related evidence and to assess depositions' genuineness and reconstruct the dynamics of a criminal event.

Several methods have been proposed in recent years to study and monitor chemical [58–63] and physical modifications [64–72] that latent fingermarks experience from the moment of deposition, with electrochemistry showing great potential in both frameworks. Indeed, most of the chemical constituents of latent fingermarks are involved in redox processes once they are deposited on a particular substrate and left to age in air. Oxidation products have been quantified to estimate the time since deposition, similarly to what has been proposed for the haemoglobin oxidation products for the age determination of blood stains [73, 74]. Van Dam and colleagues [75] measured, through fluorescence spectroscopy, the rates of degradation and generation of proteins and oxidation products. Under exposure to air, unsaturated lipids (contained in latent fingermarks) can lead to the formation of oxidized products (LipOx); in turn, these can react with tryptophan-containing proteins (Tryp, which are considered responsible for the autofluorescence of fingermarks [76]) and progress to the formation of fluorescent oxidation products (FOX). By integrating the fluorescent intensities with the two characteristic wavelength ranges of both Tryp (i.e. 313–550 nm) and FOX (i.e. 400–500 nm), the authors calculated the ratio between these two contributions and reported these values as a function of the age of fingermarks, obtaining an aging curve that allowed them to estimate the age of 55% of the fingermarks from male donors up to 3 weeks old with an uncertainty of 1.9 days.

Electrochemistry principles have been mainly applied to monitor the physical changes a fingermark experiences from the moment of its deposition on a particular substrate. This is an added value to monitoring and quantifying the changes in chemical composition of latent fingermarks.

However, most of the methodologies proposed to monitor physical changes are based on microscopy techniques; thus, significantly limiting the sensitivity of the observations to the resolution of the resulting micrograph or to the power of the microscope used. Furthermore, the results are easily influenced by the subjectivity of the researcher/operator. On the other hand, the use of electrochemical impedance spectroscopy (EIS) allows to reach a deeper understanding of the mechanisms at the basis of fingermark physical modifications, going beyond the limits imposed by microscopical observations [77]. In this case, electrochemical impedance (Z) is a measure of the ability of a circuit to resist the flow of electrical current, and it is usually measured by applying a sinusoidal AC potential to an electrochemical cell and then measuring the current through the cell. At a fixed frequency, impedance can be calculated by an expression similar to the Ohm's law (9.9) as the ratio between the applied potential (E) and the response current (I), both expressed as sinusoidal functions:

$$Z(\omega) = E(t)/I(t) = E_0 \sin(\omega t)/I_0 \sin(\omega t + \phi) = Z_0 \sin(\omega t)/\sin(\omega t + \phi) \tag{9.9}$$

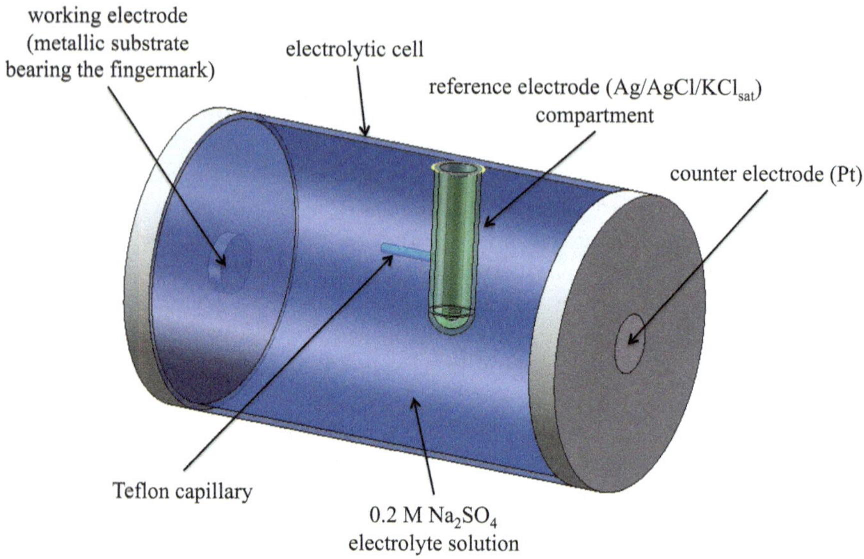

Fig. 9.7 Schematic representation of the three-electrode cell employed for EIS experiments. (Reprinted from [77], Copyright (2017), with permission from Elsevier)

Therefore, impedance is usually given in terms of a magnitude (Z_0, i.e. the ratio between the amplitudes of applied potential and resulting current) and a phase shift (ϕ, i.e. shift between potential input and the current output). In one instance, the impedance spectra were acquired in the 20 kHz–100 mHz frequency range by applying a sinusoidal potential wave (0 V bias, 10 mV rms of amplitude) in a three-electrode cell as the one depicted in Fig. 9.7. This cell included a metallic substrate bearing the fingermark as the working electrode, a Pt foil as the counter electrode, an Ag/AgCl/KCl$_{(sat)}$ as the reference electrode and 0.2 M Na_2SO_4 as the electrolytic solution.

After each measurement, the cell was disassembled and the working electrode end plate was left to dry in laboratory conditions without removing the substrate from the plate. It was found that the impedance spectra collected were significantly affected by the physical modifications associated with the aging process of latent fingermarks. A particularly useful depiction resulted by plotting the impedance phase shift as a function of frequency (Bode phase plots). Figure 9.8 shows the Bode phase plots of natural fingermarks deposited by a male donor (Caucasian, 36 years old) on AISI 316 stainless steel (Fig. 9.8a) and on commercial yellow brass (65 wt% Cu and 35 wt% Zn, Fig. 9.8b), both monitored up to 35 days. It was demonstrated that the changes observed in the Bode phase plots were not due to the electrolytic solution or to the sequence of immersion, removal and drying actions performed on the working electrode end plate.

The fitting of the experimental impedance data was performed by applying the complex non-linear least squares (CNLS) method, in an attempt to express the

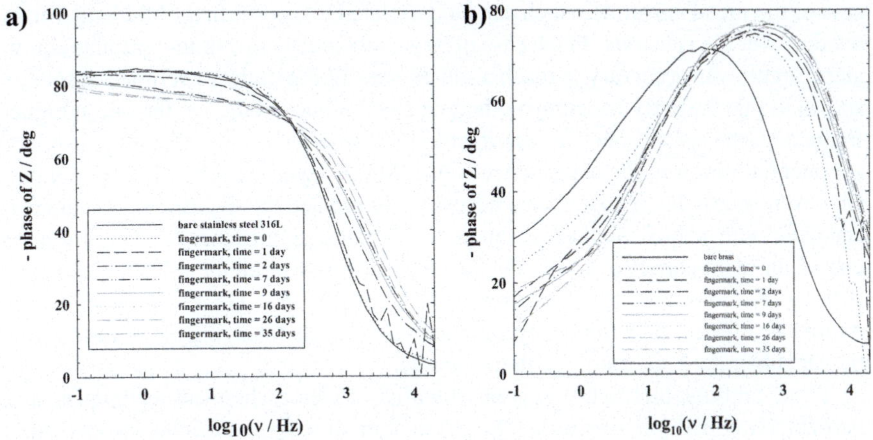

Fig. 9.8 Bode phase plots of natural fingermarks from the same donor deposited on AISI 316 stainless steel (**a**) and brass (**b**) substrates, monitored up to 35 days. (Reprinted from [77], Copyright (2017), with permission from Elsevier)

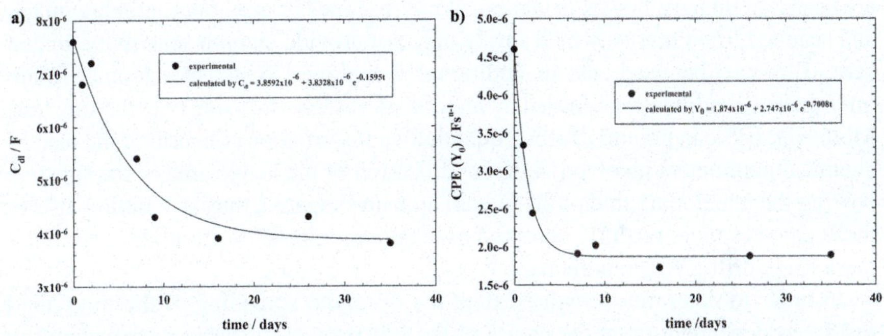

Fig. 9.9 Exponential decays of the capacitive elements of the selected equivalent circuits used to fit the experimentally obtained Bode phase plots for stainless steel (**a**) and brass (**b**) substrate, as a function of time since deposition of natural fingermarks. (Reprinted from [77], Copyright (2017), with permission from Elsevier)

observed changes in terms of quantitative and objective electrochemical parameters. Two different equivalent circuits were selected for the two different substrates studied on which natural fingermarks were deposited. Particularly for the stainless-steel substrate, a simple Randles circuit (R(CR)) was selected. It consists an Ohmic resistive element (ascribed to the resistance of the electrolyte and the electric contacts), in series with the parallel combination of the double layer capacitance (C_{dl}) and the charge transfer resistance of the faradaic reactions involved. An exponential decay in the C_{dl} values was observed as the aging of the fingermark progressed, as shown in Fig. 9.9a, that allowed calculation of the aging curve for that particular combination of donor and substrate (see the inset in Fig. 9.9a). This C_{dl} decay corresponds to a reduction in the active (i.e. not covered by the fingermark residue) area of the metal substrate which can be correlated to a spreading of the fingermark residue over the substrate. In the case of brass substrate, a two time-constants equivalent circuit was necessary to reach a satisfactory fitting of the experimental spectra shown in Fig. 9.8b. By focusing on the first mesh of the circuit, i.e. the one at higher frequency (the second one was ascribed to the corrosion of the substrate), a similar exponential decay in the capacitance of the double layer was observed (Fig. 9.9b). However, in this latter case it was necessary to replace the C_{dl} element of an ideal capacitor with a constant phase element (CPE), whose impedance can vary from the one of an ideal capacitor to the one of an ideal resistive element [77]. Therefore, only the Y_0 (constant with dimension $F{\cdot}s^{n-1}$, with $0 < n < 1$) term of the CPE was plotted as a function of frequency and used to calculate the aging curve for the particular substrate + natural fingermark sample.

These findings highlight the great potential of electrochemical techniques, and particularly by EIS, in monitoring the physical modifications of fingermarks as time since deposition progresses. The physical changes of fingermarks reported in the literature are usually limited to ridge width reductions and pores size variations, while the degree of fingermark diffusion (spreading) over the substrate is typically disregarded. This is a consequence of the techniques applied to date, most of which are based on image analysis from optical microscopes or even photographic images and scanned lifted marks, which surely may not provide enough sensitivity. In contrast, EIS can be used, as an additional technique, to detect nanometric/submicrometric modifications caused by fingermark residue diffusion over the substrate. Moreover, EIS can potentially give quantitative information concerning the electrochemical parameters involved in the modification of the active area of the substrate bearing the latent mark and of the overall area investigated, thus accounting also for those changes more recently observed as evaporation/re-deposition [43] and fingermark ridge drift [78] phenomena.

In order to determine to what extent the observed spreading of the fingermark residue is dependent upon the nature of the substrate itself, further experiments on conductive amorphous substrates, i.e. glassy carbon (GC), were performed. Similar modifications in the Bode phase plots were observed as the fingermark was aged up to 28 days (Fig. 9.10). However as already inferable from measurements at 7 days post-deposition, the phase of Z shifts back towards the original position of the bare glassy carbon substrate. This finding is also reflected by plotting the equivalent circuit parameters (determined by fitting operations) as a function of time since deposition. Indeed, also in this latter case, referred to a natural fingermark deposited on glassy carbon substrate, a CPE was necessarily used (in place of a C_{dl} element in the Randles equivalent circuit) to obtain satisfactory fitting results. The exponential decay trend of the Y_0 parameter of CPE was followed only up to 4 days of aging (Fig. 9.11), after which further substrate related phenomena probably occurred, thus affecting the extension of the active area of the working electrode. Among these phenomena, the fingermark residue absorption by the GC substrate should also be taken into consideration [79].

The use of electrochemical techniques is an optimal choice for the study of fingermarks, supported by the existence of the so-called triboelectric charging

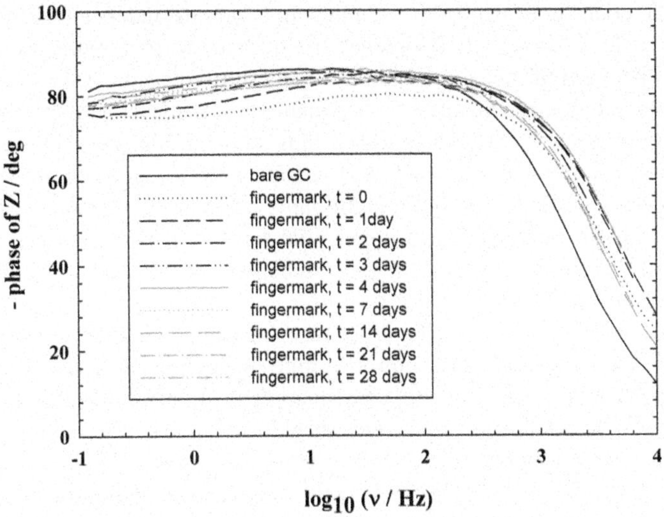

Fig. 9.10 Bode phase plots (i.e. showing the impedance phase shift with frequency) of a natural fingermark deposited on a glassy carbon (GC) substrate and monitored for 28 days

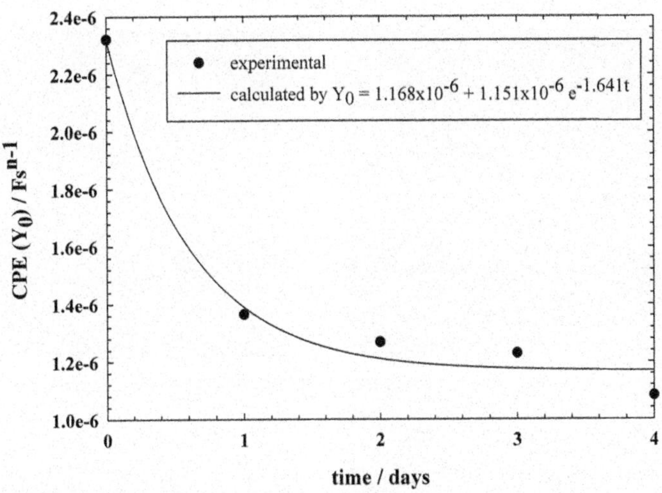

Fig. 9.11 Exponential decay of the Y_0 parameter of the CPE (constant phase element, used to replace the capacitance of the double layer) experimentally obtained, by fitting the first five impedance spectra of Fig. 9.10, and calculated for a natural fingermark deposited on glassy carbon substrate and aged up to 4 days

phenomena [80]. This consists in the formation of an invisible surface charge fingermark as the finger touches an insulating material. The amount of charge produced is mostly material dependent, and it can be inferred by the triboelectric series which rank materials on the basis of their ability in being charged upon contact [81].

In contrast, contact electrification is known to be almost independent by the donor and the transfer method [82]. Moreover, the surface charge (confined to the points of contact between the substrate and the finger) showed a slow decay during which its spatial resolution practically remains unaltered. At the same time, the absolute level of the charge is also reduced [83], thus showing important perspectives for being monitored by other electrochemical strategies in order to obtain information related to the time since deposition. In this latter perspective, Watson and colleagues [84] configured a solid-state electric potential sensor (EPS) to measure changes in spatial electric charge. Surface charge gradients were measured by the voltage induced on an electrode when moved closer to the charged surface. Although the authors proposed this method to efficiently image latent fingermarks in a non-destructive way, its potential to determine time sequence of fingermark depositions is more interesting, together with time since deposition. The charge images of a natural fingermark deposited on a PTFE sheet by a single donor and imaged immediately after deposition (a), after 5 (b) and 14 (c) days from deposition are shown in Fig. 9.12. As seen from Fig. 9.12, a good level of image definition is maintained over time.

Figure 9.12d is the charge image obtained after exposing the substrate bearing the latent fingermark to ionized air to remove the surface charge. The fingermark

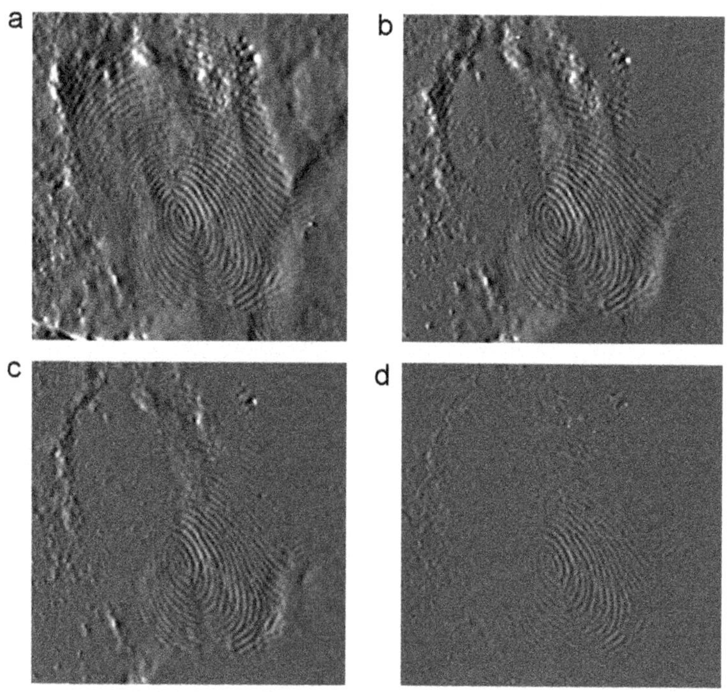

Fig. 9.12 Surface charge images of a natural latent fingermark detected immediately after deposition (**a**), 5 days after deposition (**b**), 14 days after deposition (**c**) and after discharging under ionized air (**d**). (Reprinted from [84], Copyright (2011), with permission from Elsevier)

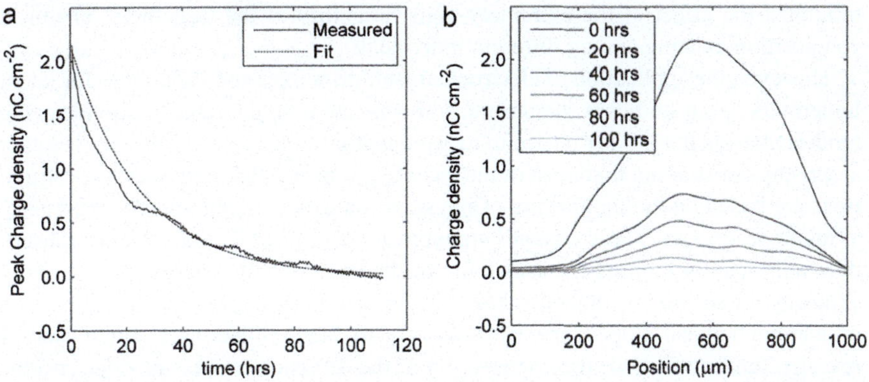

Fig. 9.13 Peak surface charge density as a function of time since deposition up to 110 h (**a**) and charge density linear profile as a function of time for the fingermark ridge considered (**b**). (Reprinted from [84], Copyright (2011), with permission from Elsevier)

still remains visible as a consequence of the presence of some residual surface charge or a different discharging involving the bare PTFE substrate and the regions where the fingermark was deposited. In order to obtain quantitative parameters for aging studies, the charge peak of a single pixel of a fingermark image is easily measured and plotted as a function of time, showing the strong time dependence reported in Fig. 9.13a. In this case, data have been fitted by a standard exponential decay equation providing the aging curve which is dependent upon the particular substrate and the environmental conditions. Figure 9.13b highlights the actual position of the charge peak considered for aging studies, and the evolution of its intensity over time.

9.4 Conclusions and Future Prospects

This chapter highlights the ubiquitous nature of electrochemistry in latent fingermark-related research. Many of the well-established, and currently in use, methodologies for latent fingermark enhancement for forensic casework exploit electrochemical principles, typically for the in situ preparation of the developer reagent. Even the more innovative techniques proposed in the scientific literature for fingermark visualization/detection are based on, or at least partially use, electrochemistry principles and electrochemical-based methodologies. The reason relates to the high degree of versatility provided by electrochemistry-based approaches. Despite being used for the visualization of positive images (i.e. by developing fingermark ridges) electrochemistry has also shown great potential in approaches aiming at targeting the regions of the substrate free from the fingermark residue (i.e. the furrows between ridges). These latter approaches have been recognized as the most appropriate to overcome intrinsic limitations of conventional techniques to minimize the effect of the extremely high variability of the fingermark chemical composition existing among different individuals.

However, the application of electrochemistry principles and techniques for latent fingermark aging studies is still scarce, with only a few studies able to provide aging models that can potentially be useful to estimate time since deposition. This is rather surprising considering that most of the recently proposed techniques for age estimations are based on the monitoring of physical changes that fingermarks experience as aging progresses. These changes are typically quantified by optical observations, thus significantly limiting the sensitivity of the detectable modifications and easily exposing the results to subjective bias. Just to mention an example, the application of electrochemical impedance spectroscopy to latent fingermark aging studies revealed (and quantified) the contribution of the diffusion process of ridge residue to the physical modifications of the fingermark on conductive substrates. Since most of the chemical changes to which a fingermark is subjected are oxidation phenomena (thus intrinsically detectable by electrochemical techniques), the disregarded use of electrochemistry in fingermark aging studies is hard to explain.

This chapter emphasizes the potential benefits of electrochemistry techniques at detecting, developing, and monitoring both physical and chemical changes a fingermark experiences from the moment of its deposition. The widely recognized advantages of electrochemistry-based techniques in terms of low-cost, miniaturization, sensitivity and selectivity and field portable instrumentation possibilities will, in all likelihood, contribute to smooth the way towards increased electrochemistry-oriented research efforts in latent fingermarks aging studies.

References

1. Bard AJ, Faulkner LR (2001) Electrochemical methods – fundamentals and applications, 2nd edn. Wiley, New York
2. Yu HA, DeTata DA, Lewis SW, Silvester DS (2017) Recent developments in the electrochemical detection of explosives: towards field-deployable devices for forensic science. Trends Anal Chem 97:374–384
3. Trejos T, Pyl CV, Menking-Hoggatt K, Alvarado AL, Arroyo LE (2018) Fast identification of inorganic and organic gunshot residues by LIBS and electrochemical methods. Forensic Chem 8:146–156
4. Florea A, de Jong M, De Wael K (2018) Electrochemical strategies for the detection of forensic drugs. Curr Opin Electrochem 11:34–40
5. Mendes LF, e Silva ARS, Bacil RP, Serrano SHP, Angnes L, Paixao TRLC, de Araujo WR (2019) Forensic electrochemistry: electrochemical study and quantification of xylazine in pharmaceutical and urine samples. Electrochim Acta 295:726–734
6. de Oliveira LP, Rocha DP, de Araujo WR, Munoz RAA, Paixao TRLC, Salles MO (2018) Forensic in hand: new trends in forensic devices (2013–2017). Anal Methods 10:5135–5163
7. Ferreira PC, Ataide VN, Chagas CLS, Angnes L, Coltro WKT, Paixao TRLC, de Araujo WR (2019) Wearable electrochemical sensors for forensic and clinical applications. Trends Anal Chem 119:115622
8. Ramotowski RS (2013, Ch. 3) Metal deposition methods. In: Ramotowski RS (ed) Lee and Gaensslen's advances in fingerprint technology, 3rd edn. CRC Press, Boca Raton, pp 55–81

9. Moenssen AA (1971) Fingerprint techniques. Chilton Book Company, Radnor. p. 120
10. Morris JR, Wells JM (1979) British Patent: 154014
11. Sodhi GS, Kaur J (2016) Physical developer method for detection of latent fingerprints: a review. Egypt J Forensic Sci 6:44–47
12. Cantu A, Johnson J (2001) Silver physical development of latent prints. In: Advances in fingerprint technology, 2nd edn. CRC Press, Boca Raton
13. de la Hunty M, Moret S, Chadwick S, Lennard C, Spindler X, Roux C (2015) Understanding physical developer (PD): part I – is PD targeting lipids? Forensic Sci Int 257:481–487
14. de la Hunty M, Moret S, Chadwick S, Lennard C, Spindler X, Roux C (2015) Understanding physical developer (PD): part II – is PD targeting eccrine constituents? Forensic Sci Int 257:488–495
15. Mohamed AA (2011) Gold is going forensic. Gold Bull 44:71–77
16. Saunders G (1989) Multimetal deposition technique for latent fingerprint development. Presented at the International Association for Identification, 74th Annual Education Conference, 1989, Pensacola, USA
17. Wuithschick M, Birnbaum A, Witte S, Sztucki M, Vainio U, Pinna N, Rademann K, Emmerling F, Kraehnert R, Polte J (2015) Turkevich in new robes: key questions answered for the most common gold nanoparticle synthesis. ACS Nano 9:7052–7071
18. Becue A, Scoundrianos A, Moret S (2012) Detection of fingermarks by colloidal gold (MMD/SMD) – beyond the pH 3 limit. Forensic Sci Int 219:39–49
19. Schnetz B, Margot P (2001) Technical note: latent fingermarks, colloidal gold and multimetal deposition (MMD) – optimisation of the method. Forensic Sci Int 118:21–28
20. Slot JW, Geuze HJ (1985) A new method of preparing gold probes for multiple-labelling cytochemistry. Eur J Cell Biol 38:87–93
21. Stauffer E, Becue A, Singh KV, Thampi KR, Champod C, Margot P (2007) Single-metal deposition (SMD) as a latent fingermark enhancement technique: an alternative to multimetal deposition (MMD). Forensic Sci Int 168:e5–e9
22. Durussel P, Stauffer E, Becue A, Champod C, Margot P (2009) Single-metal deposition: optimization of this fingermark enhancement technique. J Forensic Identif 59:80–96
23. Moret S, Becue A (2015) Single-metal deposition for fingermark detection – a simpler and more efficient protocol. J Forensic Identif 65:118–137
24. Newland TG, Moret S, Becue A, Lewis SW (2016) Further investigations into the single metal deposition (SMD II) technique for the detection of latent fingermarks. Forensic Sci Int 268:62–72
25. Moret S, Lee PLT, de la Hunty M, Spindler X, Lennard C, Roux C (2019) Single metal deposition versus physical developer: a comparison between two advanced fingermark detection techniques. Forensic Sci Int 294:103–112
26. Zhang M, Girault HH (2007) Fingerprint imaging by scanning electrochemical microscopy. Electrochem Commun 9:1778–1782
27. Zhang M, Becue A, Prudent M, Champod C, Girault HH (2007) SECM imaging of MMD-enhanced latent fingermarks. Chem Commun 38:3948–3950
28. Zhang M, Girault HH (2009) SECM for imaging and detection of latent fingerprints. Analyst 134:25–30
29. Bond JW (2008) Visualization of latent fingerprint corrosion of metallic surfaces. J Forensic Sci 53:812–822
30. Bond JW (2009) Visualization of latent fingerprint corrosion of brass. J Forensic Sci 54:1034–1041
31. Wightman G, O'Connor D (2011) The thermal visualization of latent fingermarks on metallic surfaces. Forensic Sci Int 204:88–96
32. Bond JW (2011) Optical enhancement of fingerprint deposits on brass using digital colour mapping. J Forensic Sci 56:1285–1288

33. Wightman G, Emery F, Austin C, Andersson I, Harcus L, Arju G, Steven C (2015) The interaction of fingermark deposits on metal surfaces and potential ways for visualization. Forensic Sci Int 249:241–254
34. Song DF, Sommerville D, Brown AG, Shimmon RG, Reedy BJ, Tahtouh M (2011) Thermal development of latent fingermarks on porous surfaces – further observations and refinements. Forensic Sci Int 204:97–110
35. Rosa R, Veronesi P, Leonelli C (2013) Microwave selective thermal development of latent fingerprints on porous surfaces: potentialities of the method and preliminary experimental results. J Forensic Sci 58:1314–1321
36. Choi MJ, McDonagh AM, Maynard P, Roux C (2008) Metal containing nanoparticles and nano-structured particles in fingermark detection. Forensic Sci Int 179:87–97
37. Dilag J, Kobus HJ, Ellis AV (2011) Nanotechnology as a new tool for fingermark detection: a review. Curr Nanosci 7:153–159
38. Hazarika P, Russell DA (2012) Advances in fingerprint analysis. Angew Chem Int Ed 51:3524–3531
39. Becue A, Cantu AA (2013) Fingermark detection using nanoparticles. In: Ramotowsky RS (ed) Lee and Gaensslen's advances in fingerprint technology, 3rd edn. CRC Press, Taylor & Francis Group, Boca Raton, pp 307–379
40. Spindler X, Hofstetter O, McDonagh AM, Roux C, Lennard C (2011) Enhancement of latent fingermarks on non-porous surfaces using anti-L-amino acid antibodies conjugated to gold nanoparticles. Chem Commun 47:5602–5604
41. Wood M, Maynard P, Spindler X, Lennard C, Roux C (2012) Visualization of latent fingermarks using an aptamer-based reagent. Angew Chem Int Ed 51:12272–12274
42. He Y, Xu L, Zhu Y, Wei Q, Zhang M, Su B (2014) Immunological multimetal deposition for rapid visualization of sweat fingerprints. Angew Chem Int Ed 53:12609–12612
43. Moret S, Spindler X, Lennard C, Roux C (2015) Microscopic examination of fingermark residues: opportunities for fundamentals studies. Forensic Sci Int 255:28–37
44. Jaber N, Lesniewski A, Gabizon H, Shenawi S, Mandler D, Almog J (2012) Visualization of latent fingermarks by nanotechnology: reversed development on paper – a remedy to the variation in sweat composition. Angew Chem Int Ed 51:12224–12227
45. Qin G, Zhang M, Zhang Y, Zhu Y, Liu S, Wu W, Zhang X (2013) Visualizing latent fingerprints by electrodeposition of metal nanoparticles. J Electroanal Chem 693:122–126
46. Qin G, Zhang M, Zhang Y, Zhu Y, Liu S, Wu W, Zhang X (2013) Visualization of latent fingerprints using Prussian blue thin films. Chin Chem Lett 24:173–176
47. Bersellini C, Garofano L, Giannetto M, Lusardi F, Mori G (2001) Development of latent fingerprints on metallic surfaces using electropolymerization processes. J Forensic Sci 46:871–877
48. Beresford AL, Hillman AR (2010) Electrochromic enhancement of latent fingerprints on stainless steel surfaces. Anal Chem 82:483–486
49. Brown RM, Hillman AR (2012) Electrochromic enhancement of latent fingerprints by poly(3,4-ethylenedioxythiophene). Phys Chem Chem Phys 14:8653–8661
50. Sapstead RM, Ryder KS, Fullarton C, Skoda M, Dalgliesh RM, Watkins EB, Beebee C, Barker R, Glidle A, Hillman AR (2013) Nanoscale control of interfacial processes for latent fingerprint enhancement. Faraday Discuss 164:391–410
51. Rosa R, Cannio M, Veronesi P, Leonelli C (2013) Electrophoretic deposition of nanoparticles and nano-structured particles for latent fingerprints detection on different surfaces. Proceedings of the 65th Annual Scientific Meeting of the American Academy of Forensic Sciences, February 18–23, 2013, Washington, DC, USA, vol 19, pp 30–31
52. Corni I, Ryan MP, Boccaccini AR (2008) Electrophoretic deposition: from traditional ceramics to nanotechnology. J Eur Ceram Soc 28:1353–1367
53. Zetasizer Nano User Manual, MAN0317, issue 4.0, Malvern Instruments, Worcestershire, UK, May 2008
54. Rosa R, Veronesi P, Leonelli C (2014) Strengthening research methodology for new latent fingerprints development techniques by means of design of experiments (DoE). Proceedings

of the 66th Annual Scientific Meeting of the American Academy of Forensic Sciences, Seattle, WA, USA, February 17–22, 2014, vol 20, pp 87–88
55. Girod A, Ramotowski R, Weyermann C (2012) Composition of fingermark residue: a qualitative and quantitative review. Forensic Sci Int 223:10–24
56. Sears VG, Bleay SM, Bandey HL, Bowman VJ (2012) A methodology for finger mark research. Sci Justice 52:145–160
57. van Asten AC (2014) On the added value of forensic science and grand innovation challenges for the forensic community. Sci Justice 54:170–179
58. Archer NE, Charles Y, Elliot JA, Jickells S (2005) Changes in the lipid composition of latent fingerprint residues with time after deposition on a surface. Forensic Sci Int 154:224–239
59. Weyermann C, Roux C, Champod C (2011) Initial results on the composition of fingerprints and its evolution as a function of time by GC/MS analysis. J Forensic Sci 56:102–108
60. Croxton RS, Baron MG, Butler D, Kent T, Sears VG (2006) Development of a GC-MS method for the simultaneous analysis of latent fingerprint components. J Forensic Sci 51:1329–1333
61. Croxton RS, Baron MG, Butler D, Kent T, Sears VG (2010) Variation in amino acid and lipid composition of latent fingerprints. Forensic Sci Int 199:93–102
62. Girod A, Pyratou A, Holmes D, Weyermann C (2016) Aging of target lipid parameters in fingermark residue using GC/MS: effects of influence factors and perspectives for dating purposes. Sci Justice 56:165–180
63. Girod A, Xiao L, Reedy B, Roux C, Weyermann C (2015) Fingermark initial composition and aging using Fourier transform infrared microscopy (μ-FTIR). Forensic Sci Int 254:185–196
64. Barros RM, Faria BEF, Kuckelhaus SAS (2013) Morphometry of latent palmprints as a function of time. Sci Justice 53:402–408
65. Popa G, Potorac R, Preda N (2010) Method for fingerprint age determination. Rom J Legal Med 2:149–154
66. Merkel R, Gruhn S, Dittmann J, Vielhauer C, Brautigam A (2012) On non-invasive 2D and 3D chromatic white light image sensors for age determination of latent fingerprints. Forensic Sci Int 222:52–70
67. De Alcaraz-Fossoul J, Patris CM, Muntaner AB, Feixat CB, Badia MG (2013) Determination of latent fingerprint degradation patterns – a real fieldwork study. Int J Legal Med 127:857–870
68. De Alcaraz-Fossoul J, Patris CM, Feixat CB, McGarr L, Brandelli D, Stow K, Badia MG (2016) Latent fingermark aging patterns (part I): minutiae count as one indicator of degradation. J Forensic Sci 61:322–333
69. De Alcaraz-Fossoul J, Feixat CB, Tasker J, McGarr L, Stow K, Carreras-Marin C, Oset JT, Badia MG (2016) Latent fingermark aging patterns (part II): color contrast between ridges and furrows as one indicator of degradation. J Forensic Sci 61:947–958
70. De Alcaraz-Fossoul J, Feixat CB, Carreras-Marin C, Tasker J, Zapico SC, Badia MG (2017) Latent fingermark aging patterns (part III): discontinuity index as one indicator of degradation. J Forensic Sci 62:1180–1187
71. De Alcaraz-Fossoul J, Feixat CB, Zapico SC, McGarr L, Carreras-Marin C, Tasker J, Badia MG (2019) Latent fingermark aging patterns (part IV): ridge width as one indicator of degradation. J Forensic Sci 64:1057–1066
72. De Alcaraz-Fossoul J, Mancenido M, Soignard E, Silverman N (2019) Application of 3D imaging technology to latent fingermark aging studies. J Forensic Sci 64:570–576
73. Bremmer RH, Nadort A, van Leeuwen TG, van Gemert MJC, Aalders MCG (2011) Age estimation of blood stains by hemoglobin derivative determination using reflectance spectroscopy. Forensic Sci Int 206:166–171
74. Li B, Beveridge P, O'Hare WT, Islam M (2013) The age estimation of blood stains up to 30 days old using visible wavelength hyperspectral image analysis and linear discriminant analysis. Sci Justice 53:270–277
75. van Dam A, Schwarz JCV, de Vos J, Siebes M, Sijen T, van Leeuwen TG, Aalders MCG, Lambrechts SAG (2014) Oxidation monitoring by fluorescence spectroscopy reveals the age of fingermarks. Angew Chem Int Ed 53:6272–6275

76. Lambrechts SAG, van Dam A, de Vos J, van Weert A, Sijen T, Aalders MCG (2012) On the autofluorescence of fingermarks. Forensic Sci Int 222:89–93
77. Rosa R, Giovanardi R, Bozza A, Veronesi P, Leonelli C (2017) Electrochemical impedance spectroscopy: a deeper and quantitative insight into the fingermarks physical modifications over time. Forensic Sci Int 273:144–152
78. De Alcaraz-Fossoul J, Roberts KA, Feixat CB, Hogrebe GG, Badia MG (2016) Fingermark ridge drift. Forensic Sci Int 258:26–31
79. Botelho EC, Scherbakoff N, Rezende MC (2001) Porosity control in glassy carbon by rheological study of the furfuryl resin. Carbon 39:45–52
80. Fuhrmann J (1978) Contact electrification of dielectric solids. J Electrost 4:109–118
81. Diaz AF, Felix-Navarro RM (2004) A semi-quantitative tribo-electric series for polymeric materials: the influence of chemical structure and properties. J Electrost 62:277–290
82. Lowell J, Akande AR (1988) Contact electrification-why is it variable? J Phys D Appl Phys 21:125–137
83. Beardsmore-Rust ST, Watson P, Prance RJ, Harland CJ, Prance H (2009) Imaging of charge spatial density on insulating materials. Meas Sci Technol 20:095711. (6 pp)
84. Watson P, Prance RJ, Beardsmore-Rust ST, Prance H (2011) Imaging electrostatic fingerprints with implications for a forensic timeline. Forensic Sci Int 209:e41–e45

Chapter 10
Latent Fingermarks and DNA Recovery

Sara C. Zapico

Abstract Matching a questioned fingermark from a crime scene with a suspect is one of the ways of achieving human identifications. However, if a fingermark is partial, distorted, or blurry beyond recognition, this task becomes more challenging, and sometimes cannot be attained solely based on fingermark ridge patterns. As a result, an alternative human identification approach will be necessary. Latent fingermarks are one of the sources of what has been named "touch DNA", DNA left as a result of touching objects. Thus, DNA recovered from a fingermark can be used as another source of identification to confirm the results of a fingermark comparison and vice versa. There have been several studies that have focused on "touch DNA"/ fingermark DNA in recent years. Some of these experiments investigated the source of this DNA and the factors that could affect its recovery, such as shedder status, environmental conditions, and time since deposition. Other studies concentrated on improving the techniques of DNA recovery and amplification along with other works that analyzed the potential interference of fingermark enhancement methodologies on DNA retrieval. This chapter presents an overview of the state-of-the-art technologies on DNA recovery, amplification, and profiling from "touch DNA"/ fingermark DNA.

Keywords Fingermarks · Touch DNA · Swab technique · Direct PCR · Enhancement techniques

---

S. C. Zapico (✉)
Smithsonian Institution, NMNH, Anthropology Department, Washington DC, USA

Florida International University, International Forensic Research Institute, Miami, FL, USA
e-mail: casado-zapicos@si.edu; scasadoz@fiu.edu

J. De Alcaraz-Fossoul (ed.), *Technologies for Fingermark Age Estimations: A Step Forward*, https://doi.org/10.1007/978-3-030-69337-4_10

10.1 Introduction

Since the discovery of "DNA fingerprinting" in 1985, later renamed "DNA typing", the analysis of DNA (deoxyribonucleic acid) has been a powerful tool for crime solving. The methodologies also evolved from the original use of RFLPs (restriction fragment length polymorphisms), which are differences among individuals in the lengths of DNA fragments cut by enzymes [1], to current STRs (short tandem repeats), which are DNA regions with repeat units that are two base pairs (bp) to 7 bp in length, the number of repetitions is variable among individuals [2]. This switching to STRs occurred based on (1) ease of amplification by the polymerase chain reaction (PCR), which increases the sensitivity of DNA typing, and (2) the need of as little as 100 pg (picograms) of source DNA. Moreover, it is possible to amplify and analyze multiple STRs at the same time through what is called "multiplex PCR" [3]. The Combined DNA Index System (CODIS) is the national DNA database established by the Federal Bureau of Investigation (FBI). This database requires DNA profiles to contain a minimum of 20 STRs in order to be included (full profile): https://strbase.nist.gov/fbicore.htm.

This chapter provides an overview of current studies: limitations and prospects regarding the recovery of touch/fingermark DNA, including state-of-the art methodologies to improve the DNA yield suitable for the next step of amplification and obtain a full STR profile, which includes 20 STRs, according to the current FBI standards.

10.1.1 DNA Analysis Workflow

The workflow of DNA typing starts with the collection of biological material from the crime scene, followed by its extraction and quantification (how much DNA is in the sample). Next is the amplification of STR markers by PCR and then the separation by size and detection of these markers by capillary electrophoresis (CE). Last is data analysis [2, 4]. Statistical interpretation determines the rarity of the profile on the database. This DNA analysis requires a comparison with a reference sample, which could be a potential suspect or a hit from the CODIS database (which contains profiles from convicted offenders and arrestees, as well as, DNA profiles from cold cases) [2].

After the collection of biological material, either at the crime scene, in the laboratory, or at other premises, the next step on DNA analysis is the extraction of the DNA. The recovery of DNA on FTA (Flinders Technology Associates) paper constitutes one of the main methodologies for DNA extraction. Recovered cells are lysed (broken up) and immobilized upon contact with the paper. A small punch of the paper is removed and then placed in a tube for washing and purifying to remove any inhibitors for the next step, which is the PCR reaction. Another common technique for DNA extraction is by organic (phenol-chloroform) extraction. This technique involves the use of a detergent, sodium dodecyl sulfate (SDS), and proteinase K to lyse the cells and break down the proteins. Phenol/chloroform is used to separate the DNA from proteins and cellular debris. After that, ethanol is used to precipitate the DNA, although some protocols use Microcon® 100 (Millipore) instead of ethanol to concentrate the sample. An alternative extraction methodology involves the use of a chelating-resin suspension that can be added directly to the sample, Chelex® 100 (Bio-Rad), an ion-exchange resin; its main function is chelating metal ions such as magnesium; in that way, DNA is protected of DNA-destroying nuclease enzymes. In this case, Chelex® is a protective agent, so in order to break the cells, it is required to heat the biological samples to 100 °C [2].

Although the aforementioned methods are still in use, the current tendencies are moving toward the automation of DNA extraction. In this regard, solid-phase DNA extraction methods represent a step forward. The silica-based extraction methods developed by Qiagen© are based on silica-based columns or small glass beads, which trap the DNA into the column/bead in the presence of high concentrations of chaotropic salts. This allows the DNA to be cleaned, and under alkaline conditions and low salt concentrations, the DNA can be released [2, 4, 5]. This kind of extraction has been successfully automated on several robotic platforms including EZ1, which can process between 6 and 14 samples in 20 minutes following a previous cell lysis step [2]. PrepFiler™ (Thermo Fisher Scientific) is based on magnetic beads capturing the DNA, similar to silica-based methodology described above, and also, it is possible to automate the procedure leading to the extraction of 96 samples in 2.5 hours [2].

Following the DNA isolation is the quantification of the sample. This is to determine the appropriate quantity of DNA template that needs to be included on the PCR amplification step. This quantification has to be human-specific, according to the FBI's Quality Assurance Standards. Although different techniques have been developed toward this purpose, current methodologies are based on the application of quantitative (real-time) PCR (Q-PCR), a variant of the previously referred to as PCR. Firstly, it is imperative to describe the basics of a classical PCR or commonly known as end-point PCR. A polymerase chain reaction is an enzymatic reaction which permits the replication of a DNA fragment multiple times. This process involves three steps. First is denaturation, the separation of the DNA strands at typically 94 °C. Second is the annealing step at 55–60 °C with two predesigned flanking primers (forward and reverse) to the DNA region of interest. These are short fragments of DNA (10–30 bp) complementary to each one of the DNA strands of the region (we would like to amplify, they act as "starting point" of the reaction). Finally, extension occurs at 72 °C; DNA polymerase copies the target region starting from the primers. This enzyme is similar to the one that copies the DNA in our cells. These three steps constitute one cycle of the PCR; this is repeated for approximately 25–35 cycles. This whole PCR is carried out in a thermal cycler, an instrument that heats and cools the reaction components inside a microtube.

Q-PCR analyzes the cycle-to-cycle change in fluorescence signal resulting from the amplification of a target sequence during PCR. Forensic quantifications use one type of Q-PCR called "TaqMan probes." These probes are designed specifically to hybridize with the DNA target region between the two PCR primers. The probe has two dyes: the "reporter" (R) in the 5′-end of the DNA and the "quencher" in the 3′-end. When the probe is intact, there is not fluorescence. However, during the synthesis of the DNA by the enzyme, the TaqMan probe is displaced and broken; the reporter is released and can fluoresce. The level of fluorescence is proportional to the initial DNA template quantities when compared with reference samples of known concentration of DNA. The most common assays used in forensic labs target the human telomerase reverse transcriptase gene or the human region RNU2. The kits for the latter also quantify the male chromosome, Y, and its target TSPY [2].

After quantification, the next step is the amplification of the STR regions through end-point PCR. Current commercial kits for forensic DNA typing provide the primers and all the reaction components to perform the PCR, and they are also optimized for multiplexing. This means the amplification of several STR regions at the same time in the same reaction, leading to obtain the current DNA profiles in a shorter time [2]. Other advantages are the development of direct PCR kits and the amplification of DNA directly without the steps of extraction and purification. This represents a greater advancement in terms of degraded samples. Following this line of research, the use of mini-STRs also constitutes an advantage for degraded DNA. The difference, in respect to regular STRs, is that the primers are designed closer to the region of interest than the regular STR primers, and in that way, the amplification is more attainable. However, this primer design has a major disadvantage: the multiplex is only possible with few STRs, not all, although a set of 26 new mini-STRs (some of them different than the regular STRs) has been selected at the US National Institute of Standards and Technology (NIST) to create a multiplex assay to improve the identification of degraded DNA samples [2].

Once the STRs are amplified, the next step is their detection and separation by capillary electrophoresis. For their detection, one of each pair of STR primers has to be labeled with a fluorescent dye. This separation is possible because of a capillary electrophoresis (CE) instrument. This is composed of a narrow glass capillary, two buffer vials, and two electrodes connected to a high-voltage power supply. Additionally, it also contains a laser excitation source, a fluorescence detector, an autosampler to hold the sample tubes or tray, and a computer to control the sample injection and detection. The samples are placed into a tray and injected onto the capillary by applying voltage. The capillary is filled with a polymer through which the DNA molecules travel, and the larger molecules move more slowly than the smaller, thus separating the molecules based on their size. Fluorescent dye-labeled products are excited by a laser beam and analyzed as they pass by the detector. In that way, the STRs are detected based on the fluorescent labels and the number of repetitions based on the migration speed of the DNA [2]. The final product is interpreted by genetics software that display results in the form of a diagram, called electropherogram. Figure 10.1 shows an example of an electropherogram based on the number of repetitions of an STR. These DNA profiles can then be compared with others.

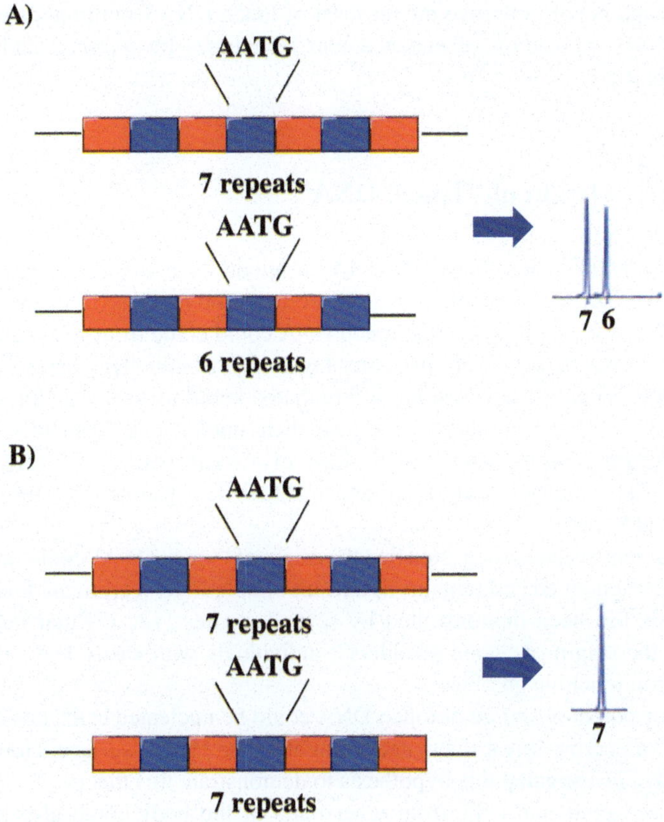

Fig. 10.1 Example of the results of DNA analysis of one STR. The sequence AATG is repeated seven times in one chromosome of the person and six times in the paired chromosome (**a**). The person is heterozygous. The electropherogram will show two peaks. (**b**) The person is homozygous, the same number of repetitions in both chromosomes. The electropherogram will show one peak

10.1.2 Fingermarks and DNA

Based on the previous workflow, the first key step in DNA recovery is to identify potential sources of DNA at crime scenes [4].

Latent fingermarks are one of the sources of what is called "touch DNA": DNA left behind as a result of touching objects [6–8]. The quantity of recovered DNA is usually low; however, since its first successful isolation by van Oorschot and Jones in 1997 from common household objects [9], other studies have demonstrated its usefulness for human identifications in forensic cases.

These studies have approached different questions regarding "touch DNA": from which type of cells is DNA originated; the quantity and quality (degradation) of DNA isolated; the effects of the surface, environment, and time on the efficiency of recovery; and how to improve the retrieval of touch DNA from fingermarks before and after the application of enhancement techniques by physical and chemical developers [6].

10.2 The Origin of "Touch DNA"

The cell types from which "touch DNA" originates are still under debate. Some studies have indicated that these cells come from the outer layer of the epidermis [10–12]. It has been proposed that touch DNA could come from shed keratinocytes or their constituent parts from the outer layers of the skin. This outer layer of the epidermis is actually composed by differentiated keratinocytes, also known as corneocytes [13]. However, these cells lose their nuclei and organelles once they mature; for this reason, some authors have dismissed them as a viable source of touch DNA [14], although others have been able to detect some DNA levels in these cell types [15, 16].

Some scientists have suggested that the origin of "touch DNA" could be "stripped nuclei" or fragmented cell remains as a result of apoptotic/necrotic pathways of cell death. This has been demonstrated by several studies [12, 17] that indicate this could be the origin of "good shedders", individuals who easily leave their DNA behind upon touching an object.

Another potential source of touch DNA could be nucleated cells, possibly originating from eccrine sweat ducts and brought to the hands' surface. However, few studies have investigated this hypothesis to demonstrate this theory [7, 18, 19].

Nucleated epithelial cells from other parts of the body could also be another source of touch DNA [6]. In addition, body fluids like saliva [20] or sebum [18], among others [21], could be another source of "touch DNA", being in this case named "transfer DNA", In fact, the study of Lacerenza et al. [21] demonstrated this transfer DNA to the hands applying RNA markers for skin and body fluids [22, 23].

Lastly, the latest trend to explain the origin of "touch DNA" is referred to as cell-free DNA (cfDNA). This is used in cancer research and prenatal screenings as a diagnostic tool [24, 25], and its origin is still controversial. It is hypothesized that cell-free DNA is released as a consequence of apoptosis/necrotic processes or, alternatively, it has an active release [6]. Although it is used as a diagnostic tool in cancer research, its presence was demonstrated in the bloodstream of healthy individuals [26] and other body fluids like saliva or semen [27, 28]. For forensic purposes, the existence of cfDNA has been demonstrated in objects that have been touched [29]. In fact, Kita et al. confirmed its presence on the topmost layer of the skin [30]. However, the study of cfDNA for forensic purposes is relatively new, so further research in this area is needed and expected.

As described above, there is no single agreement on the origin of the cell types/DNA that could generate "touch DNA." Additional studies on this area could help to improve DNA recovery techniques; DNA yields and obtains full DNA profiles from objects that have been touched. Table 10.1 provides a summary of the most

Table 10.1 Type of cells hypothesized as source of "touch DNA" and their origin

Cell type	Origin	References
Corneocytes	Outer layer epidermis	[10–13, 15, 16]
Stripped nuclei or fragmented cell remains	Apoptosis/necrosis	[12, 17]
Nucleated cells	Eccrine sweat ducts	[7, 18, 19]
Nucleated epithelial cells	Other parts of the body/body fluids	[6, 18, 20–23]
Cell-free DNA	Controversial: apoptosis/necrosis or active release	[29–31]

recent findings on this topic. Not only the quantity of DNA deposited on objects is dependent on the type of cells it originates from, but other factors such as shedder status [32], personal habits [33], handwashing [34], person's age [35], type of substrate [8, 36], environment [37], and pressure [38] influence the quantity of DNA isolated.

10.2.1 Shedder Status of the Donor

As described above, "good shedder" refers to people who easily leave their DNA behind upon contact. However, this concept is also controversial. While some authors demonstrated its consistency among individuals [32], others pointed out the opposite [39]. In general, research indicates that males are better shedders [40] by 10% [21] and that younger people are better shedders [41, 42]. However, some studies have reported high intraindividual and interindividual variation in the shedder status [34, 43, 44].

Thus, the quantity of touch DNA recovery cannot be explained solely on the shedder status. Research has repeatedly demonstrated a large variation between and within individuals and types of objects, with DNA yields ranging from 0 to 169 ng [9, 36, 37, 44]. The next factor to be considered is the substrate.

10.2.2 Type of Substrate

The first study on "touch DNA" by van Oorschot and Jones [9] demonstrated the successful DNA retrieval from common household objects such as leather briefcase handles, pens, car keys, and telephone handsets. Further research reported the recovery of DNA from different objects, including shoes, firearms, and ammunition, among others [11, 45–47]. In general, nonporous substrates are shown to be better DNA collectors than porous substrates. Among porous substrates, rough surfaces are better than smooth ones [6]. The study by Ostojic and Wurmbach [8] showed increased DNA yields and better profiles for specimens on glass, followed by plastic and paper, while no profiles were obtained from a quarter dollar coin. Other studies indicated that wood is preferable to fabrics for DNA isolation followed by glass [36], and cotton is better than plastics [44].

Along with the previously described factors, another variable to take into account is the environment.

10.2.3 The Influence of the Environment

After deposition of "touch DNA", its interaction with the environment and the elapsed time until recovery (can) affect DNA yield and retrieval of complete DNA profiles. In general, longer time intervals have a detrimental effect on DNA yield in conjunction with the environmental conditions [37]. In fact, some studies have demonstrated that the storage of "touch DNA" under controlled laboratory conditions, meaning at room temperature with low humidity, does not affect DNA recovery [48], including long-term storage of fingermarks [49]. Objects exposed to harsh and potentially damaging environments, like UV light, yielded lower quantities of DNA [37, 50]. Few studies have been developed to discover the effects of the environment on "touch DNA." Most research has focused on improving the techniques to successfully retrieve this DNA, including the time effect, as it will be described below.

10.2.4 Time After Deposition

van Oorschot and Jones [9] demonstrated that DNA from objects handled regularly by specific individuals provided DNA profiles of the user. Additional studies pointed out to the same conclusions [39, 46, 48].

As it has been referred above, the time since deposition could affect the retrieval of "touch DNA" in combination with the environment. There are several studies analyzing time as a factor. However, they are not in agreement about the best possible and adequate moment to retrieve DNA and obtain a full DNA profile. Also, it is imperative to distinguish the studies that analyze "touch DNA" or fingermarks, since the objective could be completely different as well as the results. The study of Kanokwongnuwut et al. [51] mentioned the possibility of recovering a full DNA profile up to 4 weeks after deposition from "touch DNA"; others, like the work of Ostojic and Wurmbach [8], demonstrated the retrieval of full DNA profiles from fingerprints up to 40 days. In contrast, other studies assessed the recovery after fingermark development, decreasing the time to obtain full DNA profiles [52–54].

10.2.5 Pressure

Other studies have indicated that the applied pressure against the substrate could affect the quality of "touch DNA"/fingermarks. In fact, Hefetz et al. [38] explored this possibility, demonstrating that as deposition pressure increases, the number of forensically useful DNA profiles increases, although this also depends on the type of substrate. Following this line of research, Dziak et al. [55] analyzed "touch DNA" from several samples commonly encountered in forensic casework. From these, headwear and gloves were the most successful to retrieve full DNA profiles, while in experimentally scenarios, eyeglasses, earphones, t-shirts, fabric gloves, and watches provided the highest success rates. The technique for DNA collection in this case was swabbing.

10.2.6 Personal Habits

According to Wickenheiser [33] and Phipps [34], personal habits could influence the amount of "touch DNA" recovered, especially in terms of "dominant hand." Under usual circumstances, the dominant hand will have a more frequent contact with items, and as a result, it will release more DNA than the other hand. Additionally, this hand would be more likely to come into contact with surfaces bearing the individual's own DNA (i.e., touching the face or frequently handled items like cell phones, computers, etc.), accumulating DNA on this dominant hand [34].

10.2.7 Handwashing

The study of Phipps et al. [34] and previous work from Lowe et al. [32] demonstrated that the lack of handwashing results in more DNA transfer. This increases over time although a precise pattern has not been established. However, there are not further studies analyzing this parameter individually.

10.2.8 Individual's Age

The study of Ya-Xian et al. [35] previously analyzed the influence of sex and age on shedding cells from the skin. However, they did not find any correlation among these parameters. A more recent study by Poetsch et al. [42] focused on the influence of an individual's age on the quantity of DNA and the retrieval of full DNA profiles from touched items. In children under 4 years, it was possible to obtain a full DNA profile in approximately 79% of the samples. However, in this particular

Table 10.2 Summary of the factors that affect DNA recovery. > indicates increases in DNA recovery

Factor	Effect on DNA recovery	References
Shedder status	Men > women Younger > older Intraindividual, interindividual variation	[9, 32, 36, 37, 40–44]
Substrate	Nonporous surfaces > porous Rough porous > smooth porous	[6, 8, 9, 11, 36, 44–47]
Environment	In combination with time since deposition Harsh environments decreased DNA recovery	[37, 48–50, 56]
Time	Increased yields in objects handled regularly Combination with environment Not agreement about time since deposition	[8, 9, 39, 48, 51–54, 57]
Pressure	Higher pressures increase the retrieval of useful DNA profiles Obtaining of full DNA profiles from common daily items	[38, 55]
Personal habits	Dominant hand retrieves more DNA	[33, 34]
Handwashing	Lack of handwashing retrieves more DNA	[32, 34]
Individual's age	Children (younger 4 years) > adults (21–60 years) > > elderly = teenagers	[42]

case, the authors attributed this high amount of DNA to epithelial cells and other skin cells, but also from other sources like mouth, due to childish behavior [33]. The lowest number of complete profiles was from teenagers younger than 15 years and elderly people older than 70 years, 8% in both groups. In adults between 21 and 60 years, it was possible to obtain full profiles in 25% of cases.

Table 10.2 summarizes the factors influencing the recovery of "touch DNA".

The following sections give an overview of the current state-of-the-art methodologies to improve DNA yield and obtain full DNA profiles from "touch DNA" and fingermarks.

10.3 Methodologies to Improve DNA Recovery from "Touch DNA" Samples

As in any forensic case, the workflow for DNA recovery includes its detection with different tools depending on the surface and type of biological sample; DNA extraction, again, depending on the type of biological sample the approach will be different; quantification to determine if the DNA is in enough concentration to proceed to the next step; amplification, using a multiplex PCR, amplifying 20 STRs; fragment analysis on CE; and interpretation of the results.

Some approaches have focused on the detection of "touch DNA" before extraction. The work of Kanokwongnuwut et al. [51, 58] attempted to detect touch DNA using Diamond™ Nucleic Acid Dye (DD) (Promega Corporation), which binds to the backbone of the DNA molecule as it is an external groove-binding molecule.

The authors applied this dye directly to a wide range of touched objects [51] like glass slides, credit cards, mobile phones, sim cards, ziplock bags, nickel cartridge cases, and aluminum cartridge cases. Cellular material was visualized thanks to DD and DNA and was recovered by swabbing, generating full STR profiles by direct PCR. This confirmed that DD does not interfere with DNA isolation/amplification. In addition, the authors demonstrated that DD only binds to eukaryotic DNA, not bacterial DNA, probably because of the supercoiled configuration of bacterial DNA. Following this line of research, the same authors [58] explored the possibility of applying DD to swabs as a marker of cellular presence before DNA extraction. The authors tested ten different types of swabs; however, only three swab types (black cotton swab, ultrafine dental applicator, and cylinder dental applicator) were really effective for the collection of "touch DNA," as demonstrated first by touch DNA visualization and then direct PCR profiling. In addition, the cellular material was stored for up to 4 weeks at room temperature.

Other reagents have been investigated and tested to visualize "touch DNA", like Yano et al. [59] that developed an acid-free p-dimethylaminocinnamaldehyde (DMAC) solution containing silicone oil for spraying to locate "touch DNA" samples after irradiation with a blue light emitting diode (LED). After detecting the DNA, swabbing, and extracting, they did not find any impact on DNA concentration with regard to the original DMAC containing acid. Also, in terms of fluorescence, they detected differences on the time since deposition, decreasing after 3 days despite DNA concentrations being similar. It was possible to obtain full DNA profiles, mostly 2 hours after deposition, decreasing the retrieval of these profiles after 3 days.

As mentioned above, swabbing is the common chosen methodology for "touch DNA" cases. Its efficiency in collecting DNA has also been investigated. Comte et al. [60] analyzed four different types of swabs on DNA recovery from collars, screwdrivers, and steering wheels. They studied the quality of the extracted DNA (degradation of DNA over time) and obtained full DNA profiles from these objects. As described previously [58], the type of swab influenced the quantity of DNA recovered, COPAN 4N6FLOQSwabs® (Copan) being the most suitable for this task. However, in a conservation study to determine the influence of the time since deposition and environment to the DNA recovery, the same COPAN swab showed DNA degradation over time.

With the goal to improve DNA recovery from "touch DNA" samples, Dong et al. [61] assessed four different preprocessing methods:

1. Direct cutting method, based on cutting the samples into small pieces.
2. Stubbing procedure: The stub was placed on the surface of the samples and removed until all of the area was covered, and each stub was placed on the surface of the sample approximately 20 times before it was saturated.
3. Double-swab technique: The first swab was moistened with sterile distilled water and used to scrub the entire surface of the samples; the second swab was used in the same way but was not moistened.

4. Vacuum cleaner: A 25-cm^2 white impervious membrane was placed between the one-off tip and the tip of a vacuum cleaner. Each sample was vacuumed for approximately 5 seconds on all surfaces.

These techniques were applied on porous (cloth, glove, thick rope, thin rope) and nonporous (plastic rope, glass, door handle) materials. They found that the quantity and quality (number of alleles) of DNA on porous substrates were both greater than those on nonporous substrates. The methodological approach was dependent on the material analyzed: the direct cutting method was better for porous substrates, while the vacuum cleaner method was more suitable for nonporous surfaces. Additionally, they investigated the effect of storage time from 2 to 360 days on the quantity and quality of DNA recovered from gloves and door handles by applying the stubbing procedure. In this case, the storage time did not have any effect on the retrieval of full DNA profiles from both materials, although the quantity of DNA recovered was variable.

Following this line of research, Kirgiz et al. [62] also compared the efficiency of three types of DNA recovery methods, swabbing, tape lifting, and FTA® paper (Whatman) scraping, to retrieve DNA from car steering wheels. According to this study, the three methods recovered quantifiable DNA, although the highest yield was obtained with FTA paper. In terms of quality of DNA profile, again FTA paper was the most successful technique in obtaining a full profile.

Dierig et al. [63] applied adhesive film and the staining of skin particles with Trypan blue in order to detect "touch DNA" in garments. The authors analyzed different DNA extraction methodologies, finding that direct lysis methods such as Chelex® 100 (Bio-Rad) or casework kits increased DNA yields. According to this study, the combination of both detection and extraction methodologies could improve the quantity and quality of DNA retrieved from "touch DNA" samples.

As mentioned earlier [58], another approach to increase the probability of recovering full DNA profiles from "touch DNA" could be looking at the amplification level. Kanokwongnuwut et al. [58] applied direct PCR with this goal in mind. Martin et al. [64] compared different direct PCR approaches on DNA samples recovered following standard DNA extraction methods. The substrates were common items found in forensic casework, like unfired aluminum cartridge cases, insulated wires, circuit boards, and ziplock bags. The recovery technique of choice was swabbing. They obtained a higher efficiency in recovering full DNA profiles when direct PCR was applied to these samples. This methodology was previously used by Templeton et al. [65] who demonstrated that an optimization of the swabbing technique combined with direct PCR amplification improved the retrieval of full DNA profiles. This study indicated that the time since deposition is a factor to take into consideration because the recovery of DNA decreased after 8 days. The usefulness of direct PCR amplification on "touch DNA" samples has been extensively reviewed [66].

Other authors have used mini-STRs in an attempt to increase the likelihood of obtaining meaningful DNA profiles from "touch DNA". For instance, Van Nieuwerburgh et al. [67] used "touch DNA samples" from forensic cases. The

Table 10.3 Summary of the different approaches and techniques to increase the recovery of full STR profiles from "touch DNA"

Approach	Technique	Outcome	References
Detection	Diamond nucleic acid dye/swabbing/direct PCR	Full STR profiles	[51, 58]
	DMAC + LED/swabbing/regular extraction	Full STR profiles	[59]
Recovery	Swabbing	Depending on the swab type to recover full STR profiles	[58, 60, 61]
	Direct cutting method	Full STR profiles in porous substrates	[61]
	Vacuum cleaner method	Full STR profiles in nonporous substrates	[61]
	FTA paper	Increases the probability to recover full STR profiles	[62]
	Adhesive film/direct lysis	Increases DNA yield	[63]
Amplification	Direct PCR	Increases the probability to recover full STR profiles	[58, 64, 66]
	Mini-STRs	Not really effective to recover full STR profiles	[67]

recovery technique was swabbing, followed by DNA extraction through Chelex® 100, and the amplification of DNA was performed using mini-STRs. This alternative method, however, did not significantly increase the retrieval of full DNA profiles.

To summarize this section and as shown on Table 10.3, it is likely that the detection/recovery methods influence the retrieval of full DNA profiles from "touch DNA" specimens. As a result, current research is focused on attempting to improve and develop better methods that can be specifically applied to this field.

10.4 Methodologies to Improve DNA Recovery from Fingermark Samples

Research on DNA retrieval from latent fingermarks has been mostly focused on the interaction of fingermark developers (physical and chemical enhancers) with the efficiency of DNA recovery and recovery of full STR profiles.

Ostojic and Wurmbach [8] conducted a study to analyze the efficiency of DNA recovery from latent fingermarks left on common household items: glass, plastic, paper, and metal. In this case, they did not apply any developers to visualize fingermarks. The technique chosen to collect DNA was swabbing moistened with a detergent (Triton X™-100, Sigma Aldrich), followed by extraction, quantification, and amplification with an Identifiler™ kit (Thermo Fisher Scientific). According to their results, the type of substrate had an impact on DNA retrieval, obtaining better results for glass, followed by plastic and paper, and no identifiable profiles from metals (one quarter, US dollar coin). Additionally, they tested the influence of time since deposition and were able to obtain full STR profiles up to 40 days of storage at room temperature. The effects of substrate and recovery methods were also demonstrated by Wood et al. They indicated that nonporous surfaces yielded less quantity of DNA than porous surfaces. The recovery method of choice for both substrates was swabbing, which seemed to be more appropriate for this task than using mini-tapes.

Other studies have focused on the prediction of DNA concentrations prior to its extraction. Research by Falkena et al. [68] unsuccessfully attempted to correlate the autofluorescence intensity of fingermarks when excited at 365 nm with DNA concentration and amplification of full STR profiles. This method proved to be inappropriate for predicting DNA concentrations or obtaining full DNA profiles, probably due to the concurrent detection of microbial DNA. In the same line of research and with similar negative results, the study of Haines et al. used SYBR™ Green I (Thermo Fisher Scientific) and GelGreen™ (Biotium) on fingermarks to detect DNA. However, it was not possible to distinguish between human and microbial DNA.

Methodologies for fingermark DNA detection based on imaging technologies have been extensively reviewed [69]. However, based on the proposed equipment, these techniques would not be suitable in the majority of crime labs.

Kanokwongnuwut et al. [70] used Diamond™ Nucleic Acid Dye (DD) in combination with classical methods of fingermark development. They noted that the application of aluminum or white powder developers did not interfere with the detection of cellular material by DD and the subsequent generation of STR profiles by swabbing and direct PCR. In contrast, magnetic powders and cyanoacrylate fuming negatively interfered by decreasing the quantity of DNA recovered, although it did not have any effects on generating STR profiles. In these cases, it has been recommended to first apply DD. Other research has compared the efficiency of DNA recovery after the application of latent fingermark developers. Bhoelai et al. [71] analyzed the effect of cyanoacrylate (CA) fuming, vacuum metal deposition (VMD), 1,8-diaza-9-fluorenone (DFO), ninhydrin, and physical developer on paper or plastic sheets. The sampling technique for DNA was cutting, followed by QIAamp®-based DNA (Qiagen) isolation and AmpFLSTR™ SGM™ Plus kit (Thermo Fisher Scientific). According to their results, CA and VMD did not affect STR profiling. DFO and ninhydrin presented the risk of cross-contamination, and a physical developer has deleterious effects on DNA.

A recent study by Subhani et al. [72] compared classical (powders) and modern developing and lifting fingermark techniques with regard to the efficiency of DNA retrieval and STR profiling. All the techniques tested provided meaningful STR profiles, even when traditional fingermark lifts were used, although some of them retrieved partial profiles, the majority (79.2%) gave full DNA profiles. In this line of research, Tsai et al. [73] analyzed the effects of latent fingermark enhancement techniques on DNA retrieval from porous surfaces (paper). They evaluated iodine fuming, 1,2-indanedione (in ethyl acetate or in acetic acid), silver nitrate, and Coomassie Brilliant Blue. The samples were cut out and DNA extracted, quantified, and amplified through Identifiler™ PCR Amplification Kit (Thermo Fisher). The most successful methods for STR profiling were iodine fuming and 1,2-indanedione in ethyl acetate enhancement. In contrast, 1,2-indanedione in acetic acid and silver nitrate had a negative impact, both in the quantity and quality of STR profiles. Oleiwi et al. [7] analyzed the effect of infrared fluorescent fingermark visualization powder (fpNatural® 1, Foster + Freeman) on the recovery of DNA from fingermarks deposited on glass slides. DNA was recovered by double-swabbing. This powder did not have any effect on the quantity and quality (STR profiles) of DNA recovered; therefore, it could be appropriate for developing latent fingermarks and STR analysis.

Other studies focused on the impact of cyanoacrylate (CA) on DNA retrieval. Bille et al. [52] analyzed the effects of this reagent when applied to pipe bombs. After fuming, DNA was recovered by swabbing. They did not detect any negative impact on DNA recovery after CA fuming, although the time since deposition affected the successful recovery of DNA. In this case, lower quantities of DNA were recovered on samples collected and analyzed 3 months after deflagration. Following this line of research, Khuu et al. [53] compared the effects of DNA retrieval after the application of CA treatment and one-step luminescent CA treatment, such as PolyCyano UV (Foster + Freeman Ltd) and Lumicyano™ (Sirchie) on glass slides. Their research indicated that this PolyCyano UV induced a higher degree of DNA degradation than the classical CA method. Lower quantities of DNA were isolated, and as a result, fewer full STR profiles were generated. In contrast, Lumicyano appeared to have the same effects on DNA as conventional CA techniques.

The study by Shalhoub et al. [74] developed latent fingermarks and extracted DNA using a silicone-based casting material from different substrates, such as aluminum cans, plastic bottles, coins, waxy paper cups, light bulbs, and hard plastic mobile phones. After casting the fingermarks, CA was applied, and DNA was recovered by swabbing, extracted by QIAamp® DNA Mini Kit (Qiagen), and amplified by an AmpFLSTR™ SGM Plus™ Kit (Thermo Fisher Scientific). Their results indicated that the silicone-based casting material was appropriate for nonporous surfaces but not suitable for recovering fingermarks from semi-porous surfaces. Full STR profiling was successful only in 34% of cases, so this methodology will require additional research before its application in forensic casework. The work by Plaza et al. [75] compared three different methodologies of DNA recovery from different types of paper. These included ESDA (electrostatic collection device) collection, dry swabbing, and substrate cutting on papers. Dry swabbing was the best technique in terms of quantity and retrieval of full DNA profiles, followed by ESDA. Even on porous surfaces, swabbing appeared to be more successful in terms of DNA extraction.

Other research has investigated the recovery of DNA from tape lifts [76]. The study by Steadman et al. [76] analyzed the efficiency of recovery of DNA from tape lifts from a forensic case. According to their results, it was not possible to retrieve full STR profiles from tape. The work of Feine et al. [77] indicated that acetone could facilitate and improve the DNA yield based on a cell culture-based touch DNA model.

10.4.1 Workflow Optimization

Solomon et al. [49] optimized the workflow to obtain DNA from archived latent fingermarks. Their results showed that the most efficient procedure is cutting the sample, extracting by QIAamp® DNA Investigator Kit (Qiagen) and DNA concentration by Centri-Sep™ (Thermo Fisher Scientific) followed by amplification with AmpFLSTR™ Identifiler™ Plus PCR Amplification Kit (Thermo Fisher Scientific).

In the line of protocol optimization, Sewell et al. [78] demonstrated that DNeasy® Plant Mini Kit (Qiagen) improved DNA recovery from paper compared with the classical QIAamp® DNA Mini Kit (Qiagen). In addition, certain types of paper seemed to interfere with the successful retrieval of DNA. Lin et al. [79] compared the efficiency of DNA recovery of three types of DNA kits, DNA IQ™ (Promega), QIAamp® DNA Investigator (Qiagen), and QIAsymphony® DNA investigator (Qiagen), with various fingerprint treatments. According to their results, CA fuming followed by DNA IQ extraction was more efficient in DNA profiling. Aluminum powder dusting offered the least interference in the three DNA extraction systems. As described previously by other authors [70], magnetic powder dusting strongly interfered with DNA recovery, although the worst outcome was observed with the application of physical developer as fingermark developer [73].

10.4.2 Amplification Improvements

Other research has focused on the next step, the optimization of DNA amplification after collection and extraction.

As described in Sect. 10.3., direct PCR methodologies could also be an asset to obtain full DNA profiles. Liu [80] described a workflow involving PE-Swab (a novel sample collection device) followed by direct PCR amplification. This workflow improved the generation of full DNA profiles from "touch DNA" samples. In this line of research, Templeton et al. [65] applied direct PCR after the enhancement of fingermarks with conventional powders. The introduction of direct PCR into this workflow without the previous extraction step improved the generation of full DNA profiles.

From a different experimental perspective, Fieldhouse et al. [54] analyzed the opposite. This is the quality of fingermarks after DNA recovery onto different surfaces: glass, aluminum, textured plastic, varnished wood, and photocopier paper, aged 4 hours to 4 weeks. This research demonstrated that DNA recovery methods had a destructive influence on the visual quality of latent fingermarks on glass surfaces. On the remaining surfaces, flocked swabs and gel lifts were the least destructive methods, except for aluminum sheet metal. On wood and plastic, the quality of fingermarks was less affected by swabbing, being wet swabbing and tape lifting more damaging to the fingermark.

Table 10.4 Summary of the different approaches/factors and techniques that influence the recovery of full STR profiles from fingermarks

Approach/ factor	Technique	Conclusion of research?	References
Detection	Diamond nucleic acid dye + development/swabbing/direct PCR	Full STR profiles	[51, 58, 70]
	Autofluorescence intensity of fingermarks at 365 nm	No correlation with DNA concentration/profiles	[68]
Developer	Cyanoacrylate	Not interfere with STR profile recovery	[52, 53]
	Vacuum metal deposition	Not interfere with STR profile recovery	[71]
	Ninhydrin and DFO	Risk of cross-contamination	[71]
	Physical developer	Deleterious effects on DNA	[71]
	Iodine fuming	Not interfere with STR profile recovery	[73]
	Classical powders	Not interfere with STR profile recovery	[72]
	Infrared fluorescent fingermark visualization powder	Not interfere with STR profile recovery	[7]
Amplification	Direct PCR	Increases the probability to obtain full STR profiles	[50, 56, 65, 80]
	Optimized protocol	Increases the probability to obtain full STR profiles	[49, 78, 79]

In summary (Table 10.4), "classical" techniques for latent fingermark enhancement did not significantly affect DNA recovery, with a few exceptions, like a physical developer. However, the optimization of the workflow, particularly the DNA collection technique and direct PCR amplification, improved the DNA yield and the generation of full DNA profiles.

10.5 DNA Degradation over Time

There are only a few studies investigating patterns of DNA degradation over time when exposed to different environments and substrates. For example, Goecker et al. [50] compared three different developers on glass substrates, CA fuming, black powder dusting, and the columnar-thin-film (CTF) technique, finding similar DNA degradation ratios over time with the highest at 4 weeks. None of them was detrimental to the genetic components within the fingermark. The authors also compared Quantifiler™ Trio (Thermo Fisher Scientific) and InnoQuant® (InnoGenomics) kits for detection of DNA degradation and detected that InnoQuant® detected higher degradation ratios. Samples with lower DNA concentrations showed a non-linear relationship of degradation ratios with aging, while samples with higher DNA concentrations showed a quasi-linear correlation with aging.

A recent study by Romano et al. [81] claimed the successful retrieval of DNA from fingermarks aged 14 years. These fingermarks came from a forensic case and were actually "archived" fingermarks and, thus, recovered from the crime scene 14 years before the experiment. These were enhanced by powdering, the black powders "BVDA Concentrated Black" and SIRCHIE Silk Black M101L and the silver powders "BVDA Special Silver" and "SIRCHIE Silk Gray M102L," and kept on a transparent adhesive film. DNA was extracted automatically using an EZ1® DNA Investigator Kit (Qiagen) and quantified using Quantifiler™ Trio (Thermo Fisher Scientific). STR profiles were amplified using an Amp GlobalFiler® PCR Amplification Kit (Thermo Fisher Scientific). Among 12 fingermarks, it was possible to recover quantifiable DNA from only two samples; however, these two samples did not show full DNA profiles, only partial profiles, with a variable number of typed loci between these two samples. The authors attributed this to different factors, one of them being the presence of powders that degraded the DNA and the extraction method. Here the authors directly cut out the paper where the fingermark was and extracted its DNA.

In summary, there are few studies examining patterns of "touch DNA"/fingermark DNA degradation over time. Based on these scarce works and the importance of this issue, research efforts should be focused on studying the potential degradation of DNA over time in different types of "touch DNA"/fingermark samples.

10.6 Conclusions

Several factors influence the successful retrieval and identification of DNA based on STR profiling from "touch DNA"/fingermarks samples. According to recent research, the type of substrate is one of the main influencing factors along with the recovery/amplification techniques being used. DNA swabbing followed by direct PCR appears to be the most effective combination for generating full STR profiles. Regarding fingermark enhancement techniques, the majority of the classical visualization methods (i.e., powders) do not interfere with DNA recovery; however, chemical developers, particularly physical developer, have an impact on the quantity of DNA recovered. Therefore, the selection of the most appropriate methodology for optimum fingermark development and subsequent DNA recovery will highly depend on the combination of factors that are particular to each case as well as the quality of the evidence left behind. Figure 10.2 summarizes the research described in this review.

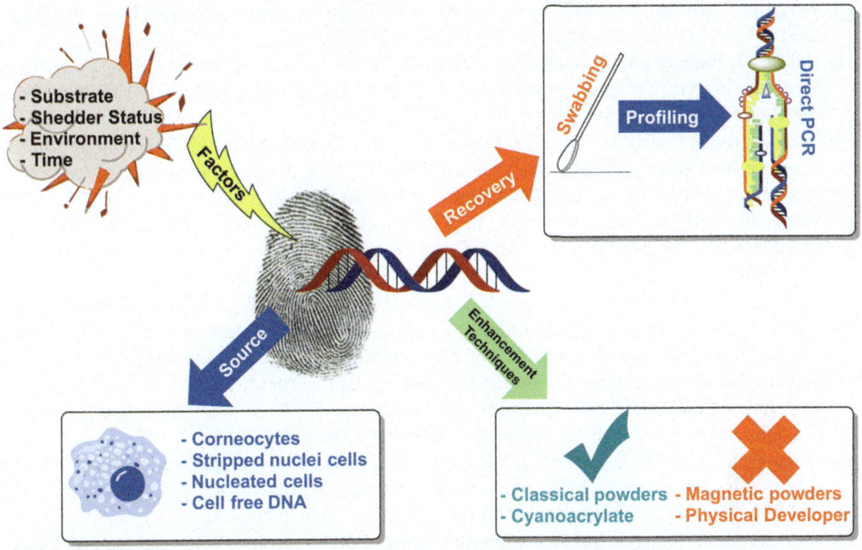

Fig. 10.2 Summary of the discoveries related to DNA recovery from latent fingermarks. Source of DNA, influencing factors, the impact of fingermark enhancement techniques, and protocol optimization for DNA profiling

References

1. Lee HC, Ladd C, Bourke MT, Pagliaro E, Tirnady F (1994) DNA typing in forensic science, I Theory and background. Am J Forensic Med Pathol 15(4):269–282. https://doi.org/10.1097/00000433-199412000-00001
2. Butler JM (2012) Advanced topics in forensic DNA typing: methodology. Elsevier/Academic Press, Walthan
3. Parson W (2018) Age estimation with DNA: from forensic DNA fingerprinting to forensic (epi) genomics: a mini-review. Gerontology 64(4):326–332. https://doi.org/10.1159/000486239
4. McCord BR, Gauthier Q, Cho S, Roig MN, Gibson-Daw GC, Young B, Taglia F, Zapico SC, Mariot RF, Lee SB, Duncan G (2019) Forensic DNA analysis. Anal Chem 91(1):673–688. https://doi.org/10.1021/acs.analchem.8b05318
5. Zapico SC, Ubelaker DH (2013) Sex determination from dentin and pulp in a medicolegal context. J Am Dent Assoc 144(12):1379–1385. https://doi.org/10.14219/jada.archive.2013.0074
6. Burrill J, Daniel B, Frascione N (2019) A review of trace "touch DNA" deposits: variability factors and an exploration of cellular composition. Forensic Sci Int Genet 39:8–18. https://doi.org/10.1016/j.fsigen.2018.11.019
7. Al Oleiwi A, Hussain I, McWhorter A, Sutton R, King RSP (2017) DNA recovery from latent fingermarks treated with an infrared fluorescent fingerprint powder. Forensic Sci Int 277:e39–e43. https://doi.org/10.1016/j.forsciint.2017.05.008
8. Ostojic L, Wurmbach E (2017) Analysis of fingerprint samples, testing various conditions, for forensic DNA identification. Sci Justice J Forensic Sci Soc 57(1):35–40. https://doi.org/10.1016/j.scijus.2016.08.009
9. van Oorschot RA, Jones MK (1997) DNA fingerprints from fingerprints. Nature 387(6635):767. https://doi.org/10.1038/42838

10. Wiegand P, Kleiber M (1997) DNA typing of epithelial cells after strangulation. Int J Legal Med 110(4):181–183. https://doi.org/10.1007/s004140050063
11. Helmus J, Bajanowski T, Poetsch M (2016) DNA transfer-a never ending story. A study on scenarios involving a second person as carrier. Int J Legal Med 130(1):121–125. https://doi.org/10.1007/s00414-015-1284-1
12. Alessandrini F, Cecati M, Pesaresi M, Turchi C, Carle F, Tagliabracci A (2003) Fingerprints as evidence for a genetic profile: morphological study on fingerprints and analysis of exogenous and individual factors affecting DNA typing. J Forensic Sci 48(3):586–592
13. Warshauer DH, Marshall P, Kelley S, King J, Budowle B (2012) An evaluation of the transfer of saliva-derived DNA. Int J Legal Med 126(6):851–861. https://doi.org/10.1007/s00414-012-0743-1
14. van den Berge M, Ozcanhan G, Zijlstra S, Lindenbergh A, Sijen T (2016) Prevalence of human cell material: DNA and RNA profiling of public and private objects and after activity scenarios. Forensic Sci Int Genet 21:81–89. https://doi.org/10.1016/j.fsigen.2015.12.012
15. Stanciu CE, Philpott MK, Kwon YJ, Bustamante EE, Ehrhardt CJ (2015) Optical characterization of epidermal cells and their relationship to DNA recovery from touch samples. F1000Research 4:1360. https://doi.org/10.12688/f1000research.7385.1
16. Wang C, Stanciu CE, Ehrhardt CJ, Yadavalli VK (2017) Nanoscale characterization of forensically relevant epithelial cells and surface associated extracellular DNA. Forensic Sci Int 277:252–258. https://doi.org/10.1016/j.forsciint.2017.06.019
17. Kamphausen T, Schadendorf D, von Wurmb-Schwark N, Bajanowski T, Poetsch M (2012) Good shedder or bad shedder--the influence of skin diseases on forensic DNA analysis from epithelial abrasions. Int J Legal Med 126(1):179–183. https://doi.org/10.1007/s00414-011-0579-0
18. Zoppis S, Muciaccia B, D'Alessio A, Ziparo E, Vecchiotti C, Filippini A (2014) DNA fingerprinting secondary transfer from different skin areas: morphological and genetic studies. Forensic Sci Int Genet 11:137–143. https://doi.org/10.1016/j.fsigen.2014.03.005
19. Oleiwi AA, Morris MR, Schmerer WM, Sutton R (2015) The relative DNA-shedding propensity of the palm and finger surfaces. Sci Justice J Forensic Sci Soc 55(5):329–334. https://doi.org/10.1016/j.scijus.2015.04.003
20. Quinque D, Kittler R, Kayser M, Stoneking M, Nasidze I (2006) Evaluation of saliva as a source of human DNA for population and association studies. Anal Biochem 353(2):272–277. https://doi.org/10.1016/j.ab.2006.03.021
21. Lacerenza D, Aneli S, Omedei M, Gino S, Pasino S, Berchialla P, Robino C (2016) A molecular exploration of human DNA/RNA co-extracted from the palmar surface of the hands and fingers. Forensic Sci Int Genet 22:44–53. https://doi.org/10.1016/j.fsigen.2016.01.012
22. Lindenbergh A, de Pagter M, Ramdayal G, Visser M, Zubakov D, Kayser M, Sijen T (2012) A multiplex (m)RNA-profiling system for the forensic identification of body fluids and contact traces. Forensic Sci Int Genet 6(5):565–577. https://doi.org/10.1016/j.fsigen.2012.01.009
23. Hanson E, Haas C, Jucker R, Ballantyne J (2012) Specific and sensitive mRNA biomarkers for the identification of skin in 'touch DNA' evidence. Forensic Sci Int Genet 6(5):548–558. https://doi.org/10.1016/j.fsigen.2012.01.004
24. Stokowski R, Wang E, White K, Batey A, Jacobsson B, Brar H, Balanarasimha M, Hollemon D, Sparks A, Nicolaides K, Musci TJ (2015) Clinical performance of non-invasive prenatal testing (NIPT) using targeted cell-free DNA analysis in maternal plasma with microarrays or next generation sequencing (NGS) is consistent across multiple controlled clinical studies. Prenat Diagn 35(12):1243–1246. https://doi.org/10.1002/pd.4686
25. Bennett CW, Berchem G, Kim YJ, El-Khoury V (2016) Cell-free DNA and next-generation sequencing in the service of personalized medicine for lung cancer. Oncotarget 7(43):71013–71035. https://doi.org/10.18632/oncotarget.11717
26. Devonshire AS, Whale AS, Gutteridge A, Jones G, Cowen S, Foy CA, Huggett JF (2014) Towards standardisation of cell-free DNA measurement in plasma: controls for extraction efficiency, fragment size bias and quantification. Anal Bioanal Chem 406(26):6499–6512. https://doi.org/10.1007/s00216-014-7835-3

27. Taki T, Kibayashi K (2015) Characterization of cellular and extracellular DNA in saliva. Legal Med 17(6):471–474. https://doi.org/10.1016/j.legalmed.2015.10.003
28. Chou JS, Jacobson JD, Patton WC, King A, Chan PJ (2004) Modified isocratic capillary electrophoresis detection of cell-free DNA in semen. J Assist Reprod Genet 21(11):397–400. https://doi.org/10.1007/s10815-004-7527-6
29. Vandewoestyne M, Van Hoofstat D, Franssen A, Van Nieuwerburgh F, Deforce D (2013) Presence and potential of cell free DNA in different types of forensic samples. Forensic Sci Int Genet 7(2):316–320. https://doi.org/10.1016/j.fsigen.2012.12.005
30. Kita T, Yamaguchi H, Yokoyama M, Tanaka T, Tanaka N (2008) Morphological study of fragmented DNA on touched objects. Forensic Sci Int Genet 3(1):32–36. https://doi.org/10.1016/j.fsigen.2008.09.002
31. Quinones I, Daniel B (2012) Cell free DNA as a component of forensic evidence recovered from touched surfaces. Forensic Sci Int Genet 6(1):26–30. https://doi.org/10.1016/j.fsigen.2011.01.004
32. Lowe A, Murray C, Whitaker J, Tully G, Gill P (2002) The propensity of individuals to deposit DNA and secondary transfer of low level DNA from individuals to inert surfaces. Forensic Sci Int 129(1):25–34. https://doi.org/10.1016/s0379-0738(02)00207-4
33. Wickenheiser RA (2002) Trace DNA: a review, discussion of theory, and application of the transfer of trace quantities of DNA through skin contact. J Forensic Sci 47(3):442–450
34. Phipps M, Petricevic S (2007) The tendency of individuals to transfer DNA to handled items. Forensic Sci Int 168(2–3):162–168. https://doi.org/10.1016/j.forsciint.2006.07.010
35. Ya-Xian Z, Suetake T, Tagami H (1999) Number of cell layers of the stratum corneum in normal skin - relationship to the anatomical location on the body, age, sex and physical parameters. Arch Dermatol Res 291(10):555–559. https://doi.org/10.1007/s004030050453
36. Daly DJ, Murphy C, McDermott SD (2012) The transfer of touch DNA from hands to glass, fabric and wood. Forensic Sci Int Genet 6(1):41–46. https://doi.org/10.1016/j.fsigen.2010.12.016
37. Raymond JJ, van Oorschot RA, Gunn PR, Walsh SJ, Roux C (2009) Trace evidence characteristics of DNA: a preliminary investigation of the persistence of DNA at crime scenes. Forensic Sci Int Genet 4(1):26–33. https://doi.org/10.1016/j.fsigen.2009.04.002
38. Hefetz I, Einot N, Faerman M, Horowitz M, Almog J (2019) Touch DNA: the effect of the deposition pressure on the quality of latent fingermarks and STR profiles. Forensic Sci Int Genet 38:105–112. https://doi.org/10.1016/j.fsigen.2018.10.016
39. Ladd C, Adamowicz MS, Bourke MT, Scherczinger CA, Lee HC (1999) A systematic analysis of secondary DNA transfer. J Forensic Sci 44(6):1270–1272
40. Fonnelop AE, Ramse M, Egeland T, Gill P (2017) The implications of shedder status and background DNA on direct and secondary transfer in an attack scenario. Forensic Sci Int Genet 29:48–60. https://doi.org/10.1016/j.fsigen.2017.03.019
41. Manoli P, Antoniou A, Bashiardes E, Xenophontos S, Photiades M, Stribley V, Mylona M, Demetriou C, Cariolou MA (2016) Sex-specific age association with primary DNA transfer. Int J Legal Med 130(1):103–112. https://doi.org/10.1007/s00414-015-1291-2
42. Poetsch M, Bajanowski T, Kamphausen T (2013) Influence of an individual's age on the amount and interpretability of DNA left on touched items. Int J Legal Med 127(6):1093–1096. https://doi.org/10.1007/s00414-013-0916-6
43. Goray M, Fowler S, Szkuta B, van Oorschot RAH (2016) Shedder status-an analysis of self and non-self DNA in multiple handprints deposited by the same individuals over time. Forensic Sci Int Genet 23:190–196. https://doi.org/10.1016/j.fsigen.2016.05.005
44. Goray M, Mitchell RJ, van Oorschot RA (2010) Investigation of secondary DNA transfer of skin cells under controlled test conditions. Legal Med 12(3):117–120. https://doi.org/10.1016/j.legalmed.2010.01.003
45. Petricevic SF, Bright JA, Cockerton SL (2006) DNA profiling of trace DNA recovered from bedding. Forensic Sci Int 159(1):21–26. https://doi.org/10.1016/j.forsciint.2005.06.004
46. Bright JA, Petricevic SF (2004) Recovery of trace DNA and its application to DNA profiling of shoe insoles. Forensic Sci Int 145(1):7–12. https://doi.org/10.1016/j.forsciint.2004.03.016

47. Horsman-Hall KM, Orihuela Y, Karczynski SL, Davis AL, Ban JD, Greenspoon SA (2009) Development of STR profiles from firearms and fired cartridge cases. Forensic Sci Int Genet 3(4):242–250. https://doi.org/10.1016/j.fsigen.2009.02.007
48. Breathnach M, Williams L, McKenna L, Moore E (2016) Probability of detection of DNA deposited by habitual wearer and/or the second individual who touched the garment. Forensic Sci Int Genet 20:53–60. https://doi.org/10.1016/j.fsigen.2015.10.001
49. Solomon AD, Hytinen ME, McClain AM, Miller MT, Dawson Cruz T (2018) An optimized DNA analysis workflow for the sampling, extraction, and concentration of DNA obtained from archived latent fingerprints. J Forensic Sci 63(1):47–57. https://doi.org/10.1111/1556-4029.13504
50. Goecker ZC, Swiontek SE, Lakhtakia A, Roy R (2016) Comparison of Quantifiler((R)) trio and InnoQuant human DNA quantification kits for detection of DNA degradation in developed and aged fingerprints. Forensic Sci Int 263:132–138. https://doi.org/10.1016/j.forsciint.2016.04.009
51. Kanokwongnuwut P, Kirkbride KP, Linacre A (2018) Detection of latent DNA. Forensic Sci Int Genet 37:95–101. https://doi.org/10.1016/j.fsigen.2018.08.004
52. Bille TW, Cromartie C, Farr M (2009) Effects of cyanoacrylate fuming, time after recovery, and location of biological material on the recovery and analysis of DNA from post-blast pipe bomb fragments*. J Forensic Sci 54(5):1059–1067. https://doi.org/10.1111/j.1556-4029.2009.01128.x
53. Khuu A, Chadwick S, Moret S, Spindler X, Gunn P, Roux C (2018) Impact of one-step luminescent cyanoacrylate treatment on subsequent DNA analysis. Forensic Sci Int 286:1–7. https://doi.org/10.1016/j.forsciint.2018.02.015
54. Fieldhouse S, Oravcova E, Walton-Williams L (2016) The effect of DNA recovery on the subsequent quality of latent fingermarks. Forensic Sci Int 267:78–88. https://doi.org/10.1016/j.forsciint.2016.08.003
55. Dziak R, Peneder A, Buetter A, Hageman C (2018) Trace DNA sampling success from evidence items commonly encountered in forensic casework. J Forensic Sci 63(3):835–841. https://doi.org/10.1111/1556-4029.13622
56. Goecker ZC, Swiontek SE, Lakhtakia A, Roy R (2016) Corrigendum to "Comparison of Quantifiler((R)) Trio and InnoQuant human DNA quantification kits for detection of DNA degradation in developed and aged fingerprints" [Forensic Sci. Int. 263C (2016) 132–138]. Forensic Sci Int 267:228. https://doi.org/10.1016/j.forsciint.2016.07.003
57. Breathnach M, Williams L, McKenna L, Moore E (2016) Response to Jamieson's and Bader's comments on Breathnach et al. probability of detection of DNA deposited by habitual wearer and/or the second individual who touched the garment. Forensic Science International Genetics 25:e3. https://doi.org/10.1016/j.fsigen.2016.10.001
58. Kanokwongnuwut P, Kirkbride P, Linacre A (2018) Visualising latent DNA on swabs. Forensic Sci Int 291:115–123. https://doi.org/10.1016/j.forsciint.2018.08.016
59. Yano R, Shimoda O, Okitsu T, Sakurada M, Ueno Y (2019) Development of a modified p-dimethylaminocinnamaldehyde solution for touch DNA analysis and its application to STR analysis. Forensic Sci Int Genet 38:86–92. https://doi.org/10.1016/j.fsigen.2018.10.007
60. Comte J, Baechler S, Gervaix J, Lock E, Milon MP, Delemont O, Castella V (2019) Touch DNA collection - performance of four different swabs. Forensic Sci Int Genet 43:102113. https://doi.org/10.1016/j.fsigen.2019.06.014
61. Dong H, Wang J, Zhang T, Ge JY, Dong YQ, Sun QF, Liu C, Li CX (2017) Comparison of preprocessing methods and storage times for touch DNA samples. Croat Med J 58(1):4–13. https://doi.org/10.3325/cmj.2017.58.4
62. Kirgiz IA, Calloway C (2017) Increased recovery of touch DNA evidence using FTA paper compared to conventional collection methods. J Forensic Legal Med 47:9–15. https://doi.org/10.1016/j.jflm.2017.01.007

63. Dierig L, Schmidt M, Wiegand P (2019) Looking for the pinpoint: optimizing identification, recovery and DNA extraction of micro traces in forensic casework. Forensic Sci Int Genet 44:102191. https://doi.org/10.1016/j.fsigen.2019.102191
64. Martin B, Blackie R, Taylor D, Linacre A (2018) DNA profiles generated from a range of touched sample types. Forensic Sci Int Genet 36:13–19. https://doi.org/10.1016/j.fsigen.2018.06.002
65. Templeton JEL, Taylor D, Handt O, Linacre A (2017) Typing DNA profiles from previously enhanced fingerprints using direct PCR. Forensic Sci Int Genet 29:276–282. https://doi.org/10.1016/j.fsigen.2017.05.006
66. Cavanaugh SE, Bathrick AS (2018) Direct PCR amplification of forensic touch and other challenging DNA samples: a review. Forensic Sci Int Genet 32:40–49. https://doi.org/10.1016/j.fsigen.2017.10.005
67. Van Nieuwerburgh F, Van Hoofstat D, Van Neste C, Deforce D (2014) Retrospective study of the impact of miniSTRs on forensic DNA profiling of touch DNA samples. Sci Justice 54(5):369–372. https://doi.org/10.1016/j.scijus.2014.05.009
68. Falkena K, Hoveling RJM, van Weert A, Lambrechts SAG, van Leeuwen TG, Aalders MCG, van Dam A (2019) Prediction of DNA concentration in fingermarks using autofluorescence properties. Forensic Sci Int 295:128–136. https://doi.org/10.1016/j.forsciint.2018.12.005
69. Wei Q, Zhang M, Ogorevc B, Zhang X (2016) Recent advances in the chemical imaging of human fingermarks (a review). Analyst 141(22):6172–6189. https://doi.org/10.1039/c6an01121g
70. Kanokwongnuwut P, Kirkbride KP, Kobus H, Linacre A (2019) Enhancement of fingermarks and visualizing DNA. Forensic Sci Int 300:99–105. https://doi.org/10.1016/j.forsciint.2019.04.035
71. Bhoelai B, de Jong BJ, de Puit M, Sijen T (2011) Effect of common fingerprint detection techniques on subsequent STR profiling. Forensic Sci Int Genet Suppl Ser 3:e429–e430
72. Subhani Z, Daniel B, Frascione N (2019) DNA profiles from fingerprint lifts-enhancing the evidential value of Fingermarks through successful DNA typing. J Forensic Sci 64(1):201–206. https://doi.org/10.1111/1556-4029.13830
73. Tsai LC, Lee CC, Chen CC, Lee JC, Wang SM, Huang NE, Linacre A, Hsieh HM (2016) The influence of selected fingerprint enhancement techniques on forensic DNA typing of epithelial cells deposited on porous surfaces. J Forensic Sci 61(Suppl 1):S221–S225. https://doi.org/10.1111/1556-4029.12893
74. Shalhoub R, Quinones I, Ames C, Multaney B, Curtis S, Seeboruth H, Moore S, Daniel B (2008) The recovery of latent fingermarks and DNA using a silicone-based casting material. Forensic Sci Int 178(2–3):199–203. https://doi.org/10.1016/j.forsciint.2008.04.001
75. Plaza DT, Mealy JL, Lane JN, Parsons MN, Bathrick AS, Slack DP (2015) ESDA(R)-Lite collection of DNA from latent fingerprints on documents. Forensic Sci Int Genet 16:8–12. https://doi.org/10.1016/j.fsigen.2014.11.011
76. Steadman SA, Hoofer SR, Geering SC, King S, Bennett MA (2015) Recovery of DNA from latent fingerprint tape lifts archived against matte acetate. J Forensic Sci 60(3):777–782. https://doi.org/10.1111/1556-4029.12716
77. Feine I, Shpitzen M, Geller B, Salmon E, Peleg T, Roth J, Gafny R (2017) Acetone facilitated DNA sampling from electrical tapes improves DNA recovery and enables latent fingerprints development. Forensic Sci Int 276:107–110. https://doi.org/10.1016/j.forsciint.2017.04.023
78. Sewell J, Quinones I, Ames C, Multaney B, Curtis S, Seeboruth H, Moore S, Daniel B (2008) Recovery of DNA and fingerprints from touched documents. Forensic Sci Int Genet 2(4):281–285. https://doi.org/10.1016/j.fsigen.2008.03.006
79. Lin SW, Ip SC, Lam TT, Tan TF, Yeung WL, Tam WM (2017) Compatibility of DNA IQ, QIAamp((R)) DNA investigator, and QIAsymphony((R)) DNA investigator((R)) with various fingerprint treatments. Int J Legal Med 131(2):293–301. https://doi.org/10.1007/s00414-016-1447-8

80. Liu JY (2015) PE-swab direct STR amplification of forensic touch DNA samples. J Forensic Sci 60(3):693–701. https://doi.org/10.1111/1556-4029.12705
81. Romano CG, Mangiaracina R, Donato L, D'Angelo R, Scimone C, Sidoti A (2019) Aged fingerprints for DNA profile: first report of successful typing. Forensic Sci Int 302:109905. https://doi.org/10.1016/j.forsciint.2019.109905

Chapter 11
Latent Fingermarks and Microbiome: Time and Community Succession

Joe Adserias-Garriga and Jesús L. Garcia-Gil

Abstract Trace evidence is a type of physical evidence that can be transferred between two objects, helping to reconstruct the sequence of events of a crime. From a forensic perspective, the human skin is a key element for, and source of, microbial trace evidence due to the constant exposure and contact with surfaces and continuous interaction with the environment. The most studied group of microorganisms in microbial forensics is bacteria. Bacteria are extremely diverse in terms of morphology, physiology, and metabolism. They have evolved to use a wide range of strategies to obtain energy, being able to survive and thrive in many different environments. This great diversity and adaptability make bacteria ubiquitous, that is, they can be found in every environment possible on Earth, a remarkable characteristic that can be of great investigative value as trace evidence. Microorganisms colonize the skin immediately after birth and remain there even after death. In healthy conditions, they are mainly located in the superficial layers of the epidermis, gland tracts, and hair follicles. The skin microbiota consists of a group of organisms that are routinely found in the human skin and can reestablish themselves after a perturbance. The human skin core microbiota includes four main phyla: *Actinobacteria*, *Firmicutes*, *Proteobacteria*, and *Bacteroides*. Recent advances in molecular analysis, especially modern sequencing techniques, have considerably increased the quantity of data available for forensic purposes. This information is valuable not only to identify the representative microbial communities but also to better understand the dynamics occurring within each community. High-throughput sequencing and alpha and beta diversity analysis have been employed to achieve a deeper understanding of the human skin microbiome and its uniqueness among individuals. This represents a means of personal identification as well as the distinctiveness of the microbiome when transferred from the skin to another surface. Several studies have reported that skin bacteria of individuals can also be found on their mobile

J. Adserias-Garriga (✉)
Department of Applied Forensic Sciences, Mercyhurst University, Erie, PA, USA
e-mail: jadseriasgarriga@mercyhurst.edu

J. L. Garcia-Gil
Department of Biology, Campus Montilivi, University of Girona, Girona, Spain

J. De Alcaraz-Fossoul (ed.), *Technologies for Fingermark Age Estimations: A Step Forward*, https://doi.org/10.1007/978-3-030-69337-4_11

phones or be traced back to a keyboard in such a way that a person could be identified based on the bacterial residue left on specific keys. The enormous metabolic and physiological diversity of bacteria allow microbial assemblages to readily evolve to the changing environments. Thus, the composition of skin microbiota has a high potential to successfully complement current forensic techniques for the reconstruction of criminal events at the scene.

Keywords Thanatomicrobiome · Molecular techniques · Bacterial succession · Next generation sequencing · Microbial forensics

Acronyms

DNA	Deoxyribonucleic acid
RNA	Ribonucleic acid
NGS	Next-generation sequencing
OTU	Operational taxonomic unit
PCR	Polymerase chain reaction
QIIME	Quantitative insights into microbial ecology
RDP	Ribosomal database project

11.1 Microbial Diversity and the Use of Microbes as Forensic Evidence

Cellular life can be classified into three major domains, *Archaea*, *Bacteria*, and *Eukarya*, according to phylogenetic analysis utilizing small ribosomal subunit sequence comparison [1]. Microscopic, unicellular life is not exclusive to *Archaea* and *Bacteria*. All *Eukarya* phylogroups, except plants, animals, and some fungi, are composed of unicellular organisms, albeit possessing a higher organization level, with a nucleus and organelles. *Archaea* and *Bacteria* came into existence after an evolutionary period of about 3.8 billion years, which is 2 billion years longer than that of eukaryotic organisms [2]. In addition to *Archaea* and *Bacteria*, viruses are also considered microorganisms; however, the main phylogenetic and evolutive features of these acellular bioentities are yet to be elucidated. In terms of the number of cells on Earth, eukaryotic cells are a small minority compared to the *Archaea* and *Bacteria*, but they are larger in size [3].

The number of prokaryotes and their cellular carbon total amounts on Earth are estimated to be 4–6 × 10^{30} cells and 350–550 Pg of C (1 Pg = 10^{15} g), respectively. The cellular production rate for all prokaryotes on Earth is estimated at 1.7 × 10^{30} cells/year and is highest in the open ocean [4]. Both the large size of the microbial population and the rapid growth rate of microorganisms provide an immense capacity for genetic diversity. The microbiota is the major component of global biodiversity and occupies a central role in ecosystem functioning from nutrient cycling to host fitness. Bacteria are ubiquitous and encompass a major portion of key nutrients, such as carbon (C), nitrogen (N), or phosphorus (P), and thus biomass on planet Earth [4].

Microorganisms are incredibly diverse at the morphological, physiological, and metabolic levels. Microbes can use a wide range of strategies to obtain energy and nutrients from their habitat, being able to survive in very different and extreme conditions. Moreover, some microorganisms can even change their metabolic strategy depending on the environmental conditions, which makes them extremely competitive at surviving in dynamic situations [3–6]. Indeed, our planet hosts a vast diversity of environments. Abiotic factors such as temperature, pH, Redox potential, pressure, water activity, radiation, or nutrient availability determine the ecological niche where individual bacterial species can thrive. Some of these parameters may reach extreme values, and yet bacteria have been able to adapt. In fact, there is a bacterial species for every ecological niche, and there are millions of possible niches in the biosphere. Therefore, we can expect extremely high bacterial diversity on Earth, as has recently been revealed through high-throughput sequencing of environmental samples, revealing previously unidentified bacterial species only known after their genomes have been sequenced. This enormous diversity and adaptability makes microbes ubiquitous, present in environments such as soil, lakes, the skin surface, and even the clean rooms at aeronautics and space facilities as well as extreme environments like hot springs or acid lakes [7, 8].

There is always a microbial community composition for any set of physical and chemical conditions. According to this ecological axiom, in a continually evolving ecosystem like a decomposing body, or an aging fingerprint, we would expect bacteria to be perfect indicators of the changing physicochemical conditions. Thus, microbes become an ideal source of physical evidence [5]. Microorganisms have been used as forensic evidence since the nineteenth century. The French pioneer criminalist, Edmond Locard, established the concept of the *exchange principle*, which states whenever two objects come in contact, a transfer of material occurs, including microorganisms. Locard used microbes as trace evidence in multiple investigations, correlating certain areas of Paris with signature microbes [6]. Since then, technical advances, especially the development of culture-independent techniques, have made it possible to deepen our understanding of microbes, highlighting the importance of the interrelationships among them and with their surrounding environment. Accordingly, current research in this field is very focused on microbiome changes [5, 6].

The term microbiota refers to the ecological community of microbes (commensal, symbiotic, and pathogenic) that occupies a specific habitat or ecosystem [7]. Its composition and the abundance of different microbial groups differ from habitat to habitat. Microbial diversity plays a central role in the habitat functioning. Understanding the microbial diversity in different environments can provide highly valuable information to comprehend any biological processes taking place. For example, the intestinal microbiota is the result of a sustained, yet immature, anaerobic ecosystem with high turnover rates [8]. Even though the techniques and methods for isolation and identification of microorganisms have become much more sophisticated, the microbial diversity has not been fully explored nor exploited [3, 6, 9].

The study of microbial communities has tremendous potential in forensic science since, as mentioned earlier, microbes are omnipresent, and certain communities are often related to specific processes and environments [6]. The application of microbiome analysis covers different areas of forensic investigations such as the cause of death [10, 11], time since death estimation [12–14], place of death, and trace evidence [15].

11.2 New Methods to Study Microbial Diversity

In studies of biodiversity and ecosystem functions [16], microbial communities have become model systems to explore how both the composition and abundance of microbial taxa can affect specific ecosystems [17–19]. Richness and functional diversity of bacteria tell us about essential features of an ecosystem's maturity and structure. Both have been hardly accessible until the advent/arrival of DNA-based analytical techniques.

Models of community assembly (i.e., bacteria clusters that are cohabiting in the same niche) have been described after the analysis of a hierarchical phylogenetic-based distribution of natural bacterial assemblages [20]. The same approach has been followed to improve our understanding of the functional role of the "rare biosphere" [21, 22]. The recent and accelerated development of quantitative molecular genetic methods to estimate bacterial representatives in ecosystems is promising for testing ecological theory. However, diversity estimates may still be dependent on technical and methodological considerations, such as the sampling and the molecular approach used.

Before the rise of molecular techniques, microbes present in natural environments had to be identified and quantified either by microscopy or cultivation [23]. These methods are now considered to only reveal a tiny fraction of the microbial taxa present [24]. Fortunately, the molecular identification of bacteria through the 16S ribosomal RNA gene (16S rRNA) has been extensively exploited over the last two decades to delve deeper into the microbial taxa of ecosystems [24, 25]. Methods based on small ribosomal subunit sequences have helped to obtain a substantially improved picture of the biotic portion of ecosystems. The growing number of genetic sequences in publicly accessible reference databases makes taxa identification from 16S rRNA more reliable than ever before.

Microbial assemblages have been approached through the analysis of 16S rRNA sequences as the standard method due, in part, to the availability of a growing number of open-source computer software tools, such as "mothur" [26], "QIIME" [27], and "RDP" [28], that allow fast and reliable sorting and comparison of massive nucleotide data. These tools ease the analysis of the large number of sequences produced by modern massive sequencing techniques.

New cutting-edge methods to characterize microbial communities through 16S rRNA sequences have developed rapidly. The classic approach consisting of creating clone libraries [29, 30] sequenced by the Sanger methods [31] has been swiftly replaced by next-generation sequencing (NGS) or massive-sequencing technologies [32], producing vast quantities of sequences in a very short time. This methodology has been widely used to study microbial communities in all sorts of environments, not only by microbial ecologists but also by disciplines such as biomedicine, agriculture, and chemical engineering. As a result, data available has exponentially grown and has significantly increased our understanding of microbial communities [33] and relevant microbiomes, both from animal [34, 35] and plant hosts [36]. Despite the clear advantages of extensively studying microbial communities, NGS presents some drawbacks. One of the main problems is the short reading length of nucleotide sequences; sequencing the complete 16S rRNA gene of entire communities is still methodologically challenging and expensive. Another problem is that NGS is a PCR-based method which may introduce some method bias to the quantitative data, especially to calculations of relative abundances.

NGS research is restricted to one, or a few, of the nine variable regions of the 16S rRNA gene, depending on the primer set designed. Different variable regions of the 16S rRNA gene show different capabilities/potential to accurately resolve microbial taxa [37–39]. The selection of the best variable region and the robustness of the derived conclusions and ecological implications are still controversial. However, common workflows are widely used which provide a better comprehensive and comparative perspective to the knowledge of microbe-inhabited ecosystems.

Even though NGS is widely employed in forensic research, its use has gained a lot of attention in the last years. Indeed, several forensic laboratories have incorporated NGS for human genetic analysis. Kits, such as the HID-Ion AmpliSeq™ Identity (SNP) Panel (Thermo Fisher Scientific) and the ForenSeq™ DNA Signature Prep Kit (Illumina®) for (X-/Y-)STRs and SNPs for human identity, biogeographical ancestry, and phenotyping applications, are commercially available and comprehensively tested by forensic researchers [40]. NGS has become affordable for most researchers in terms of costs, so it can be now implemented as a routine method for exhaustive bacterial analysis. Several companies offer high-throughput analyses of microbiota composition providing comprehensive results on the relative abundance of the main phylogenetic growth. The identification of signature taxa along the decomposition process of bodies is more accessible than ever and may help in determining the post-mortem interval [11, 12]. The use of NGS for microbial analysis in casework is still very limited, but rapid and significant advances in research will extend its application to forensic casework in the near future.

11.3 Human Skin Microbiome

The skin is the largest organ of the human body, and it is the habitat of approximately 10^7 bacterial cells that interact among themselves and with the host [41]. The human skin is composed of two layers: the epidermis and the dermis. The epidermis is the outer surface of the skin, and it includes three cell types: keratinocytes, melanocytes, and Langerhans cells. This outer layer is subjected to constant dynamics which consists of the movement of keratinocytes outward from the inner areas to the outer surface [42]. Between the epidermis and the dermis, there is the basement membrane. The basal cell layer of the epidermis and the dermis are tightly attached to each other. Both layers, together with the internal membrane, form an irregular structure, known as the dermo-epidermal junction. The dermis is comprised of a thin papillary layer and a subjacent, but much thicker, reticular layer. It includes fibroblasts, macrophages, and mast cells [43].

Microorganisms colonize the human skin immediately during and after birth, and (in healthy conditions) most of them are located in the superficial layers of the epidermis, gland tracts, and hair follicles [43]. These microbes can be classified into commensals and pathogens. Commensal microorganisms (also called resident or indigenous) are present on body surfaces without causing any harm to the human body, while pathogen microorganisms consist of those that are capable of damaging the human host. Skin microbiota significantly influences the balance between health and disease. Actually, human skin acts as the first protective barrier to prevent the invasion of harmful pathogens. In circumstances where this barrier is broken or when the balance between commensals and pathogens is disturbed, skin disease or even systemic disease can occur, which is often related to changes in the composition, structure, and diversity of the skin microbiota [44, 45].

Although the main and most studied group of microorganisms of the human skin are bacteria, other microbes such as fungi and viruses are also present. The skin microbiome includes resident and transient microorganisms. The skin resident, or core microbiota, consists of a group of organisms that are routinely found in the skin and can reestablish themselves after a perturbance. It includes four main phyla: *Actinobacteria*, *Firmicutes*, *Proteobacteria*, and *Bacteroides* [46]. Transient microbiota arise from the environment and persist for hours to days before disappearing [45].

The skin topography in different areas of the human body provides distinct ecological microenvironments, each with its own core microbiome. There are four types of environments in the human skin: moist, sebaceous, dry, and others (which include sweat glands, hair follicles, and dermal layers); for example, the palm of the hands and the fingertips are considered a dry environment [46, 47]. Nowadays, different bacteria are classified according to their similarities in the 16SRNA gene, and they are clustered in operational taxonomic units (OTUs), as proxies for species [48]. *Actinobacteria* is the most dominant bacterial phylum in the human skin; it accounts for approximately 52% to 75% of the operational taxonomic units (OTUs) in the skin [46, 47]. *Actinobacteria* skin representatives include *Propionibacterium* sp., which are abundant at sebaceous sites, and *Corynebacterium* sp. which dominates moist skin sites [49]. *Firmicutes* are the second major phylum found on the skin surface; it accounts for approximately 24% of the skin bacterial OTUs. *Staphylococcus* spp. are the dominant representative from this family, but genera such as *Streptococcus*, *Clostridium*, *Bacillus*, and *Lactobacillus* can also be identified on the skin surface. *Proteobacteria* is the third major phylum of the skin, *Betaproteobacteria* being a representative family located especially in dry areas of the skin [46].

The composition of microbial communities is primarily dependent on the physiology of the skin site, with changes in the relative abundance of bacterial taxa associated with moist, dry, and sebaceous microenvironments. Lipophilic *Propionibacterium* species dominate sebaceous sites. In contrast, bacteria that thrive in humid environments, such as *Staphylococcus* and *Corynebacterium* species, are preferentially abundant in moist areas, including the bends of the elbows and the feet [50]. The hand, which is considered a dry microenvironment in the body, harbors a high level of microbial diversity, where the main OTUs are *Streptococcus*, *Staphylococcus*, *Micrococcus*, *Enhydrobacter*, *Corynebacterium*, and *Lactobacillus* species [51–53]. The skin is a very complex organ due to the influence of its microbiota and the constant exposure to environmental factors. Metagenomic studies have shown that the microbiota on the skin is even more diverse than the gut microbiota [54].

Unlike bacterial communities, fungal community composition is similar across core body sites. Fungi of the genus *Malassezia* predominate at core body and arm sites, whereas foot sites are colonized by a more diverse combination of *Malassezia* spp., *Aspergillus* spp., *Cryptococcus* spp., *Rhodotorula* spp., *Epicoccum* spp., and others [51, 55]. In addition to bacteria and fungi, the human skin is also colonized by viruses; however, the understanding of the viral community composition associated with the human skin is still very unknown.

11.4 Latent Fingermarks, Microbiome, and Its Changes Over Time

Trace evidence is a type of physical evidence that can be transferred between two objects or materials, helping to piece together a crime investigation [5]. From the perspective of forensic science, the skin can be considered as the external interface between the human body and the environment. Therefore, it becomes a key element for the collection of, and search for, trace evidence. Several studies focused on human identification have stated the usefulness of microbiome analysis for personal identification. For example, the palm surface shares only 13% of the bacterial groups between any two individuals [52]. Therefore, the skin microbiome may be a useful tool for personal identification [6, 53, 56–58].

As mentioned earlier, modern sequencing techniques have considerably increased the amount of data available for forensic analysis. This information is valuable not only to identify the microbial community representatives but also to better understand the dynamics occurring within the community. These techniques have also been used to achieve a deeper understanding of the human skin microbiome and the uniqueness among individuals with the aim of personal identification. Moreover, the uniqueness of the microbiome transferred from the skin to another surface has already been proven. Several studies have reported that skin bacteria of individuals can also be found on their mobile phones [59, 60]. The microbial community can be traced back to a keyboard in such a way that a person could be identified based on the bacterial residue from the specific keys touched [53].

Additionally, as mentioned above, the core skin bacterial communities are relatively stable over time in healthy individuals. Sample storage under typical indoor conditions (i.e., temperature at 20 °C and no exposure to fluorescent light for about 8 h per day) has little to no influence on bacterial community composition during a maximum time interval of 14 days [53]. Studies in the succession of microbial communities over time have been conducted, focusing on the postmortem dynamics. These alterations have been studied on the body as well as on the soil where the corpse is decomposing. Bacterial communities present in the soil surrounding the cadaver dramatically change throughout the decomposition process. In the first stages of decomposition, the soil microbiota consists of indigenous soil representatives. As decomposition progresses, a decrease in soil representatives is observed due to bacteria from the decomposing body being transferred to the soil, causing a temporary niche displacement of the indigenous soil microbiota [12]. These succession changes in the postmortem microbiome can be used to estimate the time since death [6, 12, 13].

Follow-up studies of the changes in microbial community showed that *Proteobacteria*, *Firmicutes*, *Actinobacteria*, *Bacteroides*, and *Acidobacteria* as the most predominant groups. A drastic growth of *Firmicutes* and decrease of *Proteobacteria* and *Acidobacteria* were observable throughout the decomposition process in soil samples [12]. When evaluating the microbial community changes on the body surfaces, signature groups of the different stages of decomposition can be identified. The fresh stage was characterized by indigenous oral microbiome representatives, including *Lactobacillaceae*, *Staphylococcaceae*, *Gemellaceae*, *Carnobacteriaceae*, *Aerococcaceae*, *Veillonellaceae*, *Streptococcaceae*, *Campylobacteraceae*, *Micrococcaceae*, *Bifidobacteriaceae*, *Actinomycetaceae*, and *Corynebacteriaceae*. Second, the predominant families in the bloat stage were *Peptostreptococcaceae* and *Bacteroidaceae*, which are mostly oral indigenous representatives, and *Enterococcaceae*, which is a gut microbiome representative. Later, in the bloat stage, *Clostridiales* took advantage of the changes in the environmental conditions, i.e., oxygen depletion and set-up of low redox conditions. The predominant bacterial families in advanced decay were *Gammaproteobacteria*, *Pseudomonadaceae*, *Alcaligenaceae*, and *Planococcaceae*, which are soil representatives. Finally, dry remains were characterized by the presence of *Bacilli* and *Clostridia* [11].

Latent fingermarks are a natural source of microbial specimens that can be used for personal identification. However, little is known on the microbial signature of latent fingermarks, and research that focuses on changes in microbial communities on aging impressions is much needed. These data could potentially be used to estimate the time since deposition, a similar principle to time since death estimates of decomposing human bodies. These studies could adopt the techniques and methods employed for the examination of the postmortem changes of microbial communities on the substrate where a body is decaying.

11.5 Conclusions

The latest methods to study microbial communities, including through high-throughput sequencing, have allowed researchers to obtain the most complete picture to date of bacterial composition and functions. New statistical and mathematical algorithms have dramatically improved our knowledge of the structure of microbial ecosystems and the relationships among different assemblages. Additionally, taxonomic studies on the composition of communities have led to identifying several bacterial groups that act as useful indicators of changes in the abiotic conditions, no matter how tiny they are.

High-throughput sequencing has become a fabulous and valuable tool (unthinkable just a couple of decades ago) for different scientific disciplines where bacteria have always been active (or hidden) participants, especially in medicine. Forensic science is not an exception and has also benefited from the powerful possibilities offered by the new molecular techniques applied to microorganisms.

The composition of skin microbiota has a high potential to complement other current forensic techniques and to provide reliable data-driven conclusions. Microbial assemblages reflect, with high fidelity, the ever-changing environment. Forensic science is currently learning how to take advantage of this ecological feature, yet there is a long way to go. This is equally exciting and promising, especially for age estimation of fingermarks as trace evidence of microbial origin.

References

1. Woese CR, Kandler O, Wheelis ML (1990) Towards a natural system of organisms: proposal for the domains archaea, Bacteria, and Eucarya. Proc Natl Acad Sci U S A 87(12):4576–4579
2. Torsvik V, Øvreås L, Thingstad TF (2002) Prokaryotic diversity-magnitude, dynamics, and controlling factors. Science 296:1064–1066
3. Cowan MK, Smith H, Lusk J (eds) (2019) Microbiology fundamentals: a clinical approach, 3rd edn. McGraw-Hill, New York
4. Whitman WB, Coleman DC, Wiebe WJ (1998) Prokaryotes: the unseen majority. Proc Natl Acad Sci U S A 95(12):6578–6583. https://doi.org/10.1073/pnas.95.12.6578

5. Carter OD, Tomberlin JK, Benbow ME, Metcalf JL (eds) (2017) Microbial forensics. Wiley, New Jersey
6. Metcalf JL, Xu ZZ, Bouslimani A, Dorrestein P, Carter DO, Knight R (2017) Microbiome tools for forensic science. Trends Biotechnol 35(9):814–823
7. Lederberg J, McCray A (2001) The scientist: 'ome sweet' omics – a genealogical treasury of words. Science 15(7):8
8. Radjabzadeh D, Boer CG, Beth SA et al (2020) Diversity, compositional and functional differences between gut microbiota of children and adults. Sci Rep 10:1040
9. Tortora GJ, Funke BR, Case CL (eds) (2019) Microbiology: an introduction, 13th edn. Pearson, Boston
10. Kakizaki E, Kozawa S, Matsuda H, Muraoka E, Uchiyama T, Sakai M, Yukawa N (2010) Freshwater bacterioplankton cultured from liver, kidney, and lungs of a decomposed cadaver retrieved from a sandy seashore: possibility of drowning in a river and then floating out to sea. Legal Med 12(4):195–199
11. Burton JL, Saegeman V, Arribi A et al (2019) Postmortem microbiology sampling following death in hospital: an ESGFOR task force consensus statement. J Clin Pathol 72(5):329–336. https://doi.org/10.1136/jclinpath-2018-205365
12. Adserias-Garriga J, Hernandez M, Quijada NM, Rodriguez Lazaro D, Steadman D, Garcia-Gil J (2017) Daily thanatomicrobiome changes in soil as an approach of postmortem interval estimation: an ecological perspective. Forensic Sci Int 278:388–395. https://doi.org/10.1016/j.forsciint.2017.07.017
13. Adserias-Garriga J, Quijada NM, Hernandez M, Rodriguez Lazaro D, Steadman D, Garcia-Gil LJ (2017) Dynamics of the oral microbiota as a tool to estimate time since death. Mol Oral Microbiol 32(6):511–516. https://doi.org/10.1111/omi.12191.
14. Javan GT, Finley SJ, Tuomisto S, Hall A, Benbow ME, Mill DE (2019) An interdisciplinary review of the thanatomicrobiome in human decomposition. Forensic Sci Med Pathol 15(1):75–83. https://doi.org/10.1007/s12024-018-0061-0
15. Ward MJ, Goncheva M, Richardson E, McAdam PR, Raftis E, Kearns A et al (2016) Identification of source and skin populations for the emergence and global spread of the East-Asia clone of community-associated MRSA. Genome Biol 17:160
16. Cardinale BJ et al (2012) Biodiversity loss and its impact on humanity. Nature 486(7401):59–67. https://doi.org/10.1038/nature11148.
17. Petchey OL, McPhearson PT, Casey TM, Morin PJ (1999) Environmental warming alters food-web structure and ecosystem function. Nature 402:69–72. https://doi.org/10.1038/47023
18. Venail PA, Vives MJ (2013) Phylogenetic distance and species richness interactively affect the productivity of bacterial communities. Ecology 94(11):2529–2536. https://doi.org/10.1890/12-2002.1
19. Petchey OL, Gaston KJ (2006) Functional diversity: back to basics and looking forward. Ecol Lett 9(6):741–758. https://doi.org/10.1111/j.1461-0248.2006.00924.x
20. Ofiţeru ID et al (2010) Combined niche and neutral effects in a microbial wastewater treatment community. Proc Natl Acad Sci 107(35):15345–15350. https://doi.org/10.1073/pnas.1000604107.
21. Pedrós-Alió C (2012) The rare bacterial biosphere. Annu Rev Mar Sci 4:449–466
22. Besemer K et al (2012) Unraveling assembly of stream biofilm communities. ISME J 6(8):1459–1468. https://doi.org/10.1038/ismej.2011.205
23. LeChevallier MW, Seidler RJ, Evans TM (1980) Enumeration and characterization of standard plate count bacteria in chlorinated and raw water supplies. Appl Environ Microbiol 40(5):922–930. https://doi.org/10.1128/aem.40.5.922-930.1980
24. Staley JT, Konopka A (1985) Measurement of in situ activities of nonphotosynthetic microorganisms in aquatic and terrestrial habitats. Annu Rev Microbiol 39:321–346
25. Ward DM, Weller R, Bateson MM (1990) 16S rRNA sequences reveal numerous uncultured microorganisms in a natural community. Nature 345:63. https://doi.org/10.1038/345063a0

26. Schloss PD et al (2009) Introducing mothur: open-source, platform-independent, community-supported software for describing and comparing microbial communities. Appl Environ Microbiol 75:7537–7541
27. Caporaso JG et al (2010) QIIME allows analysis of high-throughput community sequencing data. Nat Methods 7(5):335–336. https://doi.org/10.1038/nmeth.f.303.
28. Cole JR et al (2014) Ribosomal database project: data and tools for high throughput rRNA analysis. Nucleic Acids Res 42(Database issue):D633–D642. https://doi.org/10.1093/nar/gkt1244
29. Larsen N et al (1993) The ribosomal database project. Nucleic Acids Res 21:3021–3023
30. Maidak JR, Lilburn TG, Parker CT Jr, Saxman RJ, Garrity GM, Olsen GJ, Schmidt TM, Tiedje JM (2001) The RDP-II (ribosomal database project). Nucleic Acids Res 29:173–174
31. Pace NR, Stahl DA, Lane DJ, Olsen GJ (1986) The analysis of natural microbial populations by ribosomal RNA sequences. In: Marshall KC (ed) Advances in microbial ecology, vol 9. Springer, Boston. https://doi.org/10.1007/978-1-4757-0611-6_1
32. Schmidt TM, DeLong EF, Pace NR (1991) Analysis of a marine picoplankton community by 16S rRNA gene cloning and sequencing. J Bacteriol 173(14):4371–4378. https://doi.org/10.1128/jb.173.14.4371-4378.1991
33. Sanger F et al (1977) Nucleotide sequence of bacteriophage φx174 DNA. Nature 265:687–695. https://doi.org/10.1038/265687a0.
34. Schuster SC (2008) Next-generation sequencing transforms today's biology. Nat Methods 5(1):16–18. https://doi.org/10.1038/nmeth1156
35. Pinto AJ, Xi C, Raskin L (2012) Bacterial community structure in the drinking water microbiome is governed by filtration processes. Environ Sci Technol 46(16):8851–8859. https://doi.org/10.1021/es302042t
36. Turnbaugh PJ et al (2006) An obesity-associated gut microbiome with increased capacity for energy harvest. Nature 444(7122):1027–1031. https://doi.org/10.1038/nature05414.
37. Arif I, Batool M, Schenk PM (2020) Plant microbiome engineering: expected benefits for improved crop growth and resilience. Trends Biotechnol 38(12):1385–1396. https://doi.org/10.1016/j.tibtech.2020.04.015
38. Vasileiadis S et al (2012) Soil bacterial diversity screening using single 16S rRNA gene V regions coupled with multi-million read generating sequencing technologies. PLoS One 7(8):e42671. https://doi.org/10.1371/journal.pone.0042671.
39. Kim M, Morrison M, Yu Z (2011) Evaluation of different partial 16S rRNA gene sequence regions for phylogenetic analysis of microbiomes. J Microbiol Methods 84(1):81–87. https://doi.org/10.1016/j.mimet.2010.10.020
40. Bruijns B, Tiggelaar R, Gardeniers R (2018) Massively parallel sequencing techniques for forensics: a review. Electrophoresis 39:2642–2654
41. Fredricks DN (2001) Microbial ecology of human skin in health and disease. J Investig Dermatol Symp Proc 6:167–169. https://doi.org/10.1046/j.0022-202x.2001.00039.
42. Weinstein GD, McCullough JL, Ross P (1984) Cell proliferation in normal epidermis. J Investig Dermatol 82:623–628
43. Tong J, Li H (2014) The human skin. In: Marchesi RJ (ed) The human microbiota and microbiome. CABI, Oxfordshire, pp 72–89
44. Leung MHY, Tong X, Wilkins D, Cheung HHL, Lee PKH (2018) Individual and household attributes influence the dynamics of the personal skin microbiota and its association network. Microbiome 6(1):26. Published 2018 Feb 2. https://doi.org/10.1186/s40168-018-0412-9
45. Dréno B, Araviiskaia E, Berardesca E, Gontijo G, Sanchez Viera M, Xiang LF, Martin R, Bieber T (2016) Microbiome in healthy skin, update for dermatologists. J Eur Acad Dermatol Venereol 30(12):2038–2047. https://doi.org/10.1111/jdv.13965
46. Grice EA, Kong HH, Conlan S, Deming CB, Davis J, Young AC et al (2009) Topographical and temporal diversity of the human skin microbiome. Science 324:1190–1192
47. Costello EK, Lauber CL, Hamady M, Fierer N, Gordon JI, Knight R (2009) Bacterial community variation in human body habitats across space and time. Science 326:1694–1697

48. Edgar RC (2018) Updating the 97% identity threshold for 16S ribosomal RNA OTUs. Bioinformatics 34(14):2371–2375. https://doi.org/10.1093/bioinformatics/bty113.
49. Fitz-Gibbon S, Tomida S, Chiu BH, Nguyen L, Du C, Liu M et al (2013) Propionibacterium acnes strain populations in the human skin microbiome associated with acne. J Investig Dermatol 133:2152–2160
50. Byrd AL, Belkaid Y, Segre JA (2018) The human skin microbiome. Nat Rev Microbiol 16(3):143–155. https://doi.org/10.1038/nrmicro.2017.157
51. Oh J, Byrd AL, Park M, NISC Comparative Sequencing Program, Kong HH, Segre JA (2016) Temporal stability of the human skin microbiome. Cell 165(4):854–866. https://doi.org/10.1016/j.cell.2016.04.008
52. Oh J, Byrd AL, Deming C et al (2014) Biogeography and individuality shape function in the human skin metagenome. Nature 514(7520):59–64. https://doi.org/10.1038/nature13786
53. Fierer N, Hamady M, Lauber CL et al (2008) The influence of sex, handedness, and washing on the diversity of hand surface bacteria. Proc Natl Acad Sci U S A 105:17994–17999
54. Bouslimani A, Porto C, Rath CM, Wang M, Guo Y, Gonzalez A et al (2015) Molecular cartography of the human skin surface in 3D. PNAS 112(17):E2120–E2129
55. Finley S, Benbow ME, Javan G (2014) Microbial communities associated with human decomposition and their potential use as postmortem clocks. Int J Legal Med 129:623–632
56. Park J, Kim SJ, Lee JA, Kim JW, Kim SB (2017) Microbial forensic analysis of human-associated bacteria inhabiting hand surface. Forensic Sci Int Genet Suppl Ser 6:e510–e512
57. Clarke TH, Gomez A, Singh H, Nelson KE, Brinkac LM (2017) Integrating the microbiome as a resource in the forensics toolkit. Forensic Sci Int Genet 30:141–147. https://doi.org/10.1016/j.fsigen.2017.06.008
58. Jain S, Kumar A, Gupta P, Prasad R (2005) Microbial forensics: a new forensic discipline. J Indian Acad Forensic Med 27(2). ISSN 0971-0973.
59. Pal P, Roy A, Moore G, Muzslay M, Lee E, Alder S, Wilson P, Powles T, Kelly J (2013) Keypad mobile phones are associated with a significant increased risk of microbial contamination compared to touch screen phones. J Infect Prev 14(2):65–68. https://doi.org/10.1177/1757177413475903
60. Meadow JF, Altrichter AE, Green JL (2014) Mobile phones carry the personal microbiome of their owners. PeerJ 2:e447. https://doi.org/10.7717/peerj.447.

Joe Adserias-Garriga is a forensic anthropologist and forensic odontologist from Spain, where she has directed and lectured several postgraduate programs in forensic science. She was an external advisor to Mossos d'Esquadra (Catalonian Police), who honored her with a Police Decoration for her contribution in forensic casework in anthropology and odontology. Dr. Adserias-Garriga worked as a forensic anthropologist in the identification of the bodies found nearby the border between Mexico and the United States in South Texas, at Forensic Anthropology Center, Texas State University, United States. She has conducted research collaborations with different entities in the United States and Europe. She is a member of the American Dental Association Standards Committee on Dental Informatics, Forensic Odontology Informatics. She is a Diplomate of the American Board of Forensic Odontology (D-ABFO); and she serves as a member of the Board Governors in the American Society of Forensic Odontology (ASFO) and as a member of the Board of Directors of the International Association of Coroners and Medical Examiners (IACME). She is a member of the INTERPOL DVI Odontology Sub-working Group and the INTERPOL DVI Pathology and Anthropology Sub-working Group. Dr. Adserias-Garriga is an assistant professor of the Department of Applied Forensic Sciences at Mercyhurst University.

Jesús L. García-Gil professor of microbiology at the UdG, has a degree in Biological Sciences from the UAB, a Master's in General and Applied Limnology (UAB), and a Doctorate in Biology (UAB). As a researcher, he started working on aquatic microbial ecology, particularly in the population dynamics of phototrophic bacteria living in anaerobic environments. He later transferred all the expertise on the microbial ecology of anaerobic ecosystems to studying the intestinal bacteria and gathered a biomedical research team on bacteria and inflammatory bowel disease (IBD). He has been leading this research group on the microbiology of intestinal disease for the latest 15 years. The current research line focuses on discovering bacterial signatures indicative of intestinal disease (basically colorectal cancer and inflammatory bowel disease) and their application to diagnosis. He has published numerous papers on bacterial indicators of IBD and supervised several theses and research works during this time. Prof. Garcia-Gil is a founding member of two SpinOffs. One of them, GoodGut SL, is now a startup dedicated to developing and marketing tools to support the diagnosis of intestinal disorders, based on bacterial indicators of health and disease, and is co-author of four patents related to bacteria use as indicators of intestinal disease.

Chapter 12
Age Estimation of Bloodstained Fingermarks

Alexis R. Weber and Igor K. Lednev

Abstract Bloodstains and fingermarks discovered at crime scenes can provide crucial information to an investigation. A DNA profile can be obtained from a bloodstain and can be used to link someone to a crime scene, corroborating or disputing a witnesses' statement. Besides DNA evidence, fingermark analysis is another common method that is also used to identify an individual. However, analysis of neither bloodstains nor fingermarks is able to answer a key question for investigators: how much time has passed since the criminal event occurred? Determining the time since deposition (TSD) of bloodstained fingermarks could be important for crime scene investigators. If crime scenes contain multiple sets of bloodstains, TSD determined for individual bloodstains should allow for selecting bloodstains relevant to the crime and as such reduce the number of samples which are collected and processed. Despite the extensive research conducted over the last century, there is still no definitive answer on how to determine the TSD of blood. This chapter summarizes the work that has been done thus far to tackle this momentous issue. However, there are still hurdles that must be overcome before methods for estimating TSD of bloodstained fingermarks become accepted by the judiciary in the court system.

Keywords Time since deposition · Bloodstained fingermarks · Hemoglobin · Trace evidence · Criminalistics · Forensic science · Biological stains

Abbreviations

ALP	alkaline phosphatase
AMF	atomic force spectroscopy
ATL	alanine transaminase

Adapted in part with permission from Weber and Lednev, *Forensic Chem*, 2020.**19**: 100248. Copyright 2020 Elsevier. https://doi.org/10.1016/j.forc.2020.100248.

A. R. Weber · I. K. Lednev (✉)
Department of Chemistry, University at Albany, Albany, NY, USA
e-mail: aweber@albany.edu; ilednev@albany.edu

J. De Alcaraz-Fossoul (ed.), *Technologies for Fingermark Age Estimations: A Step Forward*, https://doi.org/10.1007/978-3-030-69337-4_12

ATR-FTIR	attenuated total reflectance Fourier transform infrared spectroscopy
BPA	blood pattern analysis
CK	creatine kinase
CO	carbon monoxide
CMYK	cyan, magenta, yellow, and black
CN	cyanide
Ct	cycle threshold
deoxy-Hb	deoxyhemoglobin
DNA	deoxyribonucleic acid
EPR	electron paramagnetic resonance
FAD	flavin adenine dinucleotide
GOT	glutamic oxaloacetic transaminase
GPT	glutamic pyruvic transaminase
HC	hemichrome
HPLC	high-performance liquid chromatography
HSI	hyperspectral imaging
HSV	hue, saturation, and brightness
LDA	linear discriminant analysis
LDH	lactate dehydrogenase
LS-SVM	least squares support vector machine
metHb	methemoglobin
miRNA	microRNA
MLR	multiple linear regression
mRNA	messenger RNA
MSP	microspectrophotometer
NADH	reduced nicotinamide adenine dinucleotide
NADPH	reduced nicotinamide adenine dinucleotide phosphate
NIR	near-infrared
oxyHb	oxyhemoglobin
PCA	principal component analysis
PCR	principal component regression
PCR	polymerase chain reaction
PLS-DA	partial least squares discriminant analysis
PLSR	partial least squares regression
qPCR	quantitative polymerase chain reaction
RBC	red blood cell
RBG	red, blue, and green
RNA	ribonucleic acid
RT-PCR	real-time polymerase chain reaction
RT-qPCR	real-time quantitative polymerase chain reaction
TSD	time since deposition
UV-Vis	ultraviolet visible

12.1 Introduction

Blood traces are one of the most frequently found body fluids in homicide, battery, and sexual assault cases [1] and are also one of the most common contaminants of fingermarks [2–4]. Both fingermarks (friction skin ridge patterns, in general) and bloodstains can provide vital information about the crime typology and individuals involved. For example, once a stain has been identified as a bodily fluid, it can then be sent to the laboratory for DNA analysis [5, 6]. Fingermarks can be a significant source of information about the donor, such as touch DNA, the use of illicit or prescription drugs, health conditions, and more [7]. Fingermarks and DNA evidence are two evidence types that are commonly used to identify an individual.

As with all types of evidence, fingermarks can become a vital piece during investigations but are often subject to challenge during criminal trials. A common dispute is being able to demonstrate that a fingermark is relevant to the case. Therefore, to determine the probative value of evidence, it is necessary to be able to distinguish extraneous fingermarks from those that are relevant [8, 9]. While the presence of a bloody fingermark may be more of an indicator that is related to a crime, it is still circumstantial evidence and may be questioned. Thus, it is ideal to be able to temporally place evidence within the timeframe of a crime.

There are many innocuous reasons where a bloodstained fingermark may be found at any location; for example, an individual may have cut themselves shaving or while cooking. Estimating the age of a bloodstained fingermark would help investigators to determine the time period in which the crime occurred and to corroborate allegations of the event [10, 11]. In addition, if crime scenes contain multiple sets of bloodstains, determining the age of individual bloodstains can reduce the number of samples which should be collected, documented, and processed. This is why determining the age of a bloodstain or its time since deposition (TSD) is a vital task. Unfortunately, despite the years that have gone into attempting to determine the TSD of a bloodstain, there is still no unified, accurate, and validated answer between scientists so that a singular method can be introduced to the court system.

Due to the large caseload that forensic examiners are required to process, time and resource efficiency is paramount. Thus, when processing samples, it is ideal to have a rapid method of analysis which can both identify the type of bodily fluid and determine the TSD. Currently, there is no court-accepted technique that can accomplish both objectives. The existing methodology for bloodstain identification is a two-step process, requiring both presumptive tests followed by confirmatory tests [1, 6, 12]. These methods are usually time-consuming, body fluid-specific, and destructive to the sample. This is a major hindrance when analyzing bloodstained fingermarks, as the current blood identification methods destroy the ridge detail that is present, preventing further analysis of the fingermark.

To counteract these issues, new methods are emerging for the confirmatory identification of bodily fluids that are universal, noncontact, and therefore nondestructive to the sample. Vibrational spectroscopy paired with chemometric analysis has pioneered the way in this regard [13–21]. The use of spectroscopic techniques has streamlined the analysis of bloodstains, due to their rapid run time. The technology has shown potential for portable usage, which has led to the recent development of handheld instrumentation, which can be taken to crime scenes and used for on scene in situ analysis [22–24].

Spectroscopic methods are not only paving the way for the identification of bodily fluids, but they are also becoming leading techniques for estimating the TSD of bloodstains [11, 25, 26]. Scientists have attempted to discover a way to determine the TSD of bloodstains for more than a hundred years. Early techniques focused on the changes to hemoglobin in blood as it dries [27–33], and this formed the foundation for the techniques that are recently being developed. Previous approaches involved the destruction of the stain, and thus were not useful for bloodstained fingermark analysis. These methodologies included real-time reverse transcriptase-polymerase chain reaction (RT-PCR) [34–37], high-performance liquid chromatography (HPLC) [38, 39], atomic force microscopy (AFM) [40–43], and ultraviolet visible (UV-Vis) spectrophotometric analysis [44, 45]. In addition to being destructive, the problem with these techniques is they produce time estimates with large standard deviation and are unable to determine an accurate TSD [10]. As science has developed, some methods have shown more promise in assessing the age of bloodstains due to their sensitive, specific, and nondestructive analysis: vibrational and reflectance spectroscopy paired with chemometrics [2, 24, 26, 46–59].

In this chapter, we will discuss the biochemical mechanism of blood aging and how this is used to determine the age of a bloodstain. We will also briefly discuss the currently developed techniques that are used for bloodstain age determination and their limitations. We will then focus on the emerging spectroscopic techniques and technology that has shown the most promise for bloodstained fingermarks. Finally, we will discuss some of the limitations and obstacles that stand in the way of creating an accurate and reliable method for estimating the TSD of bloodstains.

12.2 Fundamentals of Blood Aging

12.2.1 Composition of Blood

Blood is a significant component within the human body, accounting for about 8% of body weight or about 12 pints (5.7 liters). Blood is composed of red blood cells (erythrocytes), white blood cells (leucocytes), and platelets (thrombocytes), which are suspended in an alkaline matrix called plasma. Other biochemical components of blood such as proteins, nucleic acids, lipid, and carbohydrate exist at low concentrations in the plasma [6, 12].

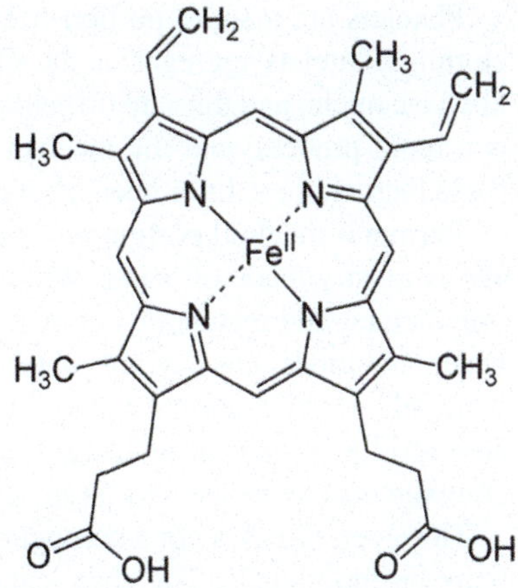

Fig. 12.1 Structure of heme. (Copyright 2010, reprinted with permission from Wikimedia Commons)

Red blood cells constitute approximately 50% of whole blood and are responsible for transporting oxygen from the lungs to the rest of the body. Red blood cells do not contain any DNA due to their lack of nuclei, and are mostly composed of hemoglobin, a protein that is responsible for the transport of oxygen [6, 10–12]. Hemoglobin (Hb) is a multiunit protein that consists of four subunit polypeptide chains, two α-chains, and two β-chains. Within each of these subunits, there is another unit called heme (Fig. 12.1), where the binding of oxygen occurs. Heme is composed of a cyclic ring called porphyrin which is made up of four pyrrole entities that are linked by a methylene bridge, forming a unit called protoporphyrin IX. When this moiety is bound with an iron in its ferrous form (Fe^{2+}), it is called ferroprotoporphyrin. The iron ion forms bonds with the four nitrogen atoms that are present in the heme compound and a fifth bond with a segment of the Hb chain [11, 12, 60]. The sixth bond that occurs with this ferrous ion is usually with O_2, so that the oxygen molecule can be transported from the lungs throughout the body. However, other chemical entities can bind to this compound, such as carbon monoxide (CO) and cyanide (CN). The binding of these alternative compounds prevents the transport of oxygen and can lead to death by chemical asphyxiation [12]. Due to red blood cells being an overwhelming contribution of blood, analysis of bloodstains commonly depends on the properties of red blood cells and specifically hemoglobin.

White blood cells only account for about 1% of whole blood (compared with 50% for red blood cells), but they play a large role in the functionality of the human body. There are three main types of white blood cells: monocytes, lymphocytes, and granulocytes [60]. White blood cells contain nuclei and thus are responsible for the genetic information within blood; they are also responsible for detecting and fighting off diseases. Common detection methods of blood utilizing white blood cells focus on the analysis of DNA and RNA. White blood cells make up a small portion of whole blood, so performing detection analysis using white blood cells can be difficult.

Platelets are a small portion (<1%) of the blood that are responsible for blood clotting. Platelets aggregate at the site of injury to stop blood from flowing. They do not have nuclei and thus are difficult to find in bloodstains. The coagulation of blood is a rapid process, and the mechanism of occurrence is intricate [10, 12]. Due to these factors, there have been no studies on determining TSD using platelets.

Plasma is the fluid portion in which the other components of blood are suspended and is mostly made of water. Within the plasma are dissolved proteins, other clotting factors, and metabolic compounds [10]. Because plasma is mostly made up of water, as bloodstains dry, the water evaporates, leaving behind the red blood cells, white blood cells, and proteins. Proteins present in blood are composed of amino acid residues, some of which are highly fluorescent. Changes to these fluorescent components can be used to estimate the TSD [61–65].

However, blood is not a homogenous substance, neither within an individual nor between individuals. The chemical makeup of blood exists in a dynamic equilibrium; for example, the structure of hemoglobin in the red blood cells varies as it moves throughout the body. This creates inherent variability to the chemical makeup of blood within individuals. As blood is expelled from the body, the starting concentration of oxygenated hemoglobin will vary, dependent on which part of the body the blood originates [12]. In addition, the chemical makeup of blood between individuals is unique. Raman spectroscopy can be used to discriminate between species [66], sexes [17, 67], and races [68] based on the variable chemical makeup of blood. There is significant variability both within and between individuals; therefore, any techniques that are created to determine the age of bloodstains must account for the variation.

12.2.2 *Biochemical Mechanism of Blood Aging*

12.2.2.1 Hemoglobin Changes

Hb is a major component of red blood cells, and as such, most methods for age determination of blood focus on measuring the changes which occur to Hb and its derivatives. Hb exists in multiple forms, depending on the chemical compound bound to the sixth coordination site of the iron and the oxidation state of the iron ion [11]. When oxygen is not bound, the Hb moieties are in the state of deoxyhemoglobin (deoxyHb). In this state, the iron ion is in the ferrous (Fe^{2+}) form, and there is no molecule present in the sixth coordination position. Here, the protoporphyrin IX ring is in a taught configuration, and the ring appears slightly bent. The spin complex of the iron atom is a high configuration, and its magnetic properties are paramagnetic. Oxygen is able to bind with Hb when it is in this state [12] (Fig. 12.2).

As oxygen interacts with the deoxyHb, it is autoxidized into the form of oxyhemoglobin (oxyHb) [6, 25, 26]. The iron ion at the center of the ring is in either the ferrous (Fe^{2+}) form or the ferric (Fe^{3+}) form; however, the ferrous form is more common. The protoporphyrin IX is in a relaxed state as the iron atom shifts into the

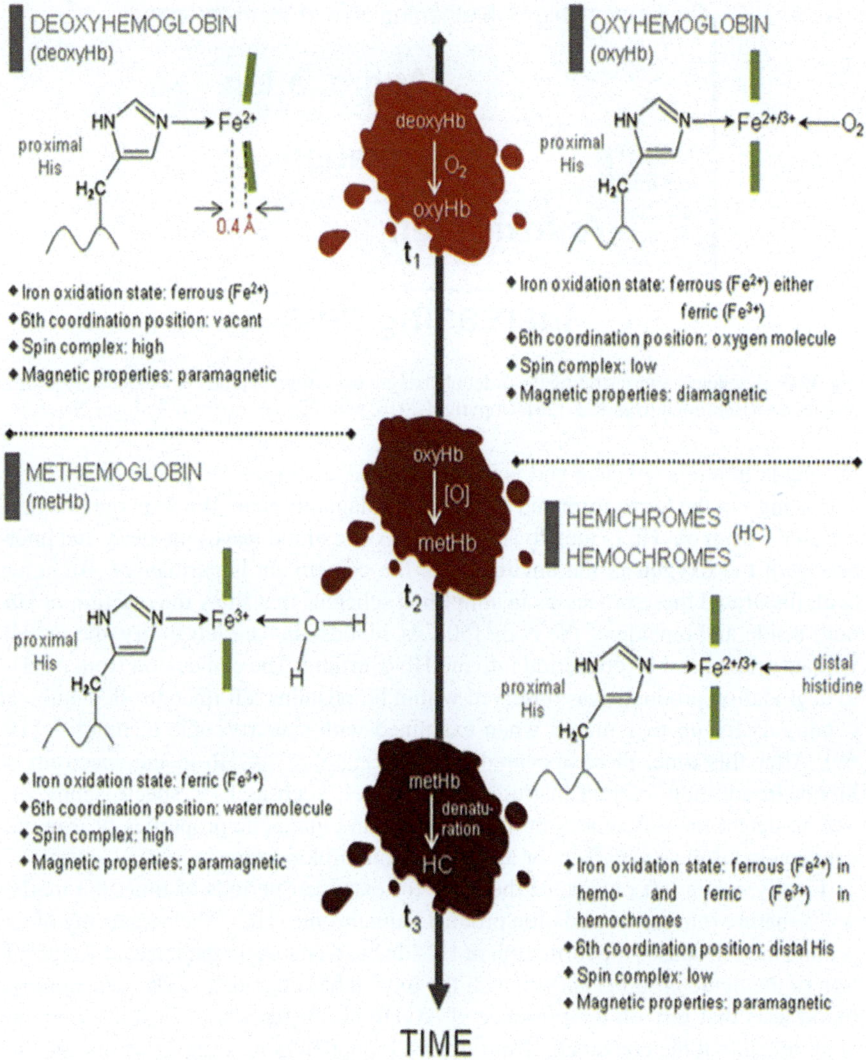

Fig. 12.2 Pathway of ex vivo hemoglobin degradation. (Reprinted from Ref. [11]. Copyright 2018, with permission from Elsevier)

plane of the ring as the oxygen is bound into the sixth coordinate position of the iron atom. The magnetic state of the iron atom is diamagnetic, and the spin state of the iron atom is low. In this state, the hemoglobin binds to the oxygen in the lungs and transports the oxygen to the rest of the body. Within the body, Hb can transform between deoxyHb and oxyHb as it moves between the lungs and the remainder of the body [11, 69].

As the Hb is continuously oxidized, oxyHb is converted into methemoglobin (metHb). In this state, the iron ion is in the ferric (Fe^{3+}) form, and the protoporphyrin

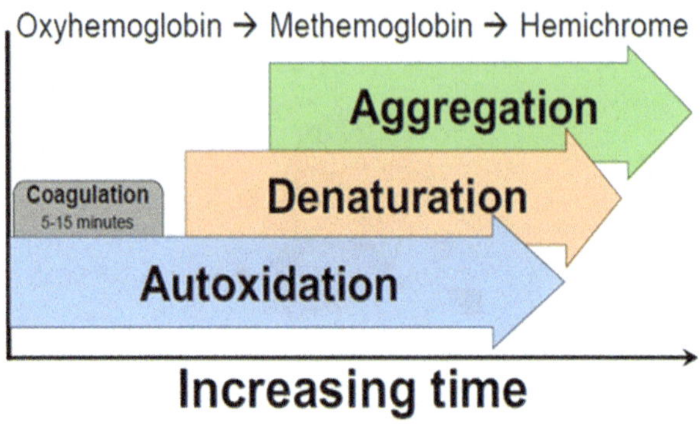

Fig. 12.3 A scheme demonstrating the order of changes occurring naturally to a bloodstain as time progresses. (Reprinted from Ref. [26]. Copyright 2016, with permission from Springer Nature)

IX ring is in a relaxed state and is unable to bind to oxygen [11]. The spin state of iron is high in the ferric form and it is in a paramagnetic state. While in the body the transition from oxyHb to metHb is reversible, once blood leaves the body and interacts with the oxygen in the environment, the conversion is permanent. Bremmer et al. illustrated these processes in simplified schemes that show the reaction of Hb, both inside and outside of the body [50]. As bloodstains are left to dry, the oxyHb will autoxidize and be converted into metHb overtime. The conversion from oxyHb to degradation products was observed within bloodstains left in controlled ambient conditions for up to 1 month when examined with spectroscopic techniques [10, 69]. After this time, there appeared to be no signs of oxyHb in the spectrum of blood. In addition, certain instrumental analyses, such as EPR spectroscopy, are able to detect the differences in the spin state and magnetic properties of iron that can then be used as a marker for age estimation of bloodstains.

Finally, when left outside of the body for extended periods of time, the metHb will denature into the degradation product hemichrome (HC). The denaturing of the protein metHb causes the formation of HC, due to a change to the internal confirmation of the heme group [11]. When all the oxyHb has degraded to HC, presumptive blood tests that rely on the presence of oxyHb no longer work [70, 71]. Bremmer et al. noted that the conversion from oxyHb to metHb is most rapid within the first 10 days of deposition; afterward, the production of HC increases, while the amount of oxyHb and metHb decreases [69], indicating that this process happens at a predictable rate in controlled conditions. Therefore, the conversion process of oxyHb → metHb → HC has commonly been used as a method to determine the TSD of bloodstains (Fig. 12.3).

As the blood becomes oxygenated and eventually denatured, there is a colorimetric change to the bloodstains over time. When blood exits the body, it is red in appearance, but as it dries, it darkens. The change from red to brown follows the oxidation of oxyHb to HC during aging. This change in color can be measured using visible spectroscopy techniques and has been used as a measurement of estimating the TSD. The variability of blood drying and the changing of colors require multiple measurements to account for the variability within and between samples [33, 50]. As the changes in color are nonlinear with time, it can be difficult to determine the age of a bloodstain based solely on colorimetric changes as it dries.

12.2.2.2 DNA, RNA, and Protein Changes Within Blood

Bloodstains are one primary source of DNA and RNA evidence. While semi-stable, these molecules are composed of biological materials which break down over time once exposed to the environment. However, due to their relatively large size and the exposure to enzymatic processes, they break down quickly and are not as useful for bloodstain age determination. The use of messenger RNA (mRNA) or microRNA (miRNA) is another possibility to determine a bloodstain's TSD [35, 36, 72]. The amount of mRNA or miRNA present is measured by the cycle threshold (Ct), which is the number of cycles that occur during a polymerase chain reaction (PCR) required to detect RNA at a quantitative level. Scientists have attempted to use the relative degradation of the mRNA or miRNA over time as a method to estimate a bloodstain's TSD. Studies have shown mRNA and miRNA are more stable against degradation because of their smaller size. They are also useful as a confirmatory test for source identification of bloodstains because they have bodily fluid-specific expression patterns. However, despite their small size, they have a high probability of degradation in heated conditions. Tests for mRNAs and miRNAs are destructive, but testing for miRNA is compatible and complementary to forensic DNA analysis because miRNA and DNA can be extracted simultaneously [34, 37].

Another component of blood is proteins, which are composed of amino acids. A large number are aromatic amino acids, including tryptophan, tyrosine, aspartic acid, and phenylalanine, which can be detected using fluorescence spectroscopy [62–64]. While the iron atom present at the center of the Hb usually quenches the fluorescence of Hb, the strength of tryptophan's fluorescence is still relatively strong. However, as with all biological materials, amino acids are still prone to break down and degrade over time [6]. Scientists have been researching methods of developing better ways to measure these changes so that they can be used for bloodstain age estimations.

A specific group of proteins in blood are enzymes, which are biologically active components that act as catalysts. These exist at varying levels within and between individuals [10, 73–75].

Some research has been conducted on the use of enzyme cascades as a measure of blood aging [75]. This technique is destructive, as utilizing an enzyme cascade to measure enzymatic activity consumes the sample. After aging the samples for 48 h, it was discovered that the enzyme alkaline phosphatase (ALP) was stable enough to estimate TSD of blood serum up to 2 days. However, this work was done using isolated blood serum that was deposited on a surface and aged. To be forensically useful, this method would need to be tested using whole blood.

Techniques that measure changes to non-hemoglobin biological components are important to research. Hb only changes so much over time, before being completely oxidized. Therefore, determining the age of bloodstains with techniques that examine the alterations to Hb can be limited. Other components of blood change at different rates and may not be as limited in their range of measurability. However, as we will describe, techniques that explore these biomarkers are usually destructive and typically rather costly. The following sections look into all of the commonly used techniques for age determination of bloodstains and compare the usability and cost-effectiveness of them for criminal investigations.

12.3 Previous Methods for Determining Time Since Deposition of Bloodstains

Violent crimes have existed for many centuries, but it was not until the start of the twentieth century that research was conducted in an attempt to create a method to temporally place a bloodstain at a scene. Original work focused mainly on the macroscopic changes to blood as it aged, while later techniques concentrated on a more microscopic level. In this section, we will review the techniques that were first developed to estimate TSD of bloodstains.

12.3.1 Techniques for the Analysis of Red Blood Cells and Hemoglobin

12.3.1.1 Techniques of the 1900s

The earliest known technique for assessing TSD of a bloodstain was established in the early 1900s. This method attempted to measure the transformations undergone by Hb as it degraded into its derivatives, by measuring the changes which occurred to its solubility and color. In 1907, Tomellini were the first to investigate TSD of bloodstains by devising a color chart. This chart acted as a reference to which a stain of unknown age could be compared and thus generated a TSD estimation [28]. Three years later, researchers explored the solubility of bloodstains for the same purpose. Leers concluded that fresh bloodstains would rapidly dissolve in water and, when the spectra were analyzed, they would show the spectrum of pure oxyHb. However, as the bloodstains aged, they would become more water-insoluble and exhibit the spectrum of metHb [29]. The solubility of bloodstains was later expanded upon by Rauschke in 1951 when attempting to correlate the solubility of bloodstains in both water and 2% KOH as a measure of TSD [76].

In 1960, the first attempt of using reflectance measurements to create a standardized color index for bloodstain age estimations was introduced. Multiple standard indexes were created in multiple environments as the ambient conditions affected the aging of the blood [33]. These results corroborated earlier work from 1930, which detected color differences between bloodstains that were aged under ultraviolet light and those that were not [77]. Kleihauer et al. later confirmed the variable effects of environmental conditions on estimating TSD in 1967 [78]. However, research into the effects of environmental conditions was not expanded on until many years later. Utilizing the spectrophotometric properties of blood as a method to determine TSD was later expanded upon by Kind et al. in 1972 and 1973 when they discovered that the absorption spectrum of blood changed in a regular manner with age. From their absorption spectra, they defined a parameter called the α-ratio, which was used to estimate TSD. They noted that this ratio was useful in calculating the age of stains up to 10–15 years old [31, 32].

12.3.1.2 High-Performance Liquid Chromatography

Other early work investigated changes to Hb and its derivatives through analysis of bloodstains using high-performance liquid chromatography (HPLC). HPLC is a technique that can separate, identify, and quantify the concentration of each chemical component in a complex solution. The proteins present in blood exhibit characteristic retention times. They are separated when passing through the HPLC column and are identified by the detector. It is important to note that this technique is destructive and cannot be used in situ because it requires the sample to be dissolved into solution before separation. However, HPLC is highly specific in its ability to discriminate between the different derivatives of hemoglobin. This technique allowed researchers to distinguish between the varying constituents of hemoglobin and compare the ratio of the peaks from the resulting chromatogram. The peaks were linearly correlated and could be used as a marker to determine the TSD of bloodstains [38, 39, 79].

12.3.1.3 Atomic Force Microscopy

As bloodstains age, there are physical changes that occur to the surface of red blood cells that can be measured using atomic force microscopy (AFM). AFM is a nondestructive technique that measures a sample's elastic properties using a high-resolution probe that scans the surface of a specimen, in this instance, bloodstains. The microscope possesses a cantilever equipped with an oscillating tip that can measure elasticity at the nanometer resolution by changes to the cantilever's resonance frequency [10, 42]. Due to a high spatial resolution of this instrument, AFM has the potential to be used with bloody fingermarks. Overall, research conducted using AFM determined that the elasticity of red blood cells (RBCs) decreases with time, even though they remain intact and do not crack. Strasser et al. were one of the first to investigate this phenomenon, measuring the morphological changes of bloodstains for up to 1 month [40]. Their results revealed the degradation that occurs to the elastic properties of blood over time. Wu et al. then expanded on analysis of AFM for the analysis of RBCs [41]. While Strasser's work measured bloodstains that were present on glass slides at room temperature, Wu et al. measured RBCs on multiple substrates and three different environmental conditions (controlled room temperature condition, uncontrolled outdoor environmental condition, and controlled low-temperature condition). They discovered that the changes in the viscoelasticity of RBCs were time-dependent, substrate-dependent, and environment-dependent [41]. Therefore, while multiple control curves would need to be created, these studies showed that AFM has the potential for estimating TSD of bloodstains.

In recent years, more work has been conducted on the applicability of AFM for age estimation of bloodstains. Smijs et al. measured the elasticity with a silicon probe tip and discovered that after four to six cell indentions, the tip of the probe became blunter, and therefore the measurements of elasticity were less precise [42]. Most recently, research by Cavalcanti et al. measured the morphological and elastic changes of RBCs on multiple surfaces: glass, metal, and ceramics. They noted that the substrate had a significant effect on changes to the average height, perimeter, area, and volume of the red blood cells. All listed characteristics of the red blood cells increased over time [43]. While their study, and the ones before it, provided promising results, more work is required to understand the changing morphological nature to bloodstains and whether these properties could be used to estimate the TSD of bloodstained fingermarks.

12.3.1.4 Ultraviolet-Visible (UV-Vis) Absorption Spectroscopy

During the pursuit to determine TSD of bloodstains for forensic cases, another requirement was added to the ideal technique, the methodology should be nondestructive and easily incorporated into the forensic workflow. The need for a technique that resulted in products for further analysis, such as for DNA evidence, led to scientists attempting to use spectroscopic analysis for TSD estimation of bloodstains. The Ballantyne group determined that the oxidative changes that Hb undergoes, induced by UV-Vis absorption spectroscopy, do not hinder the ability to extract DNA from bloodstains [44]. They then used UV-Vis absorption spectroscopy to estimate bloodstain TSD by measuring changes to the Hb Soret band, which has a max absorbance at 412 nm. They observed that this band undergoes a blue spectral shift or a shift to a lower wavelength, as the bloodstain ages as well as changes in its maximum absorbance [45]. The Ballantyne group utilized the extent of this shift as a measure of age and therefore a marker in assessing TSD of blood. However, this technique is temperature-dependent. The blue shift of the Soret band occurred faster and had a longer shift in bloodstains that were exposed to heated conditions. Analysis using UV-Vis absorption spectroscopy is still a destructive analytical technique but results in products that can still be used for DNA analysis [44]. Importantly, this was one of the first spectroscopic techniques that started the push toward nondestructive spectroscopic methods for age determination of bloodstains.

12.3.2 Techniques for the Analysis of White Blood Cells

In blood, white blood cells are the only cell type that contains a nucleus and therefore contains genetic information. Genetic information can provide highly valuable information about the bloodstain, including conclusive information about the donor. Within white blood cells, mRNA is relatively stable and was found in samples that were aged for many years [37]. However, mRNA is prone to degradation, at a slow rate, and has therefore been a subject of interest for analysis on the path of determining TSD of bloodstains. While not universal, mRNA analysis is body fluid-specific due to the discovery of tissue-specific markers. The amount of RNA present is measured by the Ct [6].

Bauer et al. attempted to use RNA degradation as an indicator of the age of bloodstains. This method utilized a semiquantitative analysis duplex paired with competitive RT-PCR to extract RNA from bloodstains. They were able to extract RNA from stains that were up to 15 years old and were able to distinguish those that had an age difference of 5 years or more [34]. While this method would not be as useful for fresh stains or fingermarks, it could potentially work well with specimens from cold cases, if samples were properly stored. Zhao et al. also tested the applicability of mRNAs for determining TSD. They analyzed the viability of seven mRNA biomarkers on blood-soaked gauze pads, including blood samples that were 30 and 50 years old. A 90% success rate in isolating mRNA was achieved from samples that were 30 years old and an 80% for 50 years old [37].

Another method for examining the validity of RNA for TSD is to compare the stability of two different species of RNA over time. This concept was explored by Anderson et al. and Alshehhi et al., who both investigated the relationship between different RNA biomarkers in blood as it aged [35, 80]. Anderson et al. examined two different regions on the same strand of RNA and discovered that they had different decay rates that allowed them to estimate the TSD. After their initial work in 2005, they acknowledge the necessity for chemometrics to analyze their data [81]. Alshehhi et al. used reverse transcription-quantitative PCR (RT-qPCR) to analyze the relative expression ratio of six mRNAs, including three target genes and three reference genes. They concluded that the individual mRNAs exhibited unique degradation profiles and that some markers were more stable than others. The ratios had a nonlinear relationship when compared with blood and were determined to be useful for blood aging [80].

Environmental conditions affect the use of mRNAs to estimate TSD of blood. Bauer et al. showed that when heated, the rate of RNA degradation increases. This acceleration in RNA loss leads to an overestimation of the blood's age [34]. In addition, analysis of mRNA in blood samples still relies on presumptive and confirmatory tests to be done before they can be tested for forensic purposes [6]. Thus, not only are these methods of determining TSD destructive, but the analysis required beforehand is also destructive. In addition, these techniques are time-consuming and costly. To test RNA, specific probes are required for analysis, which can be expensive, depending on the number of RNAs being examined. One benefit of this analysis is that RNA and DNA can be simultaneously extracted [34, 37], which does allow for an easier integration into the already existing forensic workflow.

12.3.3 Techniques for the Analysis of Blood Plasma

Like red blood cells and white blood cells, proteins that are in blood plasma undergo breakdown overtime as bloodstains age. Forensic researchers have examined this process of protein degradation as a method to determine TSD. Proteins are made up of amino acids, some of which are aromatic and produce a strong fluorescence signal. Therefore, research has focused on measuring how the fluorescence of blood samples change as they age. Fluorescence analysis is rapid, requires a minimum amount of sample preparation, and is concentration-independent. Bloodstains as small as 10μL can be measured and consistently produce reproducible results. A fluorophore is a substance that is capable of emitting light after being irradiated with energy. Blood absorbs at 274, 345, 415, 541, and 576 nm and has characteristic excitation-emission wavelength pairs at 260–630, 280–340, 340–460, and 450–520 nm [62]. These wavelength pairs were attributed to the endogenous compounds: porphyrins, tryptophan, reduced nicotinamide adenine dinucleotide (NADH) or reduced nicotinamide adenine dinucleotide phosphate (NADPH), and flavin adenine dinucleotide (FAD), respectively [61, 82].

Work by Guo et al. and Shine et al. measured changes to the fluorescence lifetime of certain fluorophores as a method to determine TSD of bloodstains. Guo et al. monitored the fluorescence lifetime of tryptophan. This amino acid is sensitive to the conformational changes of proteins, such as albumin and γ-globulins, which are the main proteins in blood. They discovered that the fluorescence lifetime/ns of tryptophan decreases over the course of 1 week, plateauing after approximately 300 h; however, the maximum wavelength remained consistent among the samples [63]. The research later conducted by Shine et al. focused on measuring the fluorescence lifetime of bloodstains, as well as blood mixed with saliva. The technique employed was useful in distinguishing fresh bloodstains from old ones and for determining TSD of blood when it was mixed with saliva. Lifetime measurements of tryptophan/ns decreased from approximately 3.0 ns to approximately 2.0 ns over the course of 91 h, after which it plateaued [64]. Shine et al. were unable to establish the specific mechanism of blood aging that led to the decrease in the fluorescence lifetime of tryptophan, indicating further research is necessary. However, this technique was promising due to the rapid performance time, the limited amount of sample preparation needed, and the concentration-independent analysis.

Enzymes are specific proteins that act as a catalyst in biochemical reactions which take place within living cells. Once these cells are no longer in the body, such as in bloodstains, the level of enzyme activity decreases. In 2015, Agudelo et al. created a biocatalytic cascade that measured the denaturation of creatine kinase (CK) and alanine transaminase (ATL) to measure their activity [74]. This cascade had two parallel subsystems that follow the activity level of each of the markers. As enzymatic activity is drastically affected by temperature, they investigated how the concentration of these markers changed when incubated at different temperatures. Isolated human serum samples that were stained on glass slides were used and how the concentration of the two markers, CK and ATL, changed with time and in various environments was examined. A multi-marker approach was employed because some biomarkers, such as CK, denature rapidly and would not be ideal for extended TSD measurements. With this method, it was possible to create a reliable method to estimate TSD of serum samples up to 5 days old [74]. Agudelo et al. continued this experiment in 2016, investigating how the concentration level of alkaline phosphatase (ALP) alters as blood serum dries [75]. The technique of utilizing an enzyme cascade to measure the activity of ALP is destructive, as it consumes the sample which can no longer be used for further analysis. This method isolated blood serum that was deposited on a surface and aged. However, to be fully comparable to other TSD estimation techniques, it would need to be tested using blood serum that is isolated from aged whole blood.

12.4 Emerging Techniques to Determine Time Since Deposition of Bloodstained Fingermarks

As technology develops, the techniques created become more rapid, specific, and most importantly nondestructive. In this section, we will first discuss what chemometric analysis is and why it is used with spectroscopic methods. Then, we will cover emerging spectroscopic techniques that, when paired with chemometrics, are able to more accurately estimate the TSD of bloodstains. In addition, due to the noncontact and nondestructive nature of these novel techniques, they are ideal for the analysis of bloodstained fingermarks. And finally, we will discuss the increased use of portable technology, which allows for the analysis of bloodstained fingermarks both in situ and on the scene.

12.4.1 Importance of Chemometrics

Previous methods described have been somewhat successful in their ability to approximate the TSD of bloodstains. However, as analytical methods improve, the techniques that are being explored are more advanced and are producing a large amount of data. Therefore, there is a necessity for improved methods that can evaluate this data, such as multivariate chemometric analysis. Chemometrics is the application of mathematical and statistical analysis to chemical data to extract meaningful information from multivariate datasets, such as spectroscopic data [83, 84]. Before discussing the emerging spectroscopic techniques for age estimation of bloodstains, we will briefly cover some of the common chemometric models that are used for TSD assessments.

Chemometric models utilized by the forensic science community are highly accepted for their ability to analyze multivariate data by statisticians. Regression-based methods are commonly used for estimating the TSD of bloodstains as many studies measure the degradation rates of different components. Multiple linear regression (MLR) analysis is the most common form of regression analysis and is used to describe the relationship between a dependent variable and multiple independent variables [83, 84]. This method is fast and provides significant results when the quantity of samples is large: however, MLR is not useful when the variables have a nonlinear relationship or have significant covariance and MLR is sensitive to outliers.

Partial least squares regression (PLSR) analysis is an extended version of the MLR model and is used when there are fewer observation variables than predictor variables [83, 84]. Other models are unable to account for variables that exceed the maximum number of predicted factors. Linear discriminant analysis (LDA) is a model used for the prediction of the group membership of an unknown variable but is not useful with a large number of variables [83, 84]. Partial least squares discriminant analysis (PLS-DA) works similarly to LDA but can account for a large number of variables. PLS-DA reduces noise and uses both a training dataset and calibration dataset. It works best with a limited number of classes of data but does not work when the classes contain samples that are not homogenous [83, 84].

Principal component regression (PCR) is a combination of principal component analysis (PCA) and MLR; it accounts for the maximum amount of variance within a multidimensional spectrum and can decompose the data into individual components that account for the most amount of variance [83, 84]. Finally, least squares support vector machine (LS-SVM) is useful for data analysis when nonlinear and linear multivariate statistics are required in tandem [83]. The most commonly used chemometric trend for estimation of TSD is regression analysis (PLSR or PCR). However, LS-SVM statistics seem to provide better results for blood aging studies [84]. This statistical model works off the data that is available and takes into account the substrate of interest. As the analysis of forensic samples becomes more robust, the data produced will also become more complex. There is a great desire within forensic science to take the subjectivity out of the analysis, which can be accomplished by implementing chemometrics with spectroscopy, as is reviewed here.

12.4.2 *Vibrational Spectroscopy Paired with Chemometrics*

Vibrational spectroscopy is a nondestructive method of analysis that measures the vibrational energy within a compound after being excited with energy. It is used to characterize compounds based on the unique vibrational energy of the bonds within a molecule (Fig. 12.4). There are two types of vibrational spectroscopy: Raman spectroscopy and infrared spectroscopy. Raman spectroscopy detects inelastically scattered radiation that has a different frequency of light than the incident beam. For a compound to be Raman active, it must exhibit a change in the polarizability of the

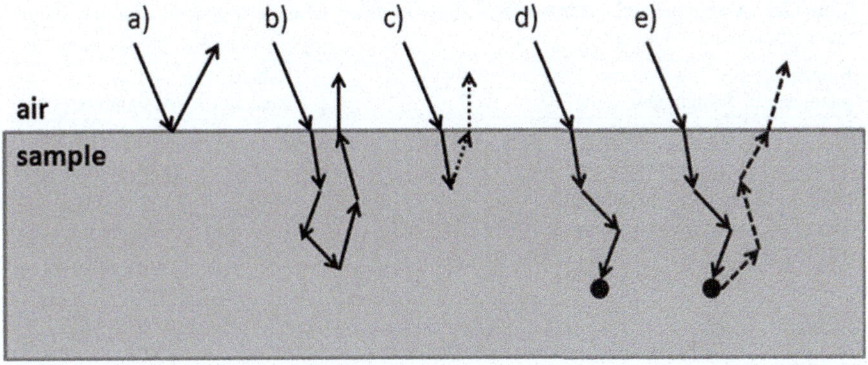

Fig. 12.4 The interaction of light with a specimen may lead to (**a**) specular reflection, (**b**) elastic scattering followed by diffuse reflection, (**c**) inelastic scattering followed by emission of Raman shifted light (dotted lines), (**d**) absorption, and (**e**) absorption followed by photoluminescence emission (dashed lines). (Reprinted from Ref. [53]. Copyright 2012, with permission from Elsevier)

compound when irradiated with light. Of the compounds present in fresh blood, Hb produces the most intense Raman signal. Therefore, the Raman spectrum of whole blood is commonly dominated by Hb.

Most of the techniques discussed thus far have focused on the analysis of hemoglobin within red blood cells. Hb is useful for determining the TSD of bloodstains, as the degradation process of hemoglobin can be measured with numerous techniques. However, aged blood samples are prone to an increase in fluorescence, which inhibits successful data acquisition via Raman spectroscopy. Thus, an alternative method is necessary when the inherent fluorescence of blood inhibits analysis. Attenuated total reflectance (ATR) Fourier transform infrared spectroscopy (FTIR) is a rapid technique that measures changes in the dipole moment of bonds after a sample is excited with electromagnetic radiation. Heme groups are not very IR active, and as such do not dominate the produced spectrum. This allows for proteins that are present in the blood to be detected nondestructively and changes which occur to them to be measured. Spectral analysis of blood with ATR-FTIR shows characteristic peaks of blood include 3300 cm^{-1} (Amide A), 2800–3000 cm^{-1} (Amide B), ~1650 cm^{-1} (Amide I), ~1540 cm^{-1} (Amide II), and 1200–1350 cm^{-1} (Amide III) [68, 85]. However, ATR-FTIR analysis requires compressing a sample on a crystal, which is destructive to bloodstained fingermarks. Therefore, only Raman spectroscopy, which is noncontact, will be the focus in this section. In this chapter, the discussion of these emerging spectroscopic techniques is an overview, and for additional information, the reader is referred to the review article by Zadora et al. which describes the most recent spectroscopic studies related to estimating TSD of bloodstains [11].

Concerning the analysis of blood, Raman spectroscopy has proven to be a rapid and nondestructive technique that can account for the inherent heterogeneous nature of biological samples. Raman mapping allows spectra to be taken from multiple points across the surface of the bloodstain instead of a single location. The Raman

Table 12.1 Assignment of spectroscopic features of blood selected by genetic algorithm coupled with artificial neural networks

Raman band, cm^{-1}	Assignment	Raman band, cm^{-1}	Assignment
715	$N^+(CH_3)$, lipids	1311	ν21
740	ν15 (pyr breathing)	1338	Some amino acids
743	ν15 (pyr breathing)	1343	Tryptophan
750	ν15 (pyr breathing)	1347	Some amino acids
754	ν15 (pyr breathing)	1367	ν (pyr half-ring) sym
790	ν46	1369	ν4
895	(C-C), glycine, glutathione, glutamine	1401	ν20
904	C–C skeletal	1447	$\delta(CH_2/CH_3)$, ν20
936	ν46	1543	ν11, ν28
942		1550	ν11, ν28
956	(C–C), leucine, lipids	1563	ν11, ν28
967	Lipids, proteins	1565	ν11, ν28
1000	ν47 (CbC1) asym phenylalanine	1575	ν(CαCm) asym, phenylalanine
1027	$\delta(=CbH_2)$ asym, Phe	1578	ν(CαCm) asym, phenylalanine
1055	$\delta(=CbH_2)$ asym	1600	ν(C=C) vinyl
1102		1604	ν(C=C)
1124	Heme, polysaccharides, ν22 (porphyrin half ring), observed in the spectra of a single human RBC	1617	Heme, phenylalanine
1170	ν (pyr half-ring) sym, tyrosine	1639	ν10
1247	ν30	1653	ν37, amide I
1248	Guanine, cytosine, proteins	1660	ν37, amide I

Reprinted and adapted with permission from Ref. [67]. Copyright 2017, with permission from American Chemical Society

spectrometer is also highly sensitive, able to collect spectra of a single red blood cell [86] as well as from bloodstains at a 1:250 dilution [87]. The leading group on the development of vibrational spectroscopy for body fluid analysis has been the Lednev laboratory. Concerning the analysis of body fluids, they have shown that vibrational spectroscopy in combination with chemometrics allows for highly discriminating analysis; the technology is capable of identifying and distinguishing between different body fluids [13, 15–19, 88], species [66, 85], sexes [67], and ethnic groups [68, 89]. Since 2016, Lednev's group has extended the analysis of bloodstains to assess TSD by continuing to pair Raman spectroscopy with advanced chemometrics. Aging of blood is commonly associated with the denaturation of Hb within RBCs and the aggregation of those degradation products. When fresh blood is excited at 785 nm, the resulting Raman spectra are dominated by the vibrational frequencies of oxyHb and its derivatives [25, 26, 55, 60, 90]. Table 12.1 shows the band position and corresponding vibrational frequencies of whole blood [90].

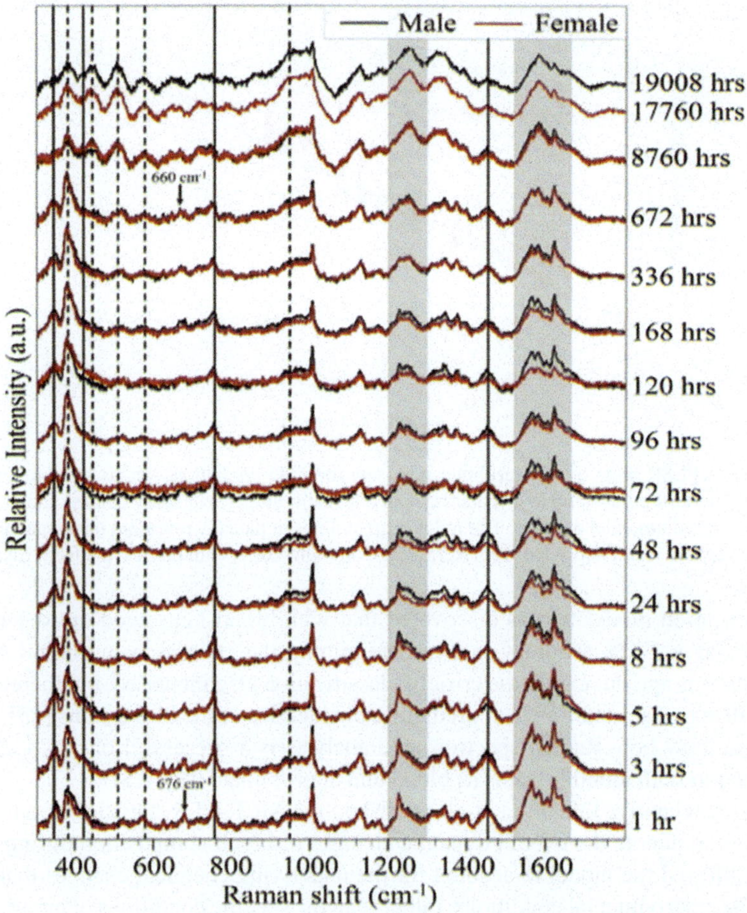

Fig. 12.5 The averaged preprocessed spectra for bloodstains from the male (black traces) and female (red traces) donors aged in an ambient environment for up to 2 years. The most prominent changes are indicated as dotted lines (increasing with time), solid lines (decreasing with time), and shaded regions. (Reprinted from Ref. [25]. Copyright 2017, with permission from Elsevier)

Doty et al. revealed that the chromatogram peaks at approximately 345, 375, 420, 440, 500, 570, 750, 970, and 1450 cm^{-1}, as well as the spectral regions 1200–1300 cm^{-1} and 1520–1670 cm^{-1}, change most noticeably as bloodstains age [26] (Fig. 12.5) Doty et al. then proved that Raman spectroscopy could estimate TSD up to 2 years using PLSR and PCR analyses. The calculated root mean squared errors were 0.29 and 0.31 for each chemometric technique, respectively; this translated to approximately a 70% accuracy for the models' ability to predict the TSD of the bloodstain [25] (Fig. 12.6) Previously by this group, a bodily fluid classification model was developed that could discriminate between five forensically relevant bodily fluids [13]. Spectra collected from the TSD study [25] were tested using this

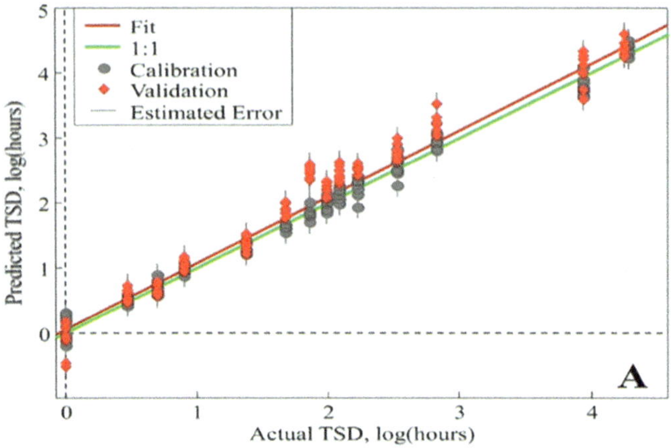

Fig. 12.6 PLSR plots on a logarithmic scale (A) for TSD predictions versus measured (actual) TSD of bloodstains aged up to 2 years showing both calibration (male, gray circles) and validation (female, red diamonds) datasets. The red and green lines in panel A represent the actual and ideal 1:1 fit, respectively. (Reprinted from Ref. [25]. Copyright 2017, with permission from Elsevier)

classification model. It was discovered that with bloodstains aged up to 1 month, there was a 100% accuracy rate in determining the stain as being blood. As the bloodstains age increased, the error of classification also increased, but there was an overall success rate of 89% in predicting the identity of the bloodstains [25]. These results show how Raman spectroscopic analysis is a successful one-step analysis method to both identify stains as blood and approximate their TSD.

More recently, Takamura et al. published the MCR-PLS statistical model in an attempt to denote the overall dynamic changes of blood composition during aging. This allowed the model to account for the underlying chemical changes that occur despite environmental conditions, rather than the specific conditions of aging. While this technique is innovative because it appears to function independently of environmental conditions, it only works for a short aging range (8.6 days) [91]. Concerning the portability of Raman spectrometers, Pereira et al. developed a method using a portable near-infrared (NIR) Raman spectrometer to identify bloodstains on various substrates in situ. They tested their instrument on four different substrates, analyzing several false positives in addition to bloodstains. Multiple chemometric techniques were examined to assess which method was best for classification of bloodstains. It was concluded that PLS-DA was the optimal chemometric method, as it was able to correctly classify the human blood as well as all the other stains on each substrate [92].

As with all methods, there are limitations: Raman spectroscopy requires irradiating the sample with a laser to induce light scattering. A high laser power has been shown to cause Hb degradation [93, 94]. When the objective of the analysis is to identify which bodily fluid the sample is, this is not a large cause for alarm. However, it has been shown that Raman spectra of bloodstains, subjected to photoinduced changes by the laser, mimic the Raman spectra of bloodstains that have been aged for an extended period [55, 90]. Wood et al. monitored the changes to Hb and discovered that five bands within the Raman spectra at 1396, 1365, 1248, 972, and 662 cm^{-1} were markers of heme aggregation. After prolonged laser exposure, Wood et al. also measured an increase in the signal-to-noise ratio, signaling an increase in the number of aggregated molecules at the spot of irradiation [46]. Research showed that prolonged exposure could induce photodegradation and thermal degradation to the sample, which could cause an overestimation of the bloodstain's age. To determine which data collection parameters produced the most informative spectra and reduced photodamage to the sample, two different research groups analyzed the effects of the laser on blood composition [93, 95]. Both groups concluded that for maximum effectiveness, the laser power must be kept low to avoid photodegradation and minimize resultant aggregation of degradation products. Their work also showed that the 785-nm laser was ideal for the analysis of bloodstains to estimate TSD [93, 95]. Another attempt to overcome the detrimental effects of laser degradation was investigated by Menżyk et al. They developed a subsampling method to reduce photodamage [59]. Current Raman acquisition modes involve either single-point measurements or mapping of the sample. A rotational collection mode was proposed which accounted for the heterogeneous nature of the samples and reduced the chance of photodamage because they are in continuous rotation. Research thus far has mainly focused on analyzing bloodstains. As Raman spectroscopy only requires a small sample size, it can also be used to analyze full or partial bloodstained fingermarks. While there is significant consideration that needs to go into the selection of collection parameters for each experiment, such as laser power, Raman spectroscopy has a high potential to be a sensitive technique for age determination of bloodstained fingermarks.

12.4.3 Reflectance Spectroscopy with Chemometrics

As blood ages, the color of the bloodstains transitions from red to brown. As discussed previously, some of the first attempts at estimating TSD of blood involved measuring changes in color. Reflectance spectroscopy is a technique that measures changes within the visible range of light (200–700 nm) and requires a white light source and a spectrometer [31, 32]. There are specific peaks present in the spectra that are indicative of blood, denoted by the Ballantyne group, including 415 nm (Soret band), 540 nm (α-band), and 576 (β-band) [44, 45]. While these peaks are indicative to blood, they are not species-specific and would require a bloodstain to be confirmed as human. Bremmer et al. in 2011 analyzed optical reflectance spectra (450–800 nm) to further measure the conversion rate of oxyHb to metHb as well as from metHb to HC at different temperatures and humidity. Research confirmed that these transitions are temperature-dependent but that only the conformational change from metHb to HC is dependent on humidity [69]. Bremmer et al. then developed a remote spectroscopic method to identify bloodstains, using a noncontact probe.

After testing 40 bloodstains and 35 non-bloodstains, they were able to successfully discriminate between the stains using their noncontact probe [51]. This noncontact probe is what has allowed on-scene analysis of bloodstains.

Bremmer et al. also demonstrated that the coefficient of determination (R^2 value) decreased as the bloodstains aged, thus showing the longer bloodstains are aged, the larger the range of uncertainty in the results. The knowledge of these decay rates was used to develop a protocol to use reflectance spectroscopy as a method for age estimation. Bloodstains on white cotton were tested to create a kinetic model to examine reflectance variations over time, based on oxidative changes in Hb. They were able to estimate TSD of bloodstains up to 55 days old with an uncertainty range of 14 days. The isosbestic points of oxyHb and HC were 525 and 585 nm, respectively [50]. The isosbestic point is a specific wavelength in the spectra where the intensity of the peak remains constant, as the concentration of the samples, oxyHb and HC in this research, change. The total sum of all fractions of Hb remained consistent, and only the quantity of each fraction changed, thus allowing to estimate the age of a bloodstain until the concentrations plateaued.

Sun et al. used visible reflectance spectroscopy paired with multiple chemometric models, each one representing a different time range. Three LS-SVM models were created to cover three different age periods from 2 to 24 h, 1 to 7 days, and 7 to 45 days [58]. This method required the use of multiple models and had the potential to be time-consuming, as testing the sample on each individual model takes time. Nonetheless, it was concluded that the creation of three separate models allowed for better predictive accuracy over an extended period and was more applicable for forensic applications. Bergmann et al. strived toward creating a TSD determination method that was substrate-independent. In this instance, pig's blood was aged on cotton, polyester, and glass for 3 weeks, collecting spectral data in the 400–640-nm range. Data was analyzed by performing dimensional reduction and k-nearest neighbor classification. During cross-validation, the model had good agreement when predicting the age of bloodstains to the observed variables ($r > 0.9$). As seen with most techniques, error was small during the earlier periods of aging (1.99-day error rate) and increased as time progressed (2.19-day error rate) [96].

12.4.4 Hyperspectral Imaging (HSI)

When analyzing bloodstained fingermarks, it is important to understand both the chemical makeup and the spatial/dimensional information of the sample. When examined in combination, TSD of fingermarks can be more accurately estimated. Hyperspectral imaging (HSI) provides the opportunity of combining both conventional imaging techniques with spectroscopy by generating a three-dimensional (x, y, λ) hyperspectral data cube for processing and analysis, where x and y represent two spatial dimensions and λ represents a spectrum acquired for each individual point on the sample surface.

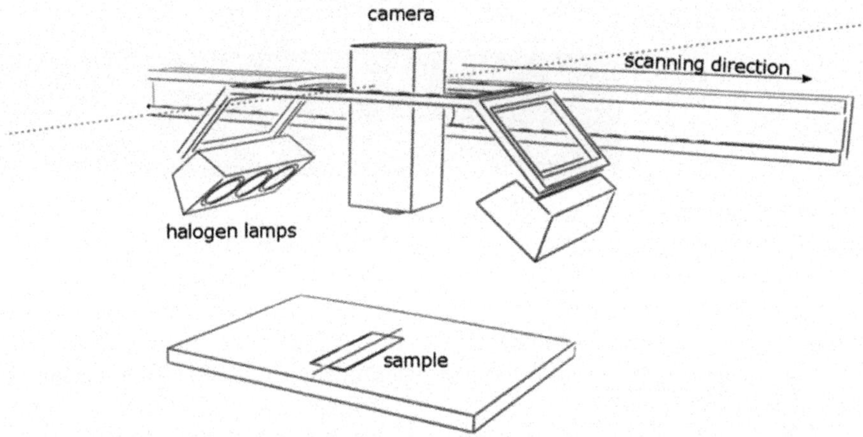

Fig. 12.7 Hyperspectral imaging setup. (Reprinted from Ref. [97]. Copyright 2018, with permission from Springer Nature)

HSI allows for reflectance spectroscopic analysis in a rapid, noncontact, and nondestructive interface (Fig. 12.7). In forensic science, HSI has been commonly used for the analysis of fingermarks, detecting both untreated and treated specimens and other traces. HSI has been combined with reflectance spectroscopy, photoluminescence spectroscopy, Raman spectroscopy, and transmission spectroscopy for the analysis of many types of forensic traces [53, 97]. In this section, the use of HSI for the analysis of bloodstains is discussed, specifically using HSI to assess TSD of bloodstains. Edelman et al. demonstrated the ability of the HSI and reflectance spectroscopy to detect latent bloodstains on dark fabrics, including washed fabrics [98]. HSI was a useful nondestructive method of visualizing bloodstains and identifying blood. From this work, it was inferred that if the stain could be identified, then the age of the stain could likely also be determined.

Edelman et al. used HSI to measure the relative and absolute TSD of bloodstains based on the relative quantities of oxyHb, metHb, and HC when compared to reference spectra. In this case, the absolute age refers to the actual TSD of a bloodstain, while the relative age is defined by the order of deposition of several samples. Edelman et al. aged blood for up to 200 days and calculated both the relative and absolute ages of their stains [54]. The absolute error of their estimations increased with the age of the stain, while the relative mean error was approximately 13.4%. For stains identified as blood, estimated value of the sample's age was automatically obtained and needed no human interference, showing that this technique has the potential to be automated. In addition, they were able to cluster the spectra of bloodstains and determine their deposition order when testing multiage samples with no reference spectra. Li et al. also used HSI paired with reflectance spectroscopy to investigate the TSD of bloodstains. They developed an LDA chemometric model to determine age up to 1 month. Within the first 24 h, the reported error rate was ±0.09 h, and within the first week, the error was ±0.27 days. This error rate was

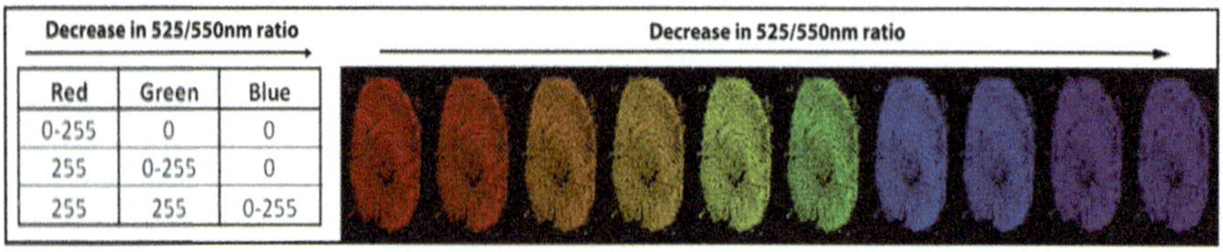

Fig. 12.8 The use of red-green-blue colors to produce false color scales based on the 525/550 nm ratio. (Reprinted from Ref. [2]. Copyright 2018, with permission from MDPI)

smaller than in other techniques; and it allowed for age estimation of stains within 1 day. After 11 days, the error rate increased to ±1.17 days [99]. However, this study showed the potential for HSI for age determination of bloodstains during an extended period.

More recently, Cadd et al. combined reflectance spectroscopy with HSI for the detection, identification, and aging of fingermarks [100]. In 2016, two scientific articles were released on their efforts to develop a novel method for the identification of bloodstained fingermarks. Their goal was to design a noncontact and nondestructive method that could identify traces of blood contaminated fingermarks. They were able to identify bloodstained fingermarks up to a 15,000-fold dilution and up to 6 months old while still obtaining ridge detail of the fingermark [4]. The pairing of HSI and reflectance spectroscopy also prevented false-positive results; Cadd et al. demonstrated that this technology was able to distinguish between blood and the tested red/brown non-blood substances [4]. In their second experiment, they showed the ability of HSI to detect and identify bloodstained fingermarks on a variety of substrates, including on light and dark ceramic tiles on both porous and nonporous substrates, but a higher resolution setup was needed for fingermarks on black tiles [3]. Cadd et al. then delved into determining TSD utilizing reflectance HSI. The spectrum of Hb between 400 and 680 nm was analyzed, using the presence of the Soret band at 415 nm for blood identification and the 525 nm (α-band) and 550 (β-band) for age estimation. A false color aging scale was created for a 24-h scale and a 30-day scale, which allowed the fingermarks to be colored differently, depending on their age (Fig. 12.8). They used the abundance ratio between the 525 and 550 nm peaks to estimate the TSD, and it was concluded that the greatest changes to the bloodstained fingermarks occurred within the first 24 h [2]. While they did not perform a blind study to test their method, this technique showed great potential for noncontact and nondestructive age estimation of bloodstains, especially bloodstained fingermarks.

12.4.5 Smartphone-Based Forensics

Being able to identify a stain as blood on the scene helps save crucial time during an investigation. This means there is a necessity for portable instrumentation that is still sensitive and specific but simple enough to be user-friendly. One computer that is constantly accessible to most individuals is the smartphone. Therefore, research has been directed toward determining the applicability of utilizing these electronic portable devices, and their ever-advancing cameras, as a means of approximating TSD of bloodstains. In 2013, Thanakiatkrai et al. was the first group to investigate the ability of smartphone cameras with a low-cost illumination system to determine TSD [24]. They compared the ability of three smartphones (Samsung Galaxy S Plus, Apple iPhone 4, and Apple iPad 2) to analyze bloodstains on multiple substrates and in different environmental conditions. Color values were extracted from the images to collect information about the stain. The RGB (red, blue, and green) colors and their counterparts the CMYK (cyan, magenta, yellow, and black) values were extracted and plotted against time to examine which color change was most related to the aging of blood. Magenta had the highest correlation ($R^2 = 0.966$) to TSD and was therefore used when estimating the bloodstain's age. Thanakiatkrai et al. discovered that the Samsung model contained the best camera for this endeavor as its camera was most equip to measure the changes to the magenta values. This research group was able to predict the age of bloodstains up to 42 days with an error rate of 12%. During their blind study, the TSD of 40 samples were correctly predicted, and 83% of mock case samples were correctly predicted [24]. This method also worked well for various donors; Thanakiatkrai et al. determined that there were no significant variations in the results of the obtained bloodstains within or between persons.

More recently, Shin et al. developed a mobile detection system called "smart forensic phone" which is an application that monitors the changes in the RGB values per pixel as a value of time. They monitored changes to bloodstains on wallpaper, fabric, glass, wood, and paper over the course of 3 days every 6 h. The observed RGB values were then converted to HSV (hue, saturation, and brightness) values. They determined that the "V" (brightness) values of the stain changed the most with the age of the bloodstain. TSD could be estimated up to 42 h; as time progressed, the error of the results increased, and the substrate affected the ability to measure the change in brightness over time [23] (Fig. 12.9). Finally, Choi et al. developed a smartphone-based colorimetric system that was used to determine the TSD of bloodstains. For this analysis, three main parameters were considered for aging: blood pool analysis, crack ratio analysis, and colorimetric analysis. The drying process of the bloodstains was then divided into six stages (coagulation, gelation, edge desiccation, center desiccation, crack propagation, and final desiccation) which linked to the estimation of TSD. Choi et al. were able to estimate TSD up to 48 h with this technique [22]. In addition, obvious changes in the physical characteristics of blood on glass were detected, but measuring the physical changes on other substrates was more difficult, due to substance-substrate interaction. The use of smartphones as an instrument for age estimation of bloodstains shows promising potential for being a rapid, inexpensive, and portable technique that requires limited training to use. While there is still research that is needed to account for environmental and substrate interferences, a smartphone application may one day be a reliable method for TSD of bloodstains at the scene of a crime.

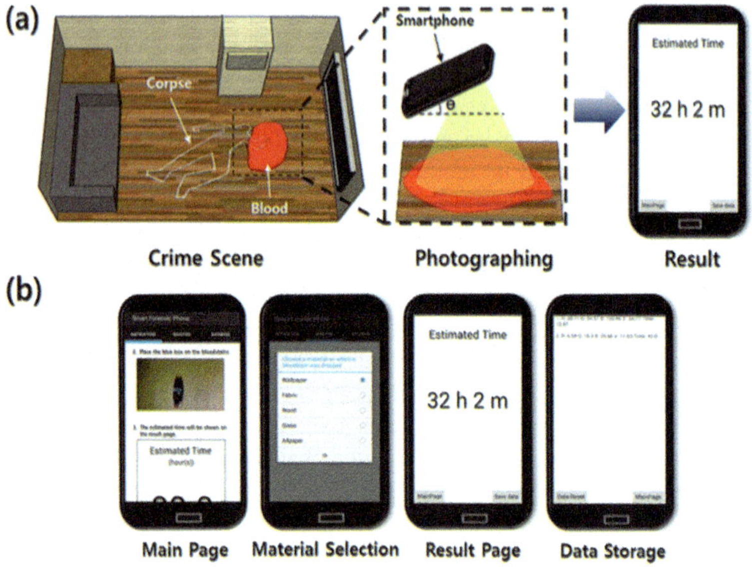

Fig. 12.9 (**a**) Schematic diagram of investigating a bloodstain's RGB variation using smart forensic phone. (**b**) Users are able to estimate the age of the bloodstain on five materials (wallpaper, fabric, glass, wood, and A4 paper) and store the data in the database. (Reprinted from Ref. [23]. Copyright 2016, with permission from Elsevier)

12.5 Limitations of Age Estimation of Bloodstained Fingermarks in Practice

12.5.1 Complexity of Fingermark Composition

Fingermarks are latent impressions that are unintentionally left on surfaces made by friction ridge skin [7]. They are composed of sweat secretions from the secretory glands, materials from the epidermis, and both intrinsic and extrinsic constituents [100]. Fingermarks are complex, containing both organic and inorganic compounds. The ability to determine the age of uncontaminated fingermarks is a complex process, but when contaminated with bloodstains, the difficulty increases. This is due to the overlap of components that are present in blood as well as fingermarks. For example, tryptophan dominates the fluorescence spectra of fingermarks [101] but is also one of the primary components of blood and is prominent in the fluorescence spectra of peripheral blood. This can cause an additive effect of the components that are found in both blood and fingermarks when analyzed with spectroscopic methods. The overlap means that the created statistical models or methodologies are not able to be used without modification. In addition, as a biological sample, the chemical composition of fingermarks exhibits significant variability within and between individuals. As it was just discussed, the additive effects of the chemical components of bloodstains and fingermarks, this variability produces another level of difficulty that must be accounted for when attempting to determine the TSD of bloodstained fingermarks. And thus, more research is needed to produce statistical models that account for the effects of fingermarks, primarily the overlap of the chemical composition of fingermarks with blood.

12.5.2 Substrate Effects

Substrates play a significant role in the analysis of bloodstains. Interaction between blood and the substrate can affect how blood clots and the bloodstain drying time [102]. Differences between porous and nonporous surfaces cause variable interactions that have yet to be fully explored. While some studies have shown that the effects of the underlying substrate do not affect the age analysis of bloodstains [45], these methods require extraction of the bloodstains from the substrate. This is not a viable approach for bloodstained fingermarks, as it would destroy the ridge detail and prevent further analysis. However, for techniques that examine bloodstains in situ, the type of substrate must be considered. For Raman spectroscopy, substrates that have inherent fluorescence are difficult to examine. Reflectance spectroscopy paired with hyperspectral imaging is inhibited by samples deposited on a dark substrate, such as black ceramic tiles [3]. Moving forward, techniques for TSD estimations are becoming more portable. Therefore, it is necessary to create a technique that is unaffected by substrate interference. McLaughlin et al. attempted to circumvent substrate interference in Raman spectroscopic analysis. By testing different laser excitation wavelengths and spectral subtraction of several substrates, they concluded that overcoming substrate effects is dependent on the nature of the substrate [103]. Smartphone cameras have recently been introduced as a novel method for age estimation of bloodstains and appear to be a promising portable technique. However, before this method can be used in forensic casework, in-depth research must be conducted to understand the scope of usability of these platforms.

12.5.3 Environmental Factors

Beyond substrate effects, environmental factors largely influence the ability to create a universal model for approximating TSD of bloodstains. Almost all research has been carried out in controlled laboratory conditions: at room temperature, controlled humidity, and either in darkness or under non-ultraviolet lights. However, crime scene conditions are not controlled nor monitored. They can occur in places of extreme temperatures, varying humidity, and outside in the direct sunlight. As shown in previously conducted research, environmental factors have a significant effect on the drying/aging process of blood. One statement that has been consistently proven is that in heated conditions, bloodstains age faster due to the denaturing and aggregation of Hb and other proteins. According to Bremmer et al., the conversion to oxyHb to metHb and the transition from metHb to HC is affected by temperature. High temperatures (above 37 °C) cause these changes to occur at an accelerated rate, but more research is needed to investigate whether the rate of acceleration is linear or nonlinear relative to the increase in temperature [10, 69]. It has also been proven that bloodstains exposed to elevated temperatures can produce false-negative results in common presumptive and confirmatory tests for bloodstain identification [70, 71]. Hanson et al. and Thanakiatkrai et al. both demonstrated that the aging process of blood stored at low temperatures (−20 °C) was halted and is thus a useful storage temperature to avoid advanced degradation of samples post-collection [24, 45].

In addition to temperature, the relative humidity is another factor that affects the decomposition of bloodstains. Studies by Bremmer et al. used reflectance spectroscopy to study how relative humidity influenced the formation rates of metHb and HC. It was discovered that humidity does not affect the transition from oxyHb to metHb by autooxidation but seems to affect the conformational change from metHb to HC [69]. Research by Hanson et al. and Thanakiatkrai et al. further explored the effect of humidity on bloodstain aging by examining the aging process of blood in different relative humidity and temperatures [24, 45]. Their findings implied that both temperature and humidity play a role in the aging process of bloodstains, where higher humidity causes the degradation process of Hb to be slowed at higher temperatures. This suggests that heat and humidity may have counteracting effects on bloodstains, which makes the understanding of environmental conditions more complex.

Due to a large number of variables that must be considered, developing a project that can cover all of these environmental conditions is difficult to conceive. Nevertheless, an in-depth study of environmental effects is needed, and once the kinetics of blood aging in different environments has been explored, perhaps a more cohesive statistical analysis model could be created. A model for determining TSD of bloodstains will not be accepted by the forensic and law communities until it is shown to be applicable in an encompassing number of situations. Current publications focus on the creation of novel methods for TSD analysis and usually discuss the effects of environmental conditions as additional work. However, research going forward should work on expanding the applicability of current techniques for a variety of different (realistic) environmental scenarios.

12.5.4 Contamination Effects

As previously discussed, violent crimes do not always happen in pristine conditions – they occur in places where evidence can be subjected to environmental contaminants such as sand, dust, and soil. The ability to identify and estimate the age of bloodstains that have been heterogeneously exposed to these conditions is a requirement for an effective standard TSD methodology. It has been shown that spatial mapping techniques, when paired with Raman spectroscopy, cannot only detect "hot spots" where the blood contribution is the highest but also identify bloodstains that have been highly contaminated [14, 104]. Besides being contaminated with extraneous conditions, the blood itself can be highly variable based on the donor's lifestyle. Individuals who are smokers have a higher amount of carboxyhemoglobin in their blood, and illnesses such as sickle cell disease can affect the morphology of RBCs, thereby interfering with analysis [105].

An ideal methodology must be highly specific, and a limitation to overcome in determining TSD of bloodstains is being able to distinguish between blood and non-blood environmental interferences, which are materials similar in appearance to blood. Rosenblatt et al. and Edelman et al. demonstrated the specificity of Raman spectroscopy to distinguish between blood and numerous red-/brown-colored samples that could be chemically or physically mistaken as blood [52, 104]. Cadd et al. established that reflectance spectroscopy paired with HSI was able to distinguish several red/brown non-blood substances from actual blood. Blood itself is also a common contaminant in fingermark analysis, and being able to estimate TSD of bloodstains while not inhibiting the further analysis of the fingermark is highly sought after. Cadd et al. have shown that HSI is useful in this regard, identifying bloodstained fingermarks nondestructively on a variety of substrates as well as estimating the age of fingermarks up to 1 month [2, 4].

12.6 Conclusion and Future Developments

The development of an accurate method for determining the time since deposition of bloodstains has been the goal of scientists for over 100 years. As we progress through the twenty-first century, there are still unanswered questions that inhibit the implementation of any reliable technique into forensic casework. Despite being one of the most common contaminants of fingermarks, much variability comes with the analysis of bloodstained fingermarks that must be taken into consideration. The composition of fingermarks, the influence of environmental conditions, and the types of substrates affect the ability to successfully estimate TSD of bloodstained fingermarks. There has been significant progress made toward the design of a rapid, inexpensive, nondestructive method for determining TSD of blood, with spectroscopy at the forefront. Infrared spectroscopy, Raman spectroscopy, and reflectance spectroscopy paired with hyperspectral imaging have shown the greatest potential for age estimation of bloodstains. In addition, the adaptation of smartphones for forensic analysis of bloodstains has opened the door to truly portable instrumentation. However, there is still much work to be done.

Many studies are showing the development of novel methods that can potentially be used to establish TSD, but there is still very limited research on validating already existing techniques. Reproducibility is one of the main components and goals of scientific (and forensic) research. Though, in the process of conducting a research project, replicates are run to produce error rates and standard deviations. However, the replicates are all analzyed within one laboratory. A methodology should be able to produce consistent results by other laboratories, showing the repoducibility of the work. Also, work conducted thus far has highly focused on the analysis of samples in laboratory conditions, briefly investigating environmental effects in realistic setups. To create a method that is usable in crime scene scenarios, the focus of research must turn from the development of a methodology to testing the full scope of those methods under "natural" conditions. Finally, as technology progresses and bloodstain analysis using smartphones is introduced, it is important to understand the limitations of smartphones compared to research-grade instrumention. The integration of smartphone technology means that nonscientifically trained individuals may be making conclusive determinations on crime evidence.

There are some requirements that should be investigated before any of the discussed techniques can be incorporated into the field: (1) estimation of error rates, (2) validation of existing TSD estimation methodologies, (3) in-depth investigation into environmental and substrate effects, and (4) understanding the applicability and limitations of smartphone technology. Being able to estimate the error rates of the proposed techniques is crucial for the method to be accepted into the court of law. As studies become more robust and create complex data, the need for comprehensive data analysis is required. The increased use of chemometrics allows for the analysis of data and the design of statistical models that can predict the TSD of bloodstains accurately. Despite the innovation in rapid identification and age determination of bloodstains, it is still imperative that evidence is examined by trained forensic scientists who understand biological sample analysis and forensic technology. Therefore, while many methods have been developed for estimating the TSD of bloodstains, there is still more research that is required before these techniques can be used on contaminated fingermarks in real crime scenarios.

Acknowledgments Nicole Ralbovsky and Alexis Barber provided valuable feedback in the preparation of this chapter.

This project was supported by Awards No. 2017-DN-BX-0135 awarded by the National Institute of Justice, Office of Justice Programs, US Department of Justice (I.K. Lednev). The opinions, findings, and conclusions or recommendations expressed here are those of the authors and do not necessarily reflect those of the US Department of Justice.

References

1. Lee HC, Palmbach T, Miller MT (2001) Henry Lee's crime scene handbook. Academic Press, London
2. Cadd S et al (2018) Age determination of blood-stained fingerprints using visible wavelength reflectance hyperspectral imaging. J Imaging 4(12):141
3. Cadd S et al (2016) The non-contact detection and identification of blood stained fingerprints using visible wavelength hyperspectral imaging: part II effectiveness on a range of substrates. Sci Justice 56(3):191–200
4. Cadd S et al (2016) The non-contact detection and identification of blood stained fingerprints using visible wavelength reflectance hyperspectral imaging: part 1. Sci Justice 56(3):181–190

5. Jeffreys AJ, Wilson V, Thein SL (1985) Hypervariable 'minisatellite' regions in human DNA. Nature 314(6006):67–73
6. Li R, Norman S, Schober J (2015) Forensic biology. CRC Press, Boca Raton
7. De Alcaraz-Fossoul J, Islam M (2019) Novel technological applications for latent and bloodstained fingermark aging studies. In: Emerging technologies for the analysis of forensic traces. Springer, Cham, pp 33–66
8. Weber AR, Lednev IK (2020) Raman spectroscopy is solving the perpetual problem of CSI: time of crime. Spectroscopy. 35(6):41-44
9. Weber AR, Lednev IK (2020) Crime clock – analytical studies for approximating time since deposition of bloodstains. Forensic Chem 19:100248
10. Bremmer RH et al (2012) Forensic quest for age determination of bloodstains. Forensic Sci Int 216(1–3):1–11
11. Zadora G, Menżyk A (2018) In the pursuit of the holy grail of forensic science – spectroscopic studies on the estimation of time since deposition of bloodstains. TrAC Trends Anal Chem 105:137–165
12. Gaensslen RE (1983) Sourcebook in forensic serology, immunology, and biochemistry. U.S. Department of Justice, National Institute of Justice, Washington, DC
13. Muro CK et al (2016) Forensic body fluid identification and differentiation by Raman spectroscopy. Forensic Chem 1:31–38
14. Sikirzhytskaya A et al (2013) Forensic identification of blood in the presence of contaminations using Raman microspectroscopy coupled with advanced statistics: effect of sand, dust, and soil. J Forensic Sci 58(5):1141–1148
15. Sikirzhytski V, Sikirzhytskaya A, Lednev IK (2012) Multidimensional Raman spectroscopic signature of sweat and its potential application to forensic body fluid identification. Anal Chim Acta 718:78–83
16. Sikirzhytski V, Sikirzhytskaya A, Lednev IK (2011) Multidimensional Raman spectroscopic signatures as a tool for forensic identification of body fluid traces: a review. Appl Spectrosc 65(11):1223–1232
17. Virkler K, Lednev IK (2010) Raman spectroscopic signature of blood and its potential application to forensic body fluid identification. Anal Bioanal Chem 396(1):525–534
18. Sikirzhytskaya A, Sikirzhytski V, Lednev IK (2012) Raman spectroscopic signature of vaginal fluid and its potential application in forensic body fluid identification. Forensic Sci Int 216(1–3):44–48
19. Sikirzhytskaya A, Sikirzhytski V, Lednev IK (2014) Raman spectroscopy coupled with advanced statistics for differentiating menstrual and peripheral blood. J Biophotonics 7(1–2):59–67
20. Mistek E et al (2019) Toward Locard's exchange principle: recent developments in forensic trace evidence analysis. Anal Chem 91(1):637–654
21. Doty KC et al (2016) What can Raman spectroscopy do for criminalistics?. J Raman Spectrosc 47(1):39–50
22. Choi W et al (2019) Highly sensitive and accurate estimation of bloodstain age using smartphone. Biosens Bioelectron 130:414–419
23. Shin J et al (2017) Smart forensic phone: colorimetric analysis of a bloodstain for age estimation using a smartphone. Sensors Actuators B Chem 243:221–225
24. Thanakiatkrai P, Yaodam A, Kitpipit T (2013) Age estimation of bloodstains using smartphones and digital image analysis. Forensic Sci Int 233(1–3):288–297
25. Doty KC, Muro CK, Lednev IK (2017) Predicting the time of the crime: bloodstain aging estimation for up to two years. Forensic Chem 5:1–7
26. Doty KC, McLaughlin G, Lednev IK (2016) A Raman "spectroscopic clock" for bloodstain age determination: the first week after deposition. Anal Bioanal Chem 408(15):3993–4001
27. Müller J (1880) Archiv für Anatomie, Physiologie und wissenschaftliche Medicin. G. Eichler, Berlin

28. Tomellini L (1907) De l'emploi d'une table chromatique pour les taches du sang. Archives d'antropologie criminelle de Criminologie 14(2)
29. Leers O (1910) Die forensische Blutuntersuchung: ein Leitfaden für Studierende, beamtete und sachverständige Ärzte und für Kriminalisten. J. Springer, Berlin
30. Schwarzacher (1930) Determination of the age of bloodstains. Am J Police Sci 1:374–380
31. Kind SS, Patterson D, Owen GW (1972) Estimation of the age of dried blood stains by a spectrophotometric method. Forensic Sci 1(1):27–54
32. Kind SS, Watson M (1973) The estimation of blood stain age from the spectrophotometric properties of ammoniacal blood stain extracts. Forensic Sci 2:325–332
33. Patterson D (1960) Use of reflectance measurements in assessing the colour changes of ageing bloodstains. Nature 187(4738):688–689
34. Bauer M, Polzin S, Patzelt D (2003) Quantification of RNA degradation by semi-quantitative duplex and competitive RT-PCR: a possible indicator of the age of bloodstains?. Forensic Sci Int 138(1–3):94–103
35. Anderson S et al (2005) A method for determining the age of a bloodstain. Forensic Sci Int 148(1):37–45
36. Lech K et al (2016) Evaluation of mRNA markers for estimating blood deposition time: towards alibi testing from human forensic stains with rhythmic biomarkers. Forensic Sci Int Genet 21:119–125
37. Zhao H et al (2017) Identification of aged bloodstains through mRNA profiling: experiments results on selected markers of 30- and 50-year-old samples. Forensic Sci Int 272:e1–e6
38. Inoue H et al (1991) Identification of fetal hemoglobin and simultaneous estimation of bloodstain age by high-performance liquid chromatography. Int J Legal Med 104(3):127–131
39. Inoue H et al (1992) A new marker for estimation of bloodstain age by high performance liquid chromatography. Forensic Sci Int 57(1):17–27
40. Strasser S et al (2007) Age determination of blood spots in forensic medicine by force spectroscopy. Forensic Sci Int 170(1):8–14
41. Wu Y et al (2009) Time-dependent surface adhesive force and morphology of RBC measured by AFM. Micron 40(3):359–364
42. Smijs T, Hosseinzoi A, Galli F (2017) Forensic application of atomic force microscopy for age determination of bloodstains. J Forensic Investig 5(1):6
43. Cavalcanti DR, Silva LP (2019) Application of atomic force microscopy in the analysis of time since deposition (TSD) of red blood cells in bloodstains: a forensic analysis. Forensic Sci Int 301:254–262
44. Marrone A, Ballantyne J (2009) Changes in dry state hemoglobin over time do not increase the potential for oxidative DNA damage in dried blood. PLoS One 4(4):e5110
45. Hanson EK, Ballantyne J (2010) A blue spectral shift of the hemoglobin Soret band correlates with the age (time since deposition) of dried bloodstains. PLoS One 5(9):e12830
46. Wood BR et al (2005) Raman microspectroscopy and imaging provides insights into heme aggregation and denaturation within human erythrocytes. J Biomed Opt 10(1):014005
47. Asghari-Khiavi M et al (2009) A resonance Raman spectroscopic investigation into the effects of fixation and dehydration on heme environment of hemoglobin. J Raman Spectrosc 40(11):1668–1674
48. Botonjic-Sehic E et al (2009) Forensic application of near-infrared spectroscopy: aging of bloodstains. Spectroscopy 24(2):42–48
49. Li B et al (2011) The estimation of the age of a blood stain using reflectance spectroscopy with a microspectrophotometer, spectral pre-processing and linear discriminant analysis. Forensic Sci Int 212(1–3):198–204
50. Bremmer RH et al (2011) Age estimation of blood stains by hemoglobin derivative determination using reflectance spectroscopy. Forensic Sci Int 206(1–3):166–171
51. Bremmer RH et al (2011) Remote spectroscopic identification of bloodstains. J Forensic Sci 56(6):1471–1475

52. Edelman G et al (2012) Identification and age estimation of blood stains on colored backgrounds by near infrared spectroscopy. Forensic Sci Int 220(1–3):239–244
53. Edelman GJ et al (2012) Hyperspectral imaging for non-contact analysis of forensic traces. Forensic Sci Int 223(1–3):28–39
54. Edelman G, Van Leeuwen TG, Aalders MC (2012) Hyperspectral imaging for the age estimation of blood stains at the crime scene. Forensic Sci Int 223(1–3):72–77
55. Lemler P et al (2014) NIR Raman spectra of whole human blood: effects of laser-induced and in vitro hemoglobin denaturation. Anal Bioanal Chem 406(1):193–200
56. Lu Z (2015) Applications of attenuated total reflectance Fourier transform infrared spectroscopy for forensic analysis. Doctoral dissertation 2015
57. Lin H et al (2017) Estimation of the age of human bloodstains under the simulated indoor and outdoor crime scene conditions by ATR-FTIR spectroscopy. Sci Rep 7(1):1-9
58. Sun H et al (2017) Accurate age estimation of bloodstains based on visible reflectance spectroscopy and chemometrics methods. IEEE Photonics J 9(1):1–14
59. Menżyk A et al (2020) Toward a novel framework for bloodstains dating by Raman spectroscopy: how to avoid sample photodamage and subsampling errors. Talanta 209:120565
60. Atkins CG et al (2017) Raman spectroscopy of blood and blood components. Appl Spectrosc 71(5):767–793
61. Masilamani V et al (2004) Cancer diagnosis by autofluorescence of blood components. J Lumin 109(3-4):143–154
62. Li BH et al (2006) Fluorescence spectral characteristics of human blood and its endogenous fluorophores. Guang Pu Xue Yu Guang Pu Fen Xi 26(7):1310–1313
63. Guo K, Achilefu S, Berezin MY (2012) Dating bloodstains with fluorescence lifetime measurements. Chemistry 18(5):1303–1305
64. Mc Shine S et al (2017) The applicability of fluorescence lifetime to determine the time since the deposition of biological stains. Anal Methods 9(13):2007–2013
65. Nagababu E, Rifkind JM (2000) Heme degradation during autoxidation of oxyhemoglobin. Biochem Biophys Res Commun 273(3):839–845
66. McLaughlin G, Doty KC, Lednev IK (2014) Raman spectroscopy of blood for species identification. Anal Chem 86(23):11628–11633
67. Sikirzhytskaya A, Sikirzhytski V, Lednev IK (2017) Determining gender by Raman spectroscopy of a bloodstain. Anal Chem 89(3):1486–1492
68. Mistek E et al (2016) Race differentiation by Raman spectroscopy of a bloodstain for forensic purposes. Anal Chem 88(15):7453–7456
69. Bremmer RH et al (2011) Biphasic oxidation of oxy-hemoglobin in bloodstains. PLoS One 6(7):e21845
70. McDonald T (2017) Investigating the effect of high temperatures and substrates on the detection of human blood using the ABAcard® Hematrace® kit. In: The school of veterinary and life sciences. Doctoral dissertation, Murdoch University, Perth
71. Katiyar K (2017) Identification of blood stains under different environmental conditions. Int J Biomed Res 8(12):707–710
72. Alrowaithi MA, McCallum NA, Watson ND (2014) A method for determining the age of a bloodstain. Forensic Sci Int 234:e30–e31
73. Tsutsumi A, Yamamoto Y, Ishizu H (1983) Determination of the age of bloodstains by enzyme activities in blood cells. Jpn J Leg Med 37(6):770–776
74. Agudelo J, Huynh C, Halámek J (2015) Forensic determination of blood sample age using a bioaffinity-based assay. Analyst 140(5):1411–1415
75. Agudelo J et al (2016) Ages at a crime scene: simultaneous estimation of the time since deposition and age of its originator. Anal Chem 88(12):6479–6484
76. Rauschke J (1951) Age determination of blood stains. Dtsch Z Gesamte Gerichtl Med 40(6):578-584
77. Schwarz F (1937) Quantitative untersuchungen der katalase und peroxydase im blutfleck. Dtsch Z Gesamte Gerichtl Med 27(1):1–34

78. Kleihauer E, Stein G, Schmidt G (1967) Beitrag zur alterbestimmung von blutflecken. Arch Kriminol 140:84–94
79. Andrasko J (1997) The estimation of age of bloodstains by HPLC analysis. J Forensic Sci 42(4):601–607
80. Alshehhi S, McCallum NA, Haddrill PR (2017) Quantification of RNA degradation of blood-specific markers to indicate the age of bloodstains. Forensic Sci Int: Gen 6:e453–e455
81. Anderson SE, Hobbs GR, Bishop CP (2011) Multivariate analysis for estimating the age of a bloodstain. J Forensic Sci 56(1):186–193
82. Lowndes S (2010) Blood interference in fluorescence spectrum- experiment, analysis and comparison with intra-operative measurements on brain tumor. Department of Biomedical Engineering, Linköping University
83. Sharma V, Kumar R (2018) Trends of chemometrics in bloodstain investigations. TrAC Trends Anal Chem 107:181–195
84. Silva CA, Braz A, Pimentel MF (2019) Vibrational spectroscopy and chemometrics in forensic chemistry: critical review, current trends and challenges. J Braz Chem Soc 30(11):2259–2290
85. Mistek E, Lednev IK (2015) Identification of species' blood by attenuated total reflection (ATR) Fourier transform infrared (FT-IR) spectroscopy. Anal Bioanal Chem 407(24):7435–7442
86. Muro CK, Lednev IK (2017) Identification of individual red blood cells by Raman micro-spectroscopy for forensic purposes: in search of a limit of detection. Anal Bioanal Chem 409(1):287–293
87. Boyd S, Bertino MF, Seashols SJ (2011) Raman spectroscopy of blood samples for forensic applications. Forensic Sci Int 208(1–3):124–128
88. Sikirzhytski V, Sikirzhytskaya A, Lednev IK (2012) Advanced statistical analysis of Raman spectroscopic data for the identification of body fluid traces: semen and blood mixtures. Forensic Sci Int 222(1–3):259–265
89. Muro CK, Lednev IK (2017) Race differentiation based on Raman spectroscopy of semen traces for forensic purposes. Anal Chem 89(8):4344–4348
90. Premasiri WR, Lee JC, Ziegler LD (2012) Surface-enhanced Raman scattering of whole human blood, blood plasma, and red blood cells: cellular processes and bioanalytical sensing. J Phys Chem B 116(31):9376–9386
91. Takamura A et al (2019) Comprehensive modeling of bloodstain aging by multivariate Raman spectral resolution with kinetics. Commun Chem 2(1):1–10
92. Pereira JFQ et al (2017) Evaluation and identification of blood stains with handheld NIR spectrometer. Microchem J 133:561–566
93. McLaughlin G, Lednev IK (2014) A modified Raman multidimensional spectroscopic signature of blood to account for the effect of laser power. Forensic Sci Int 240:88–94
94. Dasgupta R et al (2010) Hemoglobin degradation in human erythrocytes with long-duration near-infrared laser exposure in Raman optical tweezers. J Biomed Opt 15(5):055009
95. Lemler P et al (2014) NIR Raman spectra of whole human blood: effects of laser-induced and in vitro hemoglobin denaturation. Anal Bioanal Chem 406(1):193–200
96. Bergmann T, Heinke F, Labudde D (2017) Towards substrate-independent age estimation of blood stains based on dimensionality reduction and k-nearest neighbor classification of absorbance spectroscopic data. Forensic Sci Int 278:1–8
97. Majda A et al (2018) Hyperspectral imaging and multivariate analysis in the dried blood spots investigations. Appl Phys A Mater Sci Process 124(4):1-8
98. Edelman GJ, van Leeuwen TG, Aalders MC (2015) Visualization of latent blood stains using visible reflectance hyperspectral imaging and chemometrics. J Forensic Sci 60:S188–S192
99. Li B et al (2013) The age estimation of blood stains up to 30 days old using visible wavelength hyperspectral image analysis and linear discriminant analysis. Sci Justice 53(3):270–277
100. Cadd S et al (2015) Fingerprint composition and aging: a literature review. Sci Justice 55(4):219–238
101. van Dam A et al (2016) On the autofluorescence of aged fingermarks. Forensic Sci Int 258:19–25

102. Laber TL, Epstein BP (2001) Substrate effects on the clotting time of human blood. Can Soc Forensic Sci J 34(4):209–214
103. McLaughlin G, Sikirzhytski V, Lednev IK (2013) Circumventing substrate interference in the Raman spectroscopic identification of blood stains. Forensic Sci Int 231(1–3):157–166
104. Rosenblatt R et al (2019) Raman spectroscopy for forensic bloodstain identification: method validation vs. environmental interferences. Forensic Chem 16:100175
105. Edelman GJ, Aalders MCG (2017) Blood degradation and bloodstain age estimation: forensic analysis of the dead and the depositional environment. In: Taphonomy of human remains: forensic analysis of the dead and the depositional environment. Wiley, New York, pp 53–64

Chapter 13
The Application of Statistics in Fingermark Aging Research

Brooke W. Kammrath

"There are three kinds of lies: lies, damned lies, and statistics." – unknown author

Abstract Statistics play an important role in the analysis of forensic traces, including the potential dating of fingermarks. There are numerous statistical tools, both univariate and multivariate, used in fingermark source identification and aging research to study relationships and determine whether there are meaningful differences. Regardless of the statistical methods employed, it is necessary to ensure that the statistics used in forensic examinations are rigorously tested. When investigating the aging of fingermarks, both chemical and physical properties have been examined using a range of techniques and statistical approaches. While most use simple but robust statistics, a few have used sophisticated multivariate methods. Considering the complexity of fingermark aging, and specifically the number of external and internal factors, it is likely that the most successful approaches to dating a fingermark will employ multivariate statistics and possibly data fusion.

Keywords Statistics · Multivariate statistics · Chemometrics · Fingermark aging

Acronyms

ABC	Approximate Bayesian computation
AFIS	Automated fingerprint identification systems

B. W. Kammrath (✉)
Henry C. Lee College of Criminal Justice and Forensic Sciences,
Forensic Science Department, University of New Haven,
West Haven, CT, USA

Henry C. Lee Institute of Forensic Science, West Haven, CT, USA
e-mail: BKammrath@newhaven.edu

J. De Alcaraz-Fossoul (ed.), *Technologies for Fingermark Age Estimations: A Step Forward*, https://doi.org/10.1007/978-3-030-69337-4_13

ANN	Artificial neural networks
ANOVA	Analysis of variance
CLS	Classical least squares
DNA	Deoxyribonucleic acid
EIS	Electrochemical impedance spectroscopy
FID	Flame ionization detector
GC	Gas chromatography
MS	Mass spectrometer or mass spectrometry
NAS	National Academy of Sciences
NIJ	National Institute of Justice
PCA	Principal component analysis
PCR	Principal component regression
PLSR	Partial least squares regression
ROC	Receiver operating characteristics
SIMCA	Soft independent modeling of class analogy
SVM	Support vector machines

13.1 Introduction

Statistics are a valuable tool for all scientists, mathematicians, and engineers, as they can be used to help provide support for an argument or hypothesis. However, as insinuated by the above quote that is often incorrectly attributed to Mark Twain, statistics have the power to be misused and abused. In forensic science, statistics may be used for several purposes, including, but not limited to, the development of a sampling scheme, quantitative analysis, and evidence interpretation (which includes determining its probative value) [18, 86, 95–97]. Forensic science practitioners are historically quite conservative with their use of statistics and generally utilize traditional approaches that have been developed and tested for decades, if not longer. It is critical for all statistical methods to be robustly examined, which means analyses must extend beyond the laboratory testing of standards under ideal conditions before they can be applied to real case exhibits.

Although it is readily accepted that statistical methods play an important role in the analysis of forensic exhibits, there are considerable differences in opinion regarding the determination of the appropriate statistics to use for evidence interpretation. The use of Bayesian statistics and likelihood ratios for the analysis of forensic evidence has been the subject of scientific debate [18, 36, 59]. Cook et al. [18] proposed the Bayesian approach which employs the calculation of likelihood ratios for case assessment and interpretation. This has been promoted by many European forensic scientists for over two decades but has been criticized by others, particularly by scientists in the United States [11, 59, 88]. Over the last decade or so, forensic science research has begun to employ multivariate statistics, or chemometrics, for the analysis of chemical data, with Kumar and Sharma [56] and Bovens et al. [10] recently writing excellent review articles on this subject and the latter proposing how these methods can be employed in casework.

Regardless of the statistical methods employed, it is necessary to ensure that the statistics used in forensic examinations are rigorously tested. When using machine learning multivariate statistics, it is considered good practice to split the data into three different sets: a training set, a validation set, and a test set. When this is used, the training set is used to fit the model, the validation set is used to evaluate the model and tune the model's parameters, and the test set is used to provide an unbiased evaluation of the final model [85]. This is excellent practice, but is not commonly employed by those using statistics in the forensic science community. Often the argument against using this is that it requires too large of a data set to be feasibly implemented in forensic science. This may be true; thus, other methods for validating a statistical method can be successfully applied, such as cross validation methods (e.g., venetian blinds or leave-one-out) and bootstrapping.

When completing a reconstruction of past events, time is an essential factor that must be considered. In the 1999 Firth Memorial Lecture (published in *Science & Justice* in 2000) [60], Pierre Margot thoughtfully presented his philosophy on time as evidence. In this paper, when discussing a hypothetical crime where a fingermark belonging to the suspect Mr. Hugh Dunnit was recovered at the scene, Margot discusses the role time has for this evidence. He writes, "It is probabilistic when reliable statistical information exists on how fingerprints age in a well defined situation, it is conjectural otherwise." With this statement, Margot demonstrates the great value that providing statistical support to evidence statements has when evaluating traces. When there are multiple possible explanations, statistics provides numerical means of demonstrating the certainty (or uncertainty) of each outcome, which enriches the resulting logical expression.

The numerous issues with dating evidence traces are elaborated by Weyermann and Riboux in their article "Situating Forensic Traces in Time" [105]. A myriad of research studies has been published with the goal of developing techniques capable of determining the age of a trace. Some of these have focused on determining the true or absolute age, while others examine the relative age. Some of the traces that have been studied include the dating of blood deposits, inks and documents, gunshot residues, and fingermarks. These attempts at dating traces have inherent complexities which have contributed to controversial and cautionary reactions by the forensic science community [7, 47]. As a consequence of this complexity, solutions to the problem posed concerning the dating of traces has not yet been adequately solved or adopted into routine forensic science casework.

It is the goal of this author to provide the reader with a complete picture of the potential statistical analyses for the dating of (latent) fingermarks. To this purpose, this chapter is broken into three sections: (1) a discussion of statistical methods used in fingerprint research and practice with a focus on models used to support foundational statements of fingermark individuality, their identification, image analysis, and accuracy, (2) a critical review of the statistical methods used in published research on fingermark aging, and (3) a conclusion with some alternative statistical strategies that may prove useful when attempting to determine the time since deposition for fingermarks. It is not the aim here to teach the reader how to perform a statistical analysis on traces. To that end, it is recommended for readers interested in learning how to apply statistics to their own research or casework to consider reading one of the many books on statistics in science (or specifically in forensic science), or a preferred approach would be to take a university or continuing education course on this topic.

13.2 Statistics in Fingermark Research and Practice

The use of statistics in fingermark research and practice has been well covered in several other texts. The focus of the literature tends to be on using statistics to model statements about the individuality of fingerprints. Individualization is a fundamental philosophy of forensic science (or more specifically criminalistics), with the principal of individuality being introduced by one of our most notable forensic scientists, Professor Paul Leland Kirk (1902–1970). Kirk wrote that "the real aim of all forensic science is to establish individuality, or to approach it as closely as the present state of the science allows," where individuality refers to the notion that "no two objects in nature are identical" [55]. Although there have been recent criticisms of the use of the term individualization in forensic science, it truly is the best expression for the work that is done by criminalists. Individualization should be viewed as a process and not the end goal of a forensic analysis. This is eloquently described by Inman and Rudin [48] who define individualization as a reduction process of a class to one. When considered in this light, it is clear that statistical analysis can benefit this process of forensic evidence analysis by adding numerical weight to individualization determinations. It must be noted that "you cannot achieve individualization through statistics" [91]. However, statistics can provide valuable probabilities that are useful for guiding conclusions on source identifications [13, 91].

David Stoney [92] detailed the numerous models that have been proposed for determining the statistical basis for comparisons using fingerprint minutiae in his chapter "Measurement of Fingerprint Individuality." This chapter critically evaluates each individuality model, beginning with that proposed by Francis Galton in 1892 [38], and continuing to describe and evaluate nine others: Henry-Balthazard (1900–1943), Roxburgh (1933) [83], Amy (1946–1948) [3], Trauring (1963) [98], Kingston (1964) [54], Osterburg (1977–1980) [79], Stoney and Thornton (1985–1989) [90], Champod and Margot (1995–1996) [12], and Meagher, Budowle, and Ziesig (1999) [62].

Champod et al. [15] provide a valuable analysis of the use of probabilistic statements when reporting fingerprint conclusions. The authors astutely argue that statistics are "a key to transparency" that offer a "highly powerful tool for assessing fingerprint evidence." They also point out that no statistical model has yet to comprehensively include all factors of fingerprint individuality such as the general pattern, ridge flow, minutia, or pores. This does not mean that robust and thoroughly tested statistical models of fingerprint individuality should not be used in casework.

"Statistical data, even gathered through myopic models, can only help the discipline work toward more reliable and transparent methods of assessing evidential value" [15].

Neumann [69] wrote a comprehensive chapter which focused on the use of statistics and probabilities for supporting conclusions formed by fingerprint examiners. In this chapter, Neumann details the current inference process of fingermark comparisons, its limitations, and how statistics can be used to support the decision-making process.

In February of 2018, the US Department of Justice developed guidance documents for forensic scientists with regard to establishing uniform language for testimony and reports [30]. Updated in June of 2020, there are three approved conclusions for fingerprint examinations: source identification, source exclusion, and inconclusive [31]. The document provides the following definition for the first one:

> "Source identification" is an examiner's conclusion that two friction ridge skin impressions originated from the same source. This conclusion is an examiner's opinion that the observed friction ridge skin features are in sufficient correspondence such that the examiner would not expect to see the same arrangement of features repeated in an impression that came from a different source and has found insufficient friction ridge skin features in disagreement to conclude that the impressions came from different sources.
>
> The basis for a "source identification"conclusion is an examiner's opinion that the observed corresponding friction ridge skin features provide extremely strong support for the proposition that the two impressions came from the same source and extremely weak support for the proposition that the two impressions came from different sources.
>
> A "source identification" is the statement of an examiner's opinion (an inductive inference) that the probability that the two impressions were made by different sources is so small that it is negligible. [31]

It must be noted that although the definition uses the language of statistics (i.e., "probability"), no statistical analyses are required for reaching a conclusion about the source identity of a latent fingermark.

The incorporation of statistics into fingerprint identifications continues to be the subject of meaningful discussions and publications, with a growing enthusiasm for the use of match probabilities, likelihood ratios, and Bayes' theorem. A convincing discussion of the usefulness of including probabilistic reasonings and match probabilities for the determination of sufficient uniqueness of a fingerprint is presented in Champod and Evett's article "A Probabilistic Approach to Fingerprint Evidence" [14]. In subsequent research, Champod, Evett, and others demonstrate how likelihood ratios can be used to assess the value of fingerprint evidence from three minutiae [71], from the scores obtained from an automated fingerprint identification system (AFIS) which uses a proprietary algorithm to search a fingerprint database [35], during comparisons of a fingermark to ten print cards [73–75], and through spatial relationships, directions, and types of minutiae (Neumann et al. 2015). Neumann et al. [74, 75] published the results of a major project by the Forensic Science Service that developed a model based on the use of likelihood ratios for numerically evaluating fingerprints, taking into account inter-operator variability and distortion. The authors validated their model with the US national fingerprint database containing approximately 600 million fingerprints, and as of the publication of their article, the authors were performing a field study to determine the utility of the model in forensic casework in Minnesota [74, 75]. Recently, a stand-alone software program, FRStat, which uses tail probabilities rather than likelihoods, was introduced to provide statistical assessment of the strength of fingerprint evidence [93]. The FRStat program is being provided free to forensic laboratories for use in casework and has already been used by Swofford's laboratory, the US Army Criminal Investigation Laboratory at the Department of Defense, making them "the first lab in the United States to report fingerprint evidence using a statistical foundation" [104]. Although the initial work is promising, more research and independent evaluations are needed to determine the effectiveness and accuracy of this program.

Likelihood ratio-based statistical methods have also been used to study less commonly considered investigative information that may be contained in a fingerprint. Of note, Acree [1] and Kapoor and Badiye [50] used likelihood ratios to develop probability inferences to determine the biological gender of an individual from the ridge density in their fingerprint [1, 50]. Results from both of these studies, which each had relatively large datasets, demonstrated that women and men have statistically significant differences in ridge densities using t-tests and analysis of variance (ANOVA), with women having significantly greater ridge densities than men. Although more research needs to be completed before this is ready for use in criminal investigations, which should include looking at aged fingermarks and also testing the prediction capabilities with blind studies and mock evidence, these two papers are promising for gender determination from fingermarks.

Multivariate statistical methods have also been applied to the problem of fingerprint identification. Croxton et al. (2010) used gas chromatography (GC) paired with either a mass spectrometer (MS) or flame ionization detector (FID) to analyze variations in amino acid ratios in fingermarks with the aim of differentiating between donors. In addition to traditional statistical analyses (i.e., t-tests and boxplots), principal component analysis (PCA) was employed on the amino acid, fatty acid, and squalene quantitative data using to identify similarities and differences between donors and sample types (natural vs groomed prints and donor gender) (Croxton et al. 2010). Paulino et al. [80] proposed using a descriptor-based Hough transform algorithm to align and associate latent fingermark with a full fingerprint, using both minutiae and orientation field information. Daramalo [21] and Kathiravan and K.samy [51] both use support vector machines (SVM), the most well-known type of kernel method which uses a linear classifier (or classification model) to solve a nonlinear problem by transforming the data, to associate latent fingermarks with a database. Both studies demonstrated the potential for SVM to overcome issues with traditional database algorithm searched caused by fingerprint distortions and/or small areas. In 2018, Neumann and Saunders published the results of their 4-year NIJ-funded research aimed at evaluating statistical methods for pattern and trace evidence based on their validity, accuracy, and computational complexity. Ultimately, the authors concluded "that an increasingly popular method to estimate the weight of forensic evidence (i.e., the so-called score-based likelihood ratio) was incoherent and inappropriate and could not be relied upon to interpret forensic evidence." Instead of likelihood ratios, the authors proposed new models based on kernel functions. The authors created a supervised "kernel-based" model that relied on approximate Bayesian computation (ABC), a likelihood-free Bayesian inference method, paired with the use of receiver operating characteristics (ROC) curves for evaluating fingerprint evidence [43]. With the ABC-ROC curve analysis, the authors demonstrated that fingermarks can be reliably used for identification of their source and that there is a very small probability of fingermarks being misidentified for configurations composed of greater than seven minutiae.

Several studies have focused on using statistical methods to examine or improve the image quality of fingermarks. Bansal et al. [8] present a critical review on extracting minutiae from fingerprint images. The authors indicate that due to the degradation or corruption of fingermarks by several deposition or aging conditions, image enhancement techniques must be applied to create an accurate representation of the fingerprint so that minutiae can be recognized by feature-based algorithms used in automated identification systems. The authors split the methods used for minutiae extraction into two classes based on the type of input analyzed: binarized vs gray-scale images. Further, they state that lower-quality fingermark images require more processing to increase contrast and reduce noise [8]. Another review article by Kaur et al. [52] similarly evaluates several different methods of fingermark image enhancement. Sankaran et al. [84] develop automated local clarity and quality estimation algorithms, examine the correlation between the two, and determine "that local clarity assists in better estimation of local quality thus resulting in improved matching performance." Speir and Hietpas [87] address the issue of image quality by proposing a mathematical transformation called frequency filtering to process digital images to increase image clarity and quality. Frequency filtering has been widely used for decades in electrical engineering, signals, and image processing, which provides support for its use in forensic science image analysis. Frequency filtering was shown to significantly improve the clarity, number of matching minutiae, and pairwise matching minutiae for poor-quality latent fingermarks. In an article by Kalka and Hicklin [49], the distortion of fingermarks that occurs when deposited is discussed, and several metrics for their quantification and visualization are proposed. Further, the authors present software that has the potential to be used in fingerprint quality assurance procedures by aiding examiners to identify and explain distortion. Olsen et al. [77] describe ten image quality assessment algorithms, make adaptations to them, and then evaluate their performance in terms of false positives and Spearman correlation on four datasets. The authors conclude that the performance of the algorithms is heavily dataset dependent and that false-negative rates and quality features used when assessing fingermarks are contingent on the dataset, and cannot be generalized. Neumann et al. [70] use descriptive statistics, specifically a probit binary classifier which is a type of regression used to calculate a classification probability, to determine whether a latent fingermark has sufficient characteristics (quality and quantity) to be searched in AFIS. The goal of this research was to test if statistical analysis of fingermarks could increase the workflow and workload efficiency for fingerprint examinations.

Statistical analyses have been used to address the important issue of determining the accuracy or specifically the establishment of error rates in fingerprint examinations. This has been the subject of increasing scrutiny, beginning in 1993 when the US Supreme Court ruled in *Daubert v. Merrell Dow Pharmaceuticals* that judges were the gatekeepers of expert evidence admissibility [22]. One of the Daubert factors to help determine judges evaluate the reliability of scientific evidence is the determination of a "rate of error." Further, in the 2009 National Academy of Sciences (NAS) report on forensic science [68], the third recommendation stated that "research is needed to address issues of accuracy, reliability, and validity in the forensic science disciplines." Prior to the NAS report, fingerprint examiner accuracy was the subject of four studies [37, 41, 58, 103], which have been summarized by Langenburg's 2009 [57] article "A Performance Study of the ACE-V Process." A valuable black-box study which developed empirical approaches for the determination of the accuracy and reliability of latent print examiner's decisions was published by Ulery et al. [99]. In this research, 169 latent print examiners compared approximately 100 fingermarks to exemplar fingerprints, and it was determined that there was an overall 0.1% false-positive rate and a 7.5% false-negative rate. The authors also indicate that "Independent examination of the same comparisons by different participants (analogous to blind verification) was found to detect all false-positive errors and the majority of false-negative errors in this study." This research employed traditional statistical calculations for determining error rates, posterior probabilities, and trends in the data (such as the relationship of examiner skill to error rate). Subsequent studies have demonstrated similar error rates [72, 94]. Kellman et al. [53] proposed using quantitative image metrics to estimate examiner performance and error rates. This research utilized descriptive statistics, correlation measurements, regression analyses, and the creation and validation of an accuracy model. It was followed up with a 2016 NIJ report on the same topic by Mnookin et al. titled "Error Rates for Latent Fingerprinting as a Function of Visual Complexity and Cognitive Difficulty," using the same statistical metrics [65]. Most recently, Hendricks and Neumann [44] published an article that employed ABC to analyze error rates. The authors reanalyzed the data from the initial black box study by Ulery et al. [99], with similar results. The authors advocate for this new method because ABC rate estimates allow for the detection of behavioral patterns in the decision-making of examiners. This has the potential "to detect error prone examiners, enabling additional training efforts to be more tailored to each examiner, limiting the risk of errors before they occur" [44].

13.3 Statistics in Fingermark Aging Research

The dating of fingermarks has been the subject of scientific research for nearly 60 years, with the first paper published in the French journal "Revue Internationale de Criminologie et de Police Technique" in 1962 by E. Angst [4]. This research focused on the visual changes of a fingermark over time, while monitoring both the ambient temperature and humidity (which are graphically displayed). Although

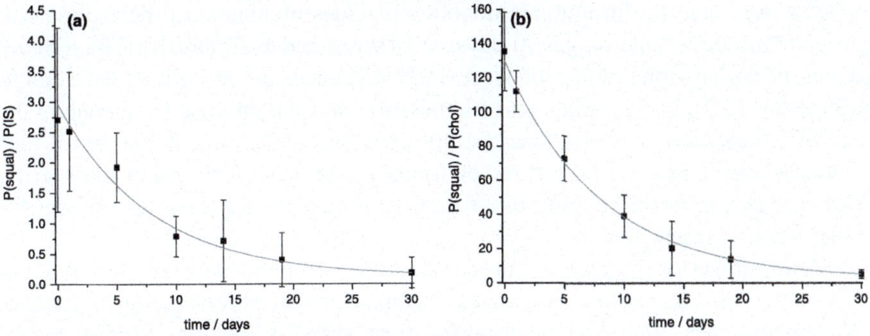

(A) FIG. 7—*GC/MS relative peak areas (P) of squalene (squal) divided by the peak areas of an (a) internal standard (IS) or (b) cholesterol (chol) as a function of time after deposition of fingerprints on microfilters: 0, 1, 5, 10, 14, 19, and 30 days.*

TABLE 4—*Exponential decay fitting of the aging curves represented in Figs 6 and 7 for the decay of squalene in fingerprint residues (t = time; y = relative peak area). The fits are better for the relative peak area (P) of squalene (squal) divided by the peak area of cholesterol (chol).*

	Glass Insert		Microfilters	
$y = a{\cdot}e^{(-t/b)}$	P(squal)/P(IS)	P(squal)/P(chol)	P(squal)/P(IS)	P(chol)
a	1.48	131.35	2.84	127.56
b	2.37	2.80	7.27	8.79
R^2	0.724	0.964	0.962	0.997

(B)

Fig. 13.1 Exponential decay fitting graphs (**a**) and results table (**b**) [106], reproduced with permission from Wiley Interscience

there were several published studies over the next 45 years, none of them employed any type of statistical data analysis [6, 16, 32, 34, 40, 46, 63].

Chemical analysis to examine aged fingermarks or determine time since deposition has been the subject of several studies using a range of statistical methods. Some published research has not employed any statistical analyses [78, 101], while others only present minimal statistical data analysis via graphs showing changes in chemical concentrations or instrument responses over time [2, 45, 66]. Mathematical modeling of the measured chemical changes has been successfully accomplished using exponential decay fitting or regression (Fig. 13.1) [67, 106]. Although these methods are appropriate for showing relationships, additional statistics (i.e., the calculation of prediction intervals) would be required for using them to predict the age of an unknown fingermark. Only one published paper, by Girod et al. [39], used advanced statistical analyses to examine the chemical changes, specifically targeted lipids identified and quantified by GC-MS, associated with the fingermark aging process. The authors used univariate exponential linear regression, PCA with soft independent modeling of class analogy (SIMCA), partial least squares regression (PLSR), and likelihood ratio calculations based on a Bayesian network for classifications and to model and predict the time since fingermark deposition. PCA with

SIMCA was able to differentiate donors regardless of time since deposition and differentiate three time ranges (0 days, < 10 days, and > 10 days) for fingermark dating of an individual donor under a single condition and in light vs dark conditions (Fig. 13.2). PLSR, using six variables, also showed promise for modeling the age of a fingermark and was successfully used to calculate likelihood ratios, thus demonstrating how it can be utilized by forensic examiners. This use of statistics by Girod et al. [39] to thoroughly interrogate their data should serve as a model for other forensic researchers.

Measurement of the physical properties or dimensions, known as morphometrics, of fingermarks has been investigated with the goal of understanding fingermark degradation over time and determining time since deposition. Similar to the chemistry-based research, one study did not employ statistical analysis [33], while others used minimal statistics in the form of graphs to show data trends [29, 102]. Fundamental but simple statistics, such as calculations of mean, standard deviation, standard error, and variance, and significance testing with t-tests or ANOVA and regression analyses were used in two studies by De Alcaraz-Fossoul and collaborators [23, 27, 28], two papers by Czech and collaborators [19, 20], and one by Colella et al. [17]. As previously stated, although these statistical methods are valuable because they are able to show relationships and demonstrate significant differences, additional statistics and testing (under non-ideal conditions) are required to capture the uncertainty associated with the prediction required for age determination of latent fingermarks.

Barros et al. [9] observed the narrowing of ridge widths during the aging of latent palmprints and studied the morphometric changes using a several statistical methods. The authors used a straightforward model hypothesis testing combined with standard procedures for para-/non-parametric data analysis, which included an examination of the homogeneity of variances, normality of the data distribution, and outliers using Bartlett's, Kolmogorov–Smirnov's and Tukey's tests, respectively. ANOVA and Kruskal–Wallis' test, followed by the Student–Newman–Keuls' or Dunn's test, were used for intrasample comparisons, paired t-tests or Wilcoxon's tests were used to compare two related samples, and unpaired t-tests or Mann–Whitney's tests were used to compare unrelated samples. The authors also evaluated deviations using linear regression analysis. This research used statistically sound methods to demonstrate the time dependence on the reduction in the width of latent palmprint ridges and in the percentage of visible ridges; however, no attempt was made to use this to predict the time since deposition of a palmprint.

Matuszewski [61] studied changes in the color contrast over time for fingermarks with different lipid content and used scatter plots and an exponential regression using a quasi-Newton method of estimation and a 5% level of significance to model the changes. This analysis showed that the changes in color contrast over time did vary based on the lipid content, and even between the three categories (high, moderate, and low lipids), there was considerable variation shown by the spread of the data. This research demonstrated the potential utility of color contrast for fingermark aging determinations and is being further investigated by De Alcaraz-Fossoul and collaborators [25].

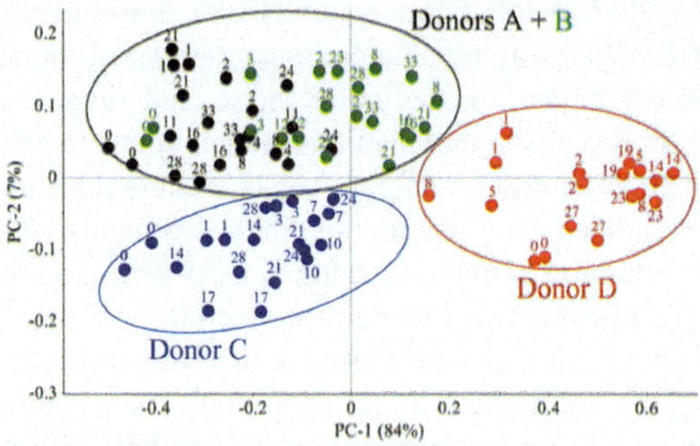

Figure 1: PCA of fresh and old fingermark specimens from different donors left on filters and aged in the fixed standard conditions. Fingermarks of "medium-rich" Donors A (black) and B (green) were aged from 0 to 33 days, fingermarks of the "medium-poor" Donor C (blue) were 0 to 28 days old and fingermarks of the "poor" Donor D (red) were aged from 0 to 27 days. Groups were observed based on the type of donor, but not on the age of the fingermarks (the numbers above the points are the fingermark age in days).

(A)

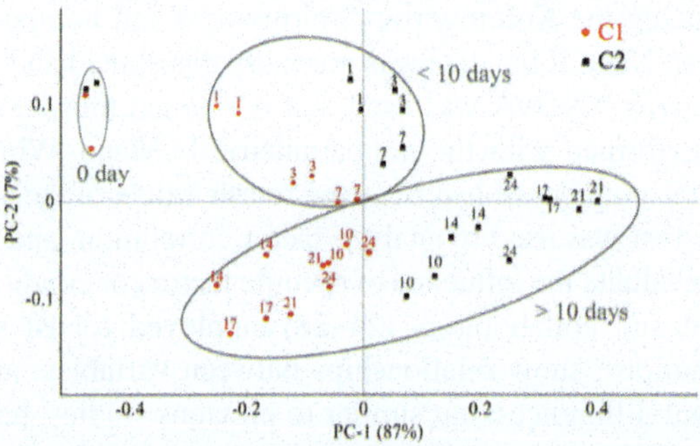

Figure 7: PCA of fresh and old fingermark specimens from Donor C ("medium-poor") left on microfiber filters in April (C1, red circles) or in July (C2, black squares) and stored in the fixed standard conditions. Fingermarks were aged from 0 to 24 days. Three groups based on fingermark age were visible: 0, < 10 or ≥ 10 days old (the numbers above the points are the fingermark age in days). However, a separation due to deposition moment was also visible, as C1 were mainly grouped on the right and C2 on the left of this PCA.

(B)

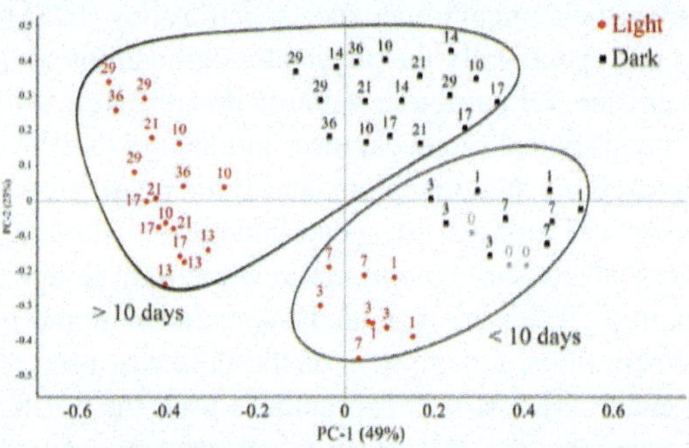

Figure 16: PCA of fresh (grey stars) and old fingermark specimens from Donor C ("Medium-poor") left on glass and stored in the light (red circles) or in the dark (black squares). Fingermarks were aged from 0 to 36 days. Lighting conditions clearly affected the specimens (fingermarks stored in the light on the right, in the dark on the left). However, two groups based on fingermark age were visible: < 10 and ≥ 10 days old (the numbers above the points are the fingermark age in days).

(C)

Fig. 13.2 PCA plots with boundaries created using SIMCA showing the ability to differentiate fingermark donors regardless of time since deposition (**a**), fingermark age ranges (0 days, < 10 days, and > 10 days) for a single donor, and fingermark ages in different lighting conditions (**c**) [39], reproduced with permission from Elsevier

De Alcaraz-Fossoul [24–28] and collaborators published a four-part study on the physical changes associated with the aging of latent fingermarks, specifically examining minutiae count, color contrast, discontinuity index, and ridge width. This group has used a range of statistically sound methods in their controlled research to thoroughly investigate the effect of different aging factors on time-dependent visual features that could be good indicators of degradation. In the first study which examined the effect of aging on minutia counts, the authors used Wilcoxon Rank-Sum tests at a 95% level of significance to evaluate reproducibility of using minutia ratios and to identify influential factors on the degradation of fingermarks, box plots to visualize trends, and regression models. When examining color contrast, histogram profiles were examined using the nonparametric Kruskal–Wallis test and Levene's test to assess the data variability and regression models were used to investigate changes over time and with different environmental parameters. In the third and fourth studies, in addition to graphical analysis, normality and homogeneity of the data were evaluated using the Kolmogorov–Smirnov test and homogeneity of variance by the Levene test. Next, if the data was normally distributed and homogeneous, then t-tests or one-way ANOVA was used, but if it failed those conditions, statistical analysis was performed with the nonparametric U-Mann–Whitney and Krustall–Wallis tests at 95% confidence levels. In the fourth article of the series, the nonparametric Spearman Test was used to analyze factor correlation, and a regression analysis was used to evaluate the influence of specific factors or conditions over time. De Alcaraz-Fossoul and collaborators [24–28] employed robust statistical methods to model distributions, show relationships between variables, and determine if there are significant differences, but similar to previous studies, have yet to use appropriate prediction statistics to test if evidence fingermark dating is possible. Further, due to the number of potential variables that may affect the aging of fingermarks, it is unknown but unlikely that these univariate statistics will prove useful for age determination.

Rosa et al. [82] used electrochemical impedance spectroscopy (EIS) to study the fingerprint aging process and specifically the physical modifications of fingermarks on metallic substrates over time. Of particular value in this research, with regard to its use of statistics, was the graphical representation and analysis of the data which employed exponential decay plots, Nyquist plots, and Bode phase plots.

One study by Merkel et al. [64] utilized advanced statistical methods and machine learning to examine the use of 2D and 3D chromatic white light image sensors for fingermark age determinations. The statistical analysis included a random selection of the data, with some restrictions, where 80% of the data was used to make the training set and 20% to make the test set. The authors used the WEKA machine learning toolbox [100] to evaluate the classification performance of five classifiers of the type Bayesian classifier (Naïve Bayes, Bayes Net) meta classifier (Rotation Forest) and tree (LMT, J48). The kappa statistic was used to calculate the classification accuracy. The authors examined the aging behavior of 17 "features" (3 binary pixel variations for 2D-intensity images, 2 binary pixel variations for 3D-topography images, and 12 statistical features) and concluded that those based on the binary pixel feature correlated best with the aging property. The authors concluded that using the binary pixel for 2D-intensity images with static normalization feature in respect to a linear property with a Naïve Bayes classifier performed the best (79.29% correct classification), with the second best being the use of the binary pixel for 2D-intensity images with dynamic normalization feature and a logarithmic curve with a Naïve Bayes classifier (76.48% correct classification). However, the authors correctly determine that the performance of the first one may be overfitted and recommend the second one going forward. The authors also attempted a feature fusion, which did not increase the performance of the classification model, and hypothesized that this was due to the dependency of the features on each other. Further, they proposed that combining the use of 2D chromatic white light imaging with additional physical traits as measured with a microscope or spectrometer could provide a robust scheme for fingermark age determination. Although these are statistically valuable, this research employed complex statistics that may be difficult for a forensic fingerprint examiner to understand and explain. Additionally, the best method had less than an 80% correct classification rate which may not be reliable enough for acceptance in court.

A combined approach using both chemical and morphometric analysis of fingermarks for age determination was proposed by Popa et al. [81]. The authors employed morphological, structural, and macro- and microscopic examinations, together with biochemical and titration DNA tests to examine the effects of the aging process. The article only includes basic graphical representation of the data. Although not explicitly stated, this research demonstrates the added information obtainable when a combined approach is taken. This can be further leveraged in the future through data fusion and multivariate statistical analysis.

13.4 Conclusion

One does not need a cannon to kill a fly, and, as such, there are a plethora of scientific problems that do not require complex multivariate statistics. However, considering the complexity of fingermark aging, and specifically the number of external and internal factors, this author is confident that the most successful approaches to dating a fingermark will employ multivariate statistics and possibly data fusion.

It is important to remember that statistical methods should be considered at the outset of a research investigation and not just after data collection has concluded. Statistical approaches must be incorporated into the experimental design. This includes correct sampling, randomizing samples, and experimental variability, such as instrumental drift to which chemometric models are very sensitive. It is not uncommon for chemometric models to classify samples based on the time since the instrument was turned on rather than real chemical differences. The classic rule for any statistical analysis will always hold true: garbage in, garbage out.

When considering alternative statistical approaches for fingermark aging research, one must take into account that the statistical methods used in forensic science must be defensible in court in order to be useful. It is the ultimate goal to be able to employ these methods to determine the age of a questioned crime scene fingermark, so the statistics must be robust and able to hold up to admissibility challenges. In light of this, it may be important to be able to trace the regression results back to the fundamental reasons that the regression is working, specifically the chemical and physical changes that are occurring in the aging fingermark. As a consequence to this, linear regression methods, such as principle component regression (PCR) and PLSR, may prove to be best suited to answer this problem. Classical least squares (CLS) is another method that has the benefits of being interpretable and offers the scientist the ability to control aspects of the regression model. It is known that when complex relationships exist, nonlinear modeling methods, such as XGBoost, SVM, artificial neural networks (ANN), and their discriminant analysis versions, are superior. Although their performance for modeling fingermark aging may prove to be better, one must recognize the significant shortcomings associated with the use of these nonlinear machine learning methods, which include the inability to interpret the mechanism or relationship of the features.

Acknowledgments I am indebted to Professor Donald Dahlberg, Ph.D. (Lebanon Valley College), for introducing me to the power of chemometrics and multivariate statistics and instilling in me the importance of rigorously testing my models and research. I am grateful to Neal Gallagher, Ph.D. (Eigenvector Research Incorporated), for our many discussions on the applications of statistics to problems of forensic interest and specifically on what methods may prove to be beneficial to solving the problem of dating fingermarks. Lastly, I would like to thank Prof. Josep De Alcaraz-Fossoul (University of New Haven) for inviting me to contribute this chapter and for many meaningful conversations regarding statistics, fingerprints, and statistics used in fingermark research.

References

1. Acree MA (1999) Is there a gender difference in fingerprint ridge density? Forensic Sci Int 102(1):35–44. https://doi.org/10.1016/S0379-0738(99)00037-7
2. Akiba N, Kuroki K, Kurosawa K, Tsuchiya K (2018) Visualization of aged fingerprints with an ultraviolet laser. J Forensic Sci 63(2):556–562. https://doi.org/10.1111/1556-4029.13588
3. Amy L (1946) Valeur de la prevue en dactyloscopie I. J Soc Stat Paris 88:80
4. Angst E (1962) Procédés pour la détermination de I'âge d'empreintes dactyloscopiques sur le papier. Revue internationale de criminologie et de police technique 16:134–146
5. Balthazard V (1911) De l'identification par les empreintes digitales. Comptes Rendus des Academies des Sciences 152:1862
6. Baniuk K (1990) Determination of age of fingerprints. Forensic Sci Int 46:133–137
7. Baniuk, K. (2000). Polemics-importance of laboratory experiment in fingerprint age determination. Problemy Kryminalistyki:69–71
8. Bansal R, Sehgal P, Bedi P (2011) Minutiae extraction from fingerprint images – a review. Int J Computer Sci 5(3):74–85
9. Barros RM, Faria BEF, Kuckelhaus SAS (2013) Morphometry of latent palmprints as a function of time. Sci Justice 53(4):402–408. https://doi.org/10.1016/j.scijus.2013.08.002
10. Bovens M, Ahrens B, Alberink I et al (2019) Chemometrics in forensic chemistry – part I: implications to the forensic workflow. Forensic Sci Int 301:82–90. https://doi.org/10.1016/j.forsciint.2019.05.030

11. Buckleton J, Robertson B, Curran J et al (2020) A review of likelihood ratios in forensic science based on a critique of Stiffelman "no longer the gold standard: probabilistic genotyping is changing the nature of DNA evidence in criminal trials". Forensic Sci Int 310:110251. https://doi.org/10.1016/j.forsciint.2020.110251
12. Champod C (1995) Reconnaissance Automatique et Analyze Statistique des Minuties sur les Emreintes Digitales. Doctoral thesis, Université de Lausanne, Institut de Police Scientifique et de Criminologie
13. Champod C (2000) Identification/individualization: overview and meaning of ID. In: Siegel J, Saukko P, Knupfer G (eds) Encyclopaedia of forensic science. Academic, London, pp 1077–1083
14. Champod C, Evett IW (2001) A probabilistic approach to fingerprint evidence. J Forensic Ident 51(2):101–122
15. Champod C, Lennard C, Margot P, Stoilovic M (2004) Fingerprints and other ridge skin impressions. CRC Press LLC, Boca Raton
16. Clements WW (1986) Latent fingerprints – one year later. Fingerprint Whorld 12(46):54
17. Colella O, Miller M, Boone E et al (2020) The effect of time and temperature on the persistence and quality of latent fingerprints recovered from 60-Watt incandescent light bulbs. J Forensic Sci 65:90–96. https://doi.org/10.1111/1556-4029.14133
18. Cook R, Evett IW, Jackson G et al (1998) A model for case assessment and interpretation. Sci Justice 38:151–156
19. Czech A, Szabelak A, Sowiński A (2019) Changes in fingerprints depending on physiological factors. J Forensic Sci 64:711–716. https://doi.org/10.1111/1556-4029.13937
20. Czech A, Gryszczyk N, Szabelak A, Sowiński A (2020) Changes in fingerprints and the quantity of material forming the print depending on hand cleanliness, gender, and ambient conditions. J Forensic Sci 65:84–89. https://doi.org/10.1111/1556-4029.14164
21. Daramola SA, Sokunbi T, Adoghe AU (2013) Fingerprint verification system using support vector machine. Int J Comput Sci Eng 5(7):678–683
22. *Daubert v. Merrell Dow Pharmaceuticals, Inc,* 509 US. 579, 593 (1993)
23. De Alcaraz-Fossoul J, Patris CM, Muntaner AB et al (2013) Determination of latent fingerprint degradation patterns – a real fieldwork study. Int J Legal Med 127:857–870. https://doi.org/10.1007/s00414-012-0797-0
24. De Alcaraz-Fossoul J, Mestres Patris C, Barrot Feixat C et al (2016a) Latent fingermark aging patterns (part I): minutiae count as one indicator of degradation. J Forensic Sci 61:322–333. https://doi.org/10.1111/1556-4029.13007
25. De Alcaraz-Fossoul J, Barrot Feixat C, Tasker J et al (2016b) Latent fingermark aging patterns (part II): color contrast between ridges and furrows as one indicator of degradation. J Forensic Sci 61:947–958. https://doi.org/10.1111/1556-4029.13099
26. De Alcaraz-Fossoul J, Barrot Feixat C, Carreras-Marin C, Tasker J, Zapico SC, Gené Badia M (2017) Latent fingermark aging patterns (part III): discontinuity index as one indicator of degradation. J Forensic Sci 62:1180–1187. https://doi.org/10.1111/1556-4029.13438
27. De Alcaraz-Fossoul J, Barrot Feixat C, Zapico SC et al (2019a) Latent fingermark aging patterns (part IV): ridge width as one indicator of degradation. J Forensic Sci 64:1057–1066. https://doi.org/10.1111/1556-4029.14018
28. De Alcaraz-Fossoul J, Mancenido M, Soignard E, Silverman N (2019b) Application of 3D imaging technology to latent fingermark aging studies. J Forensic Sci 64:570–576. https://doi.org/10.1111/1556-4029.13891
29. De Paoli G, Lewis SA, Schuette EL et al (2010) Photo- and thermal-degradation studies of select eccrine fingerprint constituents. J Forensic Sci 55:962–969. https://doi.org/10.1111/j.1556-4029.2010.01420.x
30. Department of Justice (2010) Uniform language for testimony and reports. U.S. Department of Justice. Retrieved from https://www.justice.gov/olp/uniform-language-testimony-and-reports

31. Department of Justice (2020) Uniform language for testimony and reports for the Forensic latent print discipline. U.S. Department of Justice. Retrieved from https://www.justice.gov/olp/page/file/1284786/download
32. Dikshitulu YS, Prasad L, Pal JN, Rao CVN (1986) Aging studies on fingerprint residues using thin-layer and high performance liquid chromatography. Forensic Sci Int 31:261–266
33. Dorakumbura BN, Becker T, Lewis SW (2016) Nanomechanical mapping of latent fingermarks: a preliminary investigation into the changes in surface interactions and topography over time. Forensic Sci Int 267:16–24. https://doi.org/10.1016/j.forsciint.2016.07.024
34. Duff JM, Menzel ER (1978) Laser-assisted thin-layer chromatography and luminescence of fingerprints: an approach to fingerprint age determination. J Forensic Sci 23(1):129–134
35. Egli NM, Champod C, Margot P (2007) Evidence evaluation in fingerprint comparison and automated fingerprint identification systems – modelling within finger variability. Forensic Sci Int 167(2–3):189–195. https://doi.org/10.1016/j.forsciint.2006.06.054
36. Evett IW, Jackson G, Lambert JA, McCrossan S (2000) The impact of the principles of evidence interpretation on the structure and content of statements. Sci Justice 40:233–239
37. Evett IW, Williams RL (1995) A review of the 16 point fingerprint standard in England and Wales. Fingerprint Whorld 21
38. Galton F (1892) FingerPrints. McMillan, London
39. Girod A, Spyratou A, Holmes D, Weyermann C (2016) Aging of target lipid parameters in fingermark residue using GC/MS: effects of influence factors and perspectives for dating purposes. Sci Justice 56(3):165–180. https://doi.org/10.1016/j.scijus.2015.12.004
40. Greenless D (1994) Age determination – case report. Fingerprint Whorld 20(76):50–52
41. Gutowski S (2006) Error rates in fingerprint examination: the view in 2006. Forensic Bull 2006(Autumn):18–19
42. Henry ER (1900) Classification and uses of fingerprints. Routledge & Sons, London
43. Hendricks, J. H., Neumann, C., & Saunders, C. P. (2018). Quantifying the weight of fingerprint evidence using an ROC-based approximate Bayesian computation algorithm. J Royal Stat Soc C. Retrieved from https://pdfs.semanticscholar.org/cfed/4674a51fdaf74308bb5430ed65b7d953393d.pdf
44. Hendricks JH, Neumann C (2020) A Bayesian approach for the analysis of error rate studies in forensic science. Forensic Sci Int 306:110047. https://doi.org/10.1016/j.forsciint.2019.110047
45. Hinners P, Thomas M, Young JL (2020) Determining fingerprint age with mass spectrometry imaging via Ozonolysis of Triacylglycerols. Anal Chem 92(4):3125–3132
46. Holyst B (1987) Kriminalistische Abschätzung des Spurenalters bei Fingerpapillarlinien [Criminal assessment of the age of fingerprints]. Arch Kriminol 179(3–4):94–103
47. Howorka H (1989) Question relating to the determination of the age of objects assuming relevance in criminal investigations. Fingerprint Whorld 15(57):23–28
48. Inman K, Rudin N (2000) Principles and practice of criminalistics: the profession of forensic science (protocols in forensic science). CRC Press, Boca Raton
49. Kalka ND, Hicklin RA (2014) On relative distortion in fingerprint comparison. Forensic Sci Int 244:78–84. https://doi.org/10.1016/j.forsciint.2014.08.007
50. Kapoor N, Badiye A (2015) Sex differences in the thumbprint ridge density in a central Indian population. Egyptian J Forensic Sci 5(1):23–29. https://doi.org/10.1016/j.ejfs.2014.05.001
51. Kathiravan S, K.samy A (2014) Robust latent fingerprint matching using support vector machine. Int J Res Computer App Robotics 2(4):55–62
52. Kaur N, Singh G, Vashist S (2017) Comparative analysis of various fingerprint image enhancement algorithms. Int J Eng Computer Sci 3(6):6389–6394
53. Kellman PJ, Mnookin JL, Erlikhman G, Garrigan P, Ghose T, Mettler E, Charlton E, Dror IE (2014) Forensic comparison and matching of fingerprints: using quantitative image measures for estimating error rates through understanding and predicting difficulty. PLoS One 9(5):e94617. https://doi.org/10.1371/journal.pone.0094617

54. Kingston CR (1964) Probabilistic analysis of partial fingerprint patterns. D.Crim dissertation, University of California, Berkeley
55. Kirk PL (1963) The ontogeny of criminalistics. J Criminal Law Criminol Police Sci 54(2):235–238
56. Kumar R, Sharma V (2018) Chemometrics in forensic science. Trends Anal Chem 105:191–201. https://doi.org/10.1016/j.trac.2018.05.010
57. Langenburg G (2009) A performance study of the ACE-V process. J Forensic Identification 59:219–257
58. Langenburg G, Champod P, Wertheim P (2009) Testing for potential contextual bias effects during the verification stage of the ACE-V methodology when conducting fingerprint comparisons. J Forensic Sci 54:571–582
59. Lund SP, Iyer H (2017) Likelihood ratio as weight of forensic evidence: a closer look. J Res Natl Inst Stan 122:27. https://doi.org/10.6028/jres.122.027
60. Margot P (2000) A question of time. Sci Justice 40:64–71
61. Matuszewski S (2015) Age-dependent changes of contrast in fingermarks of various content of lipids. Problems Forensic Sci 101:5–13
62. Meagher SB, Budowle B, Ziesig D (1999) 50K vs. 50K fingerprint comparison test, United States of America vs. Byron Mitchell, U.S. District Court Eastern District of Philadelphia. Government Exhibits 6–8 and 6–9 in Daubert Hearing before Judge J. Curtis Joyner, July 8–9, 1999
63. Menzel ER (1992) Fingerprint age determination by fluorescence. J Forensic Sci 37(5):1212–1213
64. Merkel R, Gruhn S, Dittmann J, Vielhauer C, Bräutigam A (2012) On non-invasive 2D and 3D chromatic white light image sensors for age determination of latent fingerprints. Forensic Sci Int 222(1–3):52–70. https://doi.org/10.1016/j.forsciint.2012.05.001
65. Mnookin JL, Kellman PJ, Dror I, Erlikhman G, Garrigan P, Ghose T, Metler E, Charlton D (2016) Error rates for latent fingerprinting as a function of visual complexity and cognitive difficulty. U.S. Department of Justice NIJ Award 2009-DN-BX-K225
66. Mountfort KA, Bronstein H, Archer N, Jickells SM (2007) Identification of oxidation products of squalene in solution and in latent fingerprints by ESI-MS and LC/APCI-MS. Anal Chem 79(7):2650–2657. https://doi.org/10.1021/ac0623944
67. Muramoto S, Sisco E (2015) Strategies for potential age dating of fingerprints through the diffusion of sebum molecules on a nonporous surface analyzed using time-of-flight secondary ion mass spectrometry. Anal Chem 87(16):8035–8038. https://doi.org/10.1021/acs.analchem.5b02018
68. National Research Council, National Academy of Sciences (2009) Strengthening forensic science in the United States: a path forward. National Academy Press, Washington, DC
69. Neumann C (2013) Chapter 15: statistics and probabilities as a means to support fingerprint examination. In: Ramotowski RS (ed) Lee and Gaensslen's advances in fingerprint technology, 3rd edn. CRC Press LLC, Boca Raton, pp 419–465
70. Neumann C, Armstrong DE, Wu T (2016) Determination of AFIS "sufficiency" in friction ridge examination. Forensic Sci Int 263:114–125. https://doi.org/10.1016/j.forsciint.2016.04.004
71. Neumann C, Champod C, Puch-Solis R et al (2006) Computation of likelihood ratios in fingerprint identification for configurations of three Minutiæ. J Forensic Sci 51:1255–1266. https://doi.org/10.1111/j.1556-4029.2006.00266.x
72. Neumann C, Champod C, Yoo M, Genessay T, Langenburg G (2013) Improving the understanding and the reliability of the concept of 'sufficiency' in friction ridge examination. U.S. Department of Justice, National Institute of Justice, Washington, DC
73. Neumann C, Evett IW, Skerrett JE, Mateos-Garcia I (2011) Quantitative assessment of evidential weight for a fingerprint comparison I. Generalisation to the comparison of a mark with set of ten prints from a suspect. Forensic Sci Int 207(1–3):101–105. https://doi.org/10.1016/j.forsciint.2010.09.006

74. Neumann C, Evett IW, Skerrett JE, Mateos-Garcia I (2012a) Quantitative assessment of evidential weight for a fingerprint comparison. Part II: a generalisation to take account of the general pattern. Forensic Sci Int 214(1–3):195–199. https://doi.org/10.1016/j.forsciint.2011.08.008
75. Neumann C, Evett IW, Skerrett J (2012b) Quantifying the weight of evidence from a forensic fingerprint comparison: a new paradigm. J Royal Stat Soc Ser A 175:371–415. https://doi.org/10.1111/j.1467-985X.2011.01027.x
76. Neumann C, Saunders CP (2019) Foundational research into the quantification of the value of forensic evidence for complex evidential forms arising from impression and pattern evidence. U.S. Department of Justice NIJ Award 2014-IJ-CX-K088
77. Olsen MA, Šmida V, Busch C (2016) Finger image quality assessment features – definitions and evaluation. IET Biometrics 5(2):47–64
78. O'Neill KC, Lee YJ (2018) Effect of aging and surface interactions on the diffusion of endogenous compounds in latent fingerprints studied by mass spectrometry imaging. J Forensic Sci 63:708–713. https://doi.org/10.1111/1556-4029.13591
79. Osterburg J et al (1977) Development of a mathematical formula for the calculation of fingerprint probabilities based on individual characteristics. J Am Stat Assoc 72:772
80. Paulino AA, Feng J, Jain AK (2013) Latent fingerprint matching using descriptor-based Hough transform. IEEE Trans Inf Forensics Secur 8(1):31–45
81. Popa G, Potorac R, Preda N (2010) Method for fingerprints age determination. RJLM 2:149–154
82. Rosa R, Giovanardi R, Bozza A et al (2017) Electrochemical impedance spectroscopy: a deeper and quantitative insight into the fingermarks physical modifications over time. Forensic Sci Int 273:144–152. https://doi.org/10.1016/j.forsciint.2017.02.016
83. Roxburgh T (1933) On the evidential value of finger prints. Sankhya: Indian J Stat 1:189
84. Sankaran A, Vatsa M, Singh R (2013) Automated clarity and quality assessment for latent fingerprints. 2013 IEEE sixth international conference on Biometrics: Theory, Applications and Systems (BTAS), Arlington, VA, pp. 1–6
85. Shah T (2017) About train, validation and test sets in machine learning. Retrieved from https://towardsdatascience.com/train-validation-and-test-sets-72cb40cba9e7
86. Slobogin C (2019) The use of statistics in criminal cases: an introduction. Behav Sci Law 37:127–132. https://doi.org/10.1002/bsl.2407
87. Speir JA, Hietpas J (2014) Frequency filtering to suppress background noise in fingerprint evidence: quantifying the fidelity of digitally enhanced fingerprint images. Forensic Sci Int 242:94–102. https://doi.org/10.1016/j.forsciint.2014.06.026
88. Stiffelman B (2019) No longer the gold standard: probabilistic genotyping is changing the nature of DNA evidence in criminal trials. Berkeley J Criminal Law 24:110–146
89. Stoney DA (1985) A quantitative assessment of fingerprint individuality. Ph.D. dissertation, University of California, Berkeley
90. Stoney DA, Thornton JI (1987) A systematic study of epidermal ridge minutiae. J Forensic Sci 32:1182
91. Stoney DA (1991) What made us ever think we could individualize using statistics? J Forensic Sci Soc 31:197–199
92. Stoney DA (2001) Measurement of fingerprint individuality. In: Lee HC, Gaensslen RE (eds) Advances in fingerprint technology, 2nd edn. CRC Press LLC, Boca Raton, pp 327–387
93. Swofford HJ, Koertner AJ, Zemp F et al (2018) A method for the statistical interpretation of friction ridge skin impression evidence: method development and validation. Forensic Sci Int 287:113–126. https://doi.org/10.1016/j.forsciint.2018.03.043
94. Tangen JM, Thompson MB, McCarthy DJ (2011) Identifying fingerprint expertise. Psychol Sci 22:995–997. https://doi.org/10.1177/0956797611414729
95. Taroni F, Aitken CGG (1998a) Probabilistic reasoning in the law. Part 1: assessment of probabilities and explanation of the value of DNA evidence. Sci Justice 38:165–177

96. Taroni F, Aitken CGG (1998b) Probabilistic reasoning in the law. Part 2: assessment of probabilities and explanation of the value of trace evidence other than DNA. Sci Justice 38:179–188
97. Tobin WA, Sheets HD, Spiegelman C (2017) Absence of statistical and scientific ethos: the common denominator in deficient forensic practices. Stat Public Policy 4(1):1–11. https://doi.org/10.1080/2330443X.2016.1270175
98. Trauring M (1963) Automatic comparison of finger-ridge patterns. Nature 197:938
99. Ulery BT, Hicklin RA, Buscaglia JA, Roberts MA (2011) Accuracy and reliability of forensic latent fingerprint decisions. PNAS 108(19):7733–7738. https://doi.org/10.1073/pnas.1018707108
100. University of Waikato Machine Learning Group: Weka Machine Learning Toolbox, Retrieved from http://www.cs.waikato.ac.nz/ml/weka/ on July 9, 2020
101. van Dam A, Aalders MCG, Todorovski T, van Leeuwen TG, Lambrechts SAG (2016) On the autofluorescence of aged fingermarks. Forensic Sci Int 258(19–25). https://doi.org/10.1016/j.forsciint.2015.11.002
102. Watson P, Prance RJ, Beardsmore-Rust ST, Prance H (2011) Imaging electrostatic fingerprints with implications for a forensic timeline. Forensic Sci Int 209(1–3):e41–e45. https://doi.org/10.1016/j.forsciint.2011.02.024
103. Wertheim K, Langenburg G, Moenssens A (2006) A report of latent print examiner accuracy during comparison training exercises. J Forensic Identification 56:55–93
104. WetsmanN(2018)Fingerprintanalysiscouldfinallygetscientific,ThankstoaNewToolGizmodo, 2018. Retrieved from https://gizmodo.com/fingerprint-analysis-could-finally-get-scientific-than-1825607912
105. Weyermann C, Ribaux O (2012) Situating forensic traces in time. Sci Justice 52:68–75
106. Weyermann C, Roux C, Champod C (2011) Initial results on the composition of fingerprints and its evolution as a function of time by GC/MS analysis. J Forensic Sci 56:102–108. https://doi.org/10.1111/j.1556-4029.2010.01523.x

The manufacturer's authorised representative in the EU is Springer Nature Customer Service Centre GmbH, Europaplatz 3, 69115 Heidelberg, Germany. If you have any concerns regarding our products, please contact ProductSafety@springernature.com

Printed and bound by CPI Group (UK) Ltd, Croydon, CR0 4YY
30/09/2026
02225491-0013